# Readings in
# Sociobiology

# Readings in
# Sociobiology

Edited by

## T. H. CLUTTON-BROCK
*King's College, Cambridge*

## PAUL H. HARVEY
*University of Sussex*

W. H. FREEMAN AND COMPANY

Reading and San Francisco

Library of congress Cataloging in Publication Data

Main entry under title:
Readings in sociobiology.
 Bibliography: p.
 Includes index.
 1. Social behavior in animals.    I. Clutton-Brock,
T. H.    II. Harvey, Paul H., 1947–
QL775.R4        591.5        77-22283

ISBN 0-7167-0191-X
ISBN 0-7167-0190-1 pbk.

Printed in the United States of America

# Contents

# Contents

## A NOTE ON SPELLING

In each paper the original American or English spelling has been adhered to.
In editorial passages, the English form is used.

# Preface

Sociobiology is the study of the biological basis of social behaviour. It is derived from two separate biological traditions: population genetics and evolutionary ecology on the one hand and ethology and animal behaviour on the other. This is the synthesis which Wilson claims, justifiably, to have achieved in *Sociobiology, the new synthesis* (1975). Though the publication of Wilson's book may represent the official birth of sociobiology as a separate discipline within biology, its gestation period has been a long one. Many of the important concepts which Wilson synthesizes have developed over the last two decades and two notable books dealing with many of the same topics in less detail were published a number of years ago (Maynard Smith 1958; Williams 1966). Though Wilson's exposition is clear and his understanding profound, we believe that many students, as well as those working in the field, will wish to return to the original papers. These have been published in a wide variety of journals and symposia and the aim of this collection is to provide easy access to some of the most important theoretical papers, supplemented by more recent work which illustrates or elaborates the principles they outline.

Selection of papers was not simple. Though sociobiology should properly include studies both of the causation and development of social behaviour, we have restricted the collection to papers concerned primarily with functional and evolutionary questions. Research on the causation and ontogeny of social behaviour has recently been reviewed in depth (Hinde 1970, 1974) and a wide variety of symposia have already devoted their attention mainly to these areas (e.g. Rosenblum 1970; Poirier 1972; Hinde and Stephenson-Hinde 1973; Bateson and Hinde 1976). Moreover, it would have been impossible to provide adequate coverage of causation and development and to deal with functional questions in the same volume. We used four criteria in selecting papers: that they should outline an important theory or provide empirical evidence supporting theoretical predictions; that they should not require a level of numeracy beyond that of the average biology student; that they should not have been recently republished; and that they should be short. Unfortunately, these criteria required us to exclude a number of papers which have played a seminal role in the field (e.g. Gilpin 1975; Hamilton 1964; Maynard Smith and Parker 1976; Selander 1966, 1972; Hinde 1956; Schoener 1968; Jarman 1974).

The papers are grouped in four sections: the first discusses the level at which natural selection operates, contrasting recent viewpoints on individual and group selection; the second introduces the idea that animals can be understood as mechanisms by which genes replicate themselves (Dawkins 1976) and discusses adaptive differences in life-history variables and breeding systems; the third section examines the evolutionary mechanisms favouring co-operative and disruptive behaviour; and the fourth (briefly) describes some attempts to explain species differences in social behaviour. The introductions to the sections are aimed at students or those new to the field. The papers have been retained in their original style and all except two are reprinted in full.

Although the introductory sections are designed to be comprehensible without recourse to further reading, we have included with each a reasonably extensive reference list for the more advanced reader who wishes to pursue a more thorough literature search.

We are grateful for help, advice or criticism from R. M. Gibson, P. J. Greenwood, P. W. Greig-Smith, A. M. Harvey, M. Leverton, G. Mace, J. Maynard Smith and G. A. Parker. The choice of papers included in the volume is our own.

We would especially like to thank Dr. John Gillman for his help throughout the preparation of this book, and Dr. M. Kavanagh for editorial assistance.

Finally, we must thank Priscilla Barrett for the five beautiful illustrations which appear at the beginning of each section and on the cover.

T. H. CLUTTON-BROCK
*King's College Research Centre, Cambridge, U.K.*

PAUL H. HARVEY
*School of Biology, University of Sussex, Brighton, U.K.*

*February 1977*

# The Authors

B. C. R. Bertram  *King's College Research Centre, Cambridge*

T. H. Clutton-Brock  *King's College Research Centre, Cambridge*

W. D. Hamilton  *Department of Zoology, Imperial College, London*

P. H. Harvey  *School of Biological Sciences, University of Sussex, Brighton*

D. Lack  *Late, of Edward Grey Institute, University of Oxford*

J. Maynard Smith  *School of Biological Sciences, University of Sussex, Brighton*

G. H. Orians  *Department of Zoology, University of Washington, Seattle*

C. Packer  *School of Biological Sciences, University of Sussex, Brighton*

G. A. Parker  *Department of Zoology, University of Liverpool, Liverpool*

E. R. Pianka  *Department of Zoology, University of Texas, Austin, Texas*

R. L. Trivers  *Museum Comparative Zoology, Harvard University*

E. O. Wilson  *Museum Comparative Zoology, Harvard University*

V. C. Wynne-Edwards  *Department of Natural History, University of Aberdeen*

# Section 1

# Group benefit or individual advantage?

# Section 1

# Group benefit or individual advantage?

## Introduction

'In October 1838, that is, fifteen months after I had begun my systematic enquiry, I happened to read for amusement Malthus on "Population", and being well prepared to appreciate the struggle for existence which everywhere goes on from long-continued observation of the habits of animals and plants, it at once struck me that under these circumstances favourable variations would tend to be preserved, and unfavourable ones to be destroyed.'

Charles Darwin, *Autobiography* (1876)

'Owing to this struggle [for existence] variations, however slight and from whatever cause proceeding, if they be in any degree profitable to the individuals of a species, in their infinitely complex relations to other organic beings and to their physical conditions of life, will tend to the preservation of such individuals, and will generally be inherited by the offspring. The offspring, also, will thus have a better chance of surviving, for, of the many individuals of a species which are periodically born, but a small number can survive. I have called this principle by which each slight variation, if useful, is preserved, by the term Natural Selection, in order to mark its relation to man's power of selection.'

Charles Darwin, *The origin of species* (1859)

Superabundance of all necessary resources is, in most animal societies, a rare and transitory state. In winter, many insectivorous birds are forced to feed intensively from dawn to dusk to obtain sufficient food to cover their basic metabolic expenses (Gibb 1956) and starvation is probably the commonest factor controlling population density among vertebrates (Lack 1954) though, in some cases, social behaviour imposes proximal limits (see Lack 1966; Watson and Jenkins 1968; LeBoeuf 1974). The nub of Darwin's theory of natural selection was the realization that morphological and behavioural differences between individuals can affect their ability to compete for resources, and he later commended Spencer's alternative phrase, the 'survival of the fittest' as more accurate and more convenient. Subsequent generations of evolutionary theoreticians have reduced the emphasis on direct competition for resources, realizing that reproductive success (usually measured in terms of the number of offspring reared to reproductive age) is a better yardstick of evolutionary fitness; but, otherwise, the original formulation of natural selection remains the same.

During the first half of the twentieth century, the power of the theory became clear. Variation in anatomy and physiology and, later, in embryology, immunology, and behaviour which had previously puzzled scientists was interpreted. It became widely accepted that most interspecific differences must have a functional basis (Mayr 1963). By and large, this remains a reasonable assumption: the biological literature contains many cases where the presence or distribution of traits was ascribed to chance or phylogenetic inertia (the constraints imposed on genetic change by a species' evolutionary history) and which subsequently proved to be directly functional. However, the same period saw a relaxation of the rigour with which arguments were examined. Precisely because Darwin's concept of natural selection was rooted in Malthusian theory, leading him to see the process as a competitive struggle for existence, he was insistent on linking functional explanations to individual reproductive success. Later theorists, particularly when attempting to explain complex social behaviour, lacked his discipline and proposed explanations which, though apparently sensible, are difficult to state in terms of individual advantage. For example, it has been commonly argued that the function of dominance hierarchies is to restrict resource access to superior individuals and to channel mortality into physically inferior specimens:

'The social hierarchy, or "peck order" and its equivalent, is a common and important product of conventional competititon, and its function is to differentiate automatically, whenever such a situation arises, between the haves and the have-nots. For those high enough in the scale the rewards— space, food, mates—are forthcoming, but when food, for instance, is already being exploited up to the optimum level, the surplus individuals must abide by the conventional code and not remain to contest the issue if necessary to the death. It is in the interest of survival of the stock and the species that this should be so, but it ruthlessly suppresses the temporary interests of the

rejected individual who may be condemned to starve while food still abounds.'
(Wynne-Edwards 1962, p. 18. See also Chance and Jolly 1970; Kummer 1971)

Such arguments rely on the assumption that some individuals act, to their own detriment, for the good of the population or species (an assumption which Darwin specifically avoided). The necessity for formalizing arguments of this kind was realized by Wynne-Edwards (1962) though it had, in fact, been previously outlined by Carr-Saunders (1922). Wynne-Edwards was primarily concerned with attempting to explain how population density is regulated. Populations, he argued, should not over-exploit their food supplies, for this was likely to lead to reduced food yield and, thus, to lower reproductive success. As an example, he described how the potential yield of many fisheries became drastically reduced when they were over-exploited but could recover if catches were voluntarily restricted. Similar processes should apply, he argued, to animal populations, which should restrict their population density and rate of reproduction rather than endanger their food supply:

'Ideally, the habitat should be made to carry everywhere an optimum density, related to its productivity or capacity, without making any parts so crowded as to subject the inhabitants to privation or leaving other parts needlessly empty.'                                                          (Wynne-Edwards 1962, p. 4)

His book goes on to suggest that delayed reproduction and slow breeding, small clutch or litter size, sociality, dominance hierarchies, and territoriality represent adaptations to restricting population density in order to prevent over-exploitation of the food supply.

Since Wynne-Edwards' theory depended on the assumption that individuals did *not* always maximize their own reproductive success, it was necessary to extend evolutionary theory to account for this. The first paper in this section summarizes Wynne-Edwards' hypothesis: that natural selection operates between competing groups or populations of animals (as well as between individuals) and that groups whose members behave 'altruistically', suppressing their individual interests for those of the group, are likely to have higher reproductive success overall and are thus going to contribute more to the next generation than groups whose members behave selfishly, increasing their own fitness at the cost of other individuals in the group. Consider, for example, two troops of monkeys occupying separate home ranges. In troop A, all individuals breed as fast as possible, and the size of the group increases to a level where the impact of the animals on their food supplies is so heavy that it reduces food regeneration, which eventually leads to a permanent reduction of troop size. In troop B, troop members assess the availability of food and produce only as many offspring as the home range can support, maintaining the maximum population density for the environment. Monkeys belonging to troop B will contribute more offspring to subsequent generations and, assuming that such behaviour is heritable, their genotype will spread.

There is no doubt that differential survival of groups (or species) occurs and that this will have some effect on the genetic structure of populations (Maynard Smith 1976). However, it is most unlikely that group selection is as important as Wynne-Edwards originally suggested, or that it can account for the evolution of the various cases where individuals apparently act to the detriment of their own reproductive success. Attempts to explain such traits by group selection face a fundamental theoretical problem (Maynard Smith 1964). Consider the original pair of monkey troops and suppose that, in the one that over-exploited its food supply, a mutation occured in one animal which caused it to limit its reproductive output, thereby benefiting the rest of the group and increasing the average reproductive success of group members. Even though such a group might be more likely to reproduce successfully, the 'altruistic' individuals would (by definition) produce fewer offspring than other group members and selection operating *within the group* would eradicate the trait. This difficulty can be overcome if one assumes that whole groups of altruists can arise (for example, by genetic drift or the 'founder principle') thus sheltering altruists from competition with selfish individuals. However, within such groups, selfish mutants or migrants will still be at an advantage and the position is unlikely to be evolutionarily stable. This criticism strikes at the roots of Wynne-Edwards' formulation of group selection as an evolutionary force which can produce traits detrimental to their carriers but beneficient to the population or species. The view that group selection cannot do this has become orthodox and, as Maynard Smith's paper (1976) points out, invites attack. A range of models of group selection have recently been produced and they are critically reviewed in this paper: most of these models apply to extremely specialized conditions and provide no explanation for the evolution of altruistic traits in vertebrates, except in groups of related individuals (see below).

This does not totally rule out the possibility that differential survival of groups will have some evolutionary effect. For example, where traits evolve which are to the individual's *advantage*, selection at the group level may have some additive effect on selection at the individual level, though this is likely to be unimportant since group extinction is less common than individual death.

Although we have stressed that selection acts primarily on individual differences in reproductive success, this is not strictly accurate since individuals may ensure the successful replication of their genes by means other than reproduction. In particular, they may assist close relatives (who share a large proportion of their genes) to reproduce. Though this idea was anticipated by Darwin (1859) and realized by both Fisher (1930) and Haldane (1953), it was first developed by W. D. Hamilton in 1963, in the paper included in this section.

To an individual, the selective advantage of assisting a relative will depend on three factors: the cost to its own fitness, the benefit to the relative, and the average proportion of genes shared by the two animals (the degree of relatedness ($r$)). For the action to be selectively favoured, the cost to the individual ($c$) must be less than the increment to the recipient's fitness (i.e. the benefit $b$) multiplied by the degree of relatedness. In his paper, Hamilton formulates the same proposition in a different order by stating that the ratio of benefit to cost ($k$) must exceed the reciprocal of the coefficient of relatedness ($1/r$). Methods for calculating co-efficients of relatedness are outlined by Dawkins (1976) and discussed in detail by Jacquard (1974). Modifications to the mathematical basis of Hamilton's theory have been discussed in several recent papers (e.g. Orlove 1975; West-Eberhard 1975), but no major developments have occured.

Hamilton used the principle in further papers (Hamilton 1964, 1972; see also Trivers and Hare 1976) to explain why facultative sterility has evolved in many species of social Hymenoptera (ants, bees, and wasps). In these animals, unfertilized eggs develop into males, fertilized eggs into females. Since the former are unfertilized, they are haploid, having only one set of chromosomes all of which they inherit from their mother. In contrast, females are diploid, inheriting one set of chromosomes from their mother and another from their father. Sibling females (workers) will all inherit the same genes from their father and consequently 50% of their genes will be identical. The other half of their genes will be inherited from their mother. Since she is diploid they inherit one half of her genes, and will (on average) have half of these in common. Thus they will be related to each other by 75% (50% from having the same father and 25% from having the same mother), in contrast to most sexual animals where both parents are diploid and full siblings (on average) have only 50% of their genes in common. To the individual workers, female siblings thus contain a higher proportion of their own genes (75%) than potential offspring (50%) and for this reason, Hamilton argues, it should be selectively advantageous for workers to help their mother produce more siblings than to reproduce themselves. (See also Trivers and Hare, 1976). One weakness of this argument (Alexander 1974) is that facultative sterility also occurs in some groups of termites where males are diploid and workers are related to each other by 50%.

Hamilton called his extension of fitness to include that of relatives 'inclusive fitness' but it is more widely referred to as 'kin selection', a name suggested by Maynard Smith in 1964. The use of this term has the danger that the process might be considered as an alternative to natural selection. This is a fallacy, as Dawkins (1976) has pointed out, offering the alternative term 'gene selection' to emphasize that the product of natural selection is the differential survival of genes. This, in turn, has the disadvantage that it implies that selection operates directly on genes when it really operates on individuals, producing differences in gene frequencies. 'Inclusive fitness' is probably still the best term, though it is often convenient to use 'kin selection' to refer to increments in fitness gained by assisting relatives.

In the third section, we shall consider a variety of cases of altruism in animal societies which, like the evolution of facultative sterility in the social Hymenoptera, probably represent cases where it is to an individual's advantage to assist its relatives to breed. One final point concerning inclusive fitness needs to be made here. As we have already mentioned, if individuals are related to each other, a gene responsible for altruistic behaviour can spread through a population, eventually producing inter-group differences in reproductive success. Several workers (see D. S. Wilson 1974; E. O. Wilson 1975) have produced valid models of 'group selection' which depend on group members being closely related to each other. Should such a process be called group selection or kin selection? E. O. Wilson (1975) argues that 'pure kin and pure interdemic (group) selection are two poles at the end of a gradient of selection on ever enlarging nested sets of related individuals', while Maynard Smith (p. 23) emphasizes that group selection must depend on differential extinction of groups and that evolutionary processes dependent on individuals being related to each other should be regarded as kin selection. Since it is important to distinguish clearly between explanations of altruistic behaviour which require an individual to act to the detriment of its own inclusive fitness and those which require it to act to the detriment of its personal reproductive success while not reducing its inclusive fitness, Maynard Smith's position is preferable.

# Intergroup selection in the evolution of social systems

V. C. WYNNE-EDWARDS†

In a recent book[1] I advanced a general proposition which may be summarized in the following way. (1) Animals, especially in the higher phyla, are variously adapted to control their own population densities. (2) The mechanisms involved work homeostatically, adjusting the population density in relation to fluctuating levels of resources; where the limiting resource is food, as it most frequently is, the homeostatic system prevents the population from increasing to densities that would cause over-exploitation and the depletion of future yields. (3) The mechanisms depend in part on the substitution of conventional prizes, namely, the possession of territories, homes, living space and similar real property, or of social status as the proximate objects of competition among the members of the group concerned, in place of the actual food itself. (4) Any group of individuals engaged together in such conventional competitition automatically constitutes a society, all social behaviour having sprung originally from this source.

In developing the theme it soon became apparent that the greatest benefits of sociality arise from its capacity to override the advantage of the individual members in the interests of the survival of the group as a whole. The kind of adaptations which make this possible, as explained more fully here, belong to and characterize social groups as entities, rather than their members individually. This in turn seems to entail that natural selection has occurred between social groups as evolutionary units in their own right, favouring the more efficient variants among social systems wherever they have appeared, and furthering their progressive development and adaptation.

The general concept of intergroup selection is not new. It has been widely accepted in the field of evolutionary genetics, largely as a result of the classical analysis of Sewall Wright.[2-4] He has expressed the view that 'selection between the genetic systems of

† Reprinted with permission from the author and *Nature, Lond.* **200**,(1963). Copyright © 1963 by Macmillan Journals Ltd.

local populations of a species . . . has been perhaps the greatest creative factor of all in making possible selection of genetic systems as wholes in place of mere selection according to the net effects of alleles'.[4] Intergroup selection has been invoked also to explain the special case of colonial evolution in the social insects.[5-7]

In the context of the social group a difficulty appears, with selection acting simultaneously at the two levels of the group and the individual. It is that the homeostatic control of population density frequently demands sacrifices of the individual; and while population control is essential to the long-term survival of the group, the sacrifices impair fertility and survivorship in the individual. One may legitimately ask how two kinds of selection can act simultaneously when on fundamental issues they are working at cross purposes. At first sight there seems to be no easy way of reconciling this clash of interests; and to some people consequently the whole idea of intergroup selection is unacceptable.

Before attempting to resolve the problem it is necessary to fill in some of the background and give it clearer definition. The survival of a local group or population naturally depends among other things on the continuing annual yield of its food resources. Typically, where the tissues of animals or plants are consumed as food, persistent excessive pressure of exploitation can rather quickly overtax the resource and reduce its productivity, with the result that yields are diminished in subsequent years. This effect can be seen in the overfishing of commercial fisheries which is occurring now in different parts of the world, and in the over-grazing of pastoral land, which in dry climates can eventually turn good grassland into desert. Damage to the crop precedes the onset and spread of starvation in the exploiting population, and may be further aggravated by it. The general net effect of over-population is thus to diminish the carrying-capacity of the habitat.

In natural environments undisturbed by man this kind of degradation is rare and exceptional: the normal evolutionary trend is in the other direction, towards building up and sustaining the productivity of the habitat at the highest attainable level. Predatory animals do not in these conditions chronically depress their stocks of prey, nor do herbivores impair the regeneration of their food-plants. Many animals, especially among the larger vertebrates (including man), have themselves virtually no predators or parasites automatically capable of disposing of a population surplus if it arises.

It is the absence or undependability of external destructive agencies that makes it valuable, if not in some cases mandatory, that animals should be adapted to regulate their own numbers. By so doing, the population density can be balanced around the optimum level, at which the highest sustainable use is made of food resources.

Only an extraordinary circumstance could have concealed this elementary conclusion and prevented our taking it immediately for granted. Some eight or more thousand years ago, as neolithic man began to achieve new and greatly enhanced levels of production from the land through the agricultural revolution, the homeostatic conventions of his hunting ancestors, developed there as in other primates to keep population density in balance with carrying-capacity, were slowly and imperceptibly allowed to decay. We can tell this from the centrally important place always occupied by fertility-limiting and functionally similar conventions in the numerous stoneage cultures which persisted into modern times.[8] Since these conventions disappeared nothing has been acquired in their place: growing skills in resource development have, except momentarily, always outstripped the demands of a progressively increasing population; there has consequently been no effective natural selection against a freely expanding economy. So far as the regulation of numbers is concerned, the human race provides a spectacular exception to the general rule.

A secondary factor, tending to obscure the almost universal powers possessed by animals for controlling their numbers, is our everyday familiarity with insect and other pests which appear to undergo uncontrolled and sometimes violent fluctuations in abundance. In fact, human and land-use practices seldom leave natural processes alone for any length of time: vegetation is gathered, the ground is tilled, treated or irrigated, single-species crops are planted and rotated, predators and competitors are destroyed, and the animals' regulating mechanisms are thereby, understandably, often defeated. In the comparably drastic environmental fluctuations of the polar and desert regions, similar population fluctuations occur without human intervention.

The methods by which natural animal populations curb their own increase and promote the efficient exploitation of food resources include the control of recruitment and, when necessary, the expulsion and elimination of unwanted surpluses. The individual member has to be governed by the homeostatic system even when, as commonly happens, this means his exclusion from food in the midst of apparent plenty, or detention from reproduction when others are breeding. The recruitment rate must be determined by the contemporary relation between population density and resources; under average conditions, therefore, only part of the potential fecundity of the group needs to be realized in a given year or generation.

This is a conclusion amply supported by the results of experiments on fecundity versus density in laboratory populations, in a wide variety of animals including Crustacea, insects, fish, and mammals; and also by field data from natural populations.[1] But it conflicts with the assumption, still rather widely made, that under natural selection there can be no alternative to promoting the fecundity of the individual, provided this results in his leaving a larger contribution of surviving progeny to posterity.[9] This assumption is the chief obstacle to accepting the principle of intergroup selection.

One of the most important premises of intergroup selection is that animal populations are typically self-perpetuating, tending to be strongly localized and persistent on the same ground. This is illustrated by the widespread use of traditional breeding sites by birds, fishes, and animals of many other kinds; and by the subsequent return of the great majority of experimentally marked young to breed in their native neighbourhood. It is true of non-migratory species, for example the more primitive communities of man; and all the long-distance two-way migrants that have so far been experimentally tagged, whether they are birds, bats, seals or salmon, have developed parallel and equally remarkable navigating powers that enable them to return precisely to the same point, and consequently preserve the integrity of their particular local stock. Isolation is normally not quite complete, however. Provision is made for an element of pioneering, and infiltration into other areas; but the gene-flow that results is not commonly fast enough to prevent the population from accumulating heritable characteristics of its own. Partly genetic and partly traditional, these differentiate it from other similar groups.

Local groups are the smallest racial units capable of continuous existence for long enough to undergo evolutionary differentiation. In the course of generations some die out; others survive, and have the opportunity to spread out new or vacated ground as it becomes available, themselves subdividing as they grow. In so far as the successful ones take over the habitat left vacant by the unsuccessful, the groups are in a relation of passive competititon. Their survival or extinction is partly a matter of chance, arising from various forms of *force majeure*, including secular changes in the environment; for the rest it is determined, in general terms, by heritable qualities of fitness.

Gene-frequencies within the group may alter as time passes, through gene-flow, drift (the Sewall Wright effect), and selection at the individual level. Through the latter, adaptations to local conditions may accumulate. Population fitness, however, depends on something over and above the heritable basis that determines the success as individuals of a continuing stream of independent members. It becomes particularly clear in relation to population homeostatis that social groups have highly important adaptive characteristics in their own right.

When the balance of a self-regulating population is disturbed, for example by heavy accidental mortality, or by a change in food-resource yields, a restorative reaction is set in motion. If the density has dropped below the optimum, the recruitment rate may be increased in a variety of ways, most simply by drawing on the reserve of potential fecundity referred to earlier, and so raising the reproductive output. Immigrants appearing from surrounding areas can be allowed to remain as recruits also. If the density has risen too high, aggression between individuals may build up to the point of expelling the surplus as emigrants; the reproductive rate may drop; and mortality due to social stress (and in some species cannibalism) may rise. These are typical examples of density-dependent homeostatic responses.

Seven years of investigation of the population ecology of the red grouse (*Lagopus scoticus*) near Aberdeen, by the Nature Conservancy Unit of Grouse and Moorland Ecology, have revealed many of these processes at work.[10] Their operation in this case depends to a great extent on the fact that individual members of a grouse population living together on a moor, even of the same sex, are not equal in social status. Some of the cock birds are sufficiently dominant to establish themselves as territory owners, parcelling out the ground among them and holding sway over it, with a varying intensity of possessiveness, almost the whole year round. During February-June their mates enjoy the same established status.

In the early dawn of August and September mornings, after a short and almost complete recess, the shape of a new territorial pattern begins to he hammered out. In most years this quickly identifies a large surplus of males, old and young, which are not successful in securing any part of the ground, and consequently assume a socially inferior status. They are grouped with the hens at this stage as unestablished birds; and day by day their security is so disturbed by the dawn aggressive stress that almost at once some begin to get forced out, never to return. By about 8 a.m. each day the passion subsides in all but the most refractory territorial cocks, after which the moor reverts to communal ground on which the whole population can feed freely for the rest of the day. As autumn wears on and turns to winter, the daily period of aggression becomes fiercer and lasts longer; birds with no property rights have to feed at least part of the time on territorial ground defended by owners that may at any moment chase them off. More and more are driven out altogether; and since they can rarely find a safe nook to occupy elsewhere in the neighbourhood, they become outcasts, and are easily picked up by hawks and foxes, or succumb to malnutrition. Females are included among those expelled; but about February the remaining ones begin to establish marital attachments; and at the same time, quite suddenly, territories are vigorously defended all day. Of the unestablished birds still present in late winter, some achieve promotion by filling the gaps caused by casualties in the establishment. Some may persist occasionally until spring; but unless a cock holds a territory exceeding a minimum threshold capacity, or a hen becomes accepted by a territorially qualified mate, breeding is inhibited.

Territories are not all of uniform size, and on average the largest are held by the most dominating cocks. More important still, the average territory size changes from year to year, thus varying the basic population density, apparently in direct response to changes in productivity of the staple food-plant, heather (*Calluna vulgaris*). As yet this productivity has been estimated only by subjective methods; but significant mutual correlations have been established between annual average values for body-weight of adults, adult survival, clutch-size, hatching success, survival of young, and, finally, breeding density the following year. As would be expected with changing densities, the size of the autumn surplus, measured by the proportions of unestablished to established birds, also varies from year to year.

There are increasing grounds for concluding that this is quite a typical organization, so far as birds are concerned, and that social stratification into established and unestablished members, particularly in the breeding season, is common to many other species, and other classes of animals. In different circumstances the social hierarchy may take the form of a more or less linear series or peck-order. Hierarchies commonly play a leading part in regulating animal populations; not only can they be made to cut off any required proportion of the population from breeding, but also they have exactly the same effect in respect of food when it is in short supply. According to circumstances, the surplus tail of the hierarchy may either be disposed of or retained as a non-participating reserve if resources permit.

It is not necessary here to explore in detail the elaborate patterns of behaviour by which the social hierarchy takes effect. The processes are infinitely varied and complex, though the results are simple and functionally always the same. The hierarchy is essentially an overflow mechanism, continuously variable in terms of population pressure on one hand, and habitat capacity on the other. In operation it is purely conventional, prescribing a code of behaviour. When a more dominant individual exerts sufficient aggressive pressure, usually expressed as threat although frequently in some more subtle and sophisticated form, his subordinates yield, characteristically without physical resistance or even demur. It may cost them their sole chance of reproduction to do so, if not their lives. The survival of the group depends on their compliance.

This has been taken as an example to illustrate one type of adaptation possessed by the group, transcending the individuality of its members. It subordinates the advantage of particular members to the advantage of the group; its survival value to the latter is clearly very great. The hierarchy as a system of behaviour has innumerable variants in different species and different phyla, analogous to those of a somatic unit like the nervous or vascular system. Like them, it must have been subject to adaptive evolutionary change through natural selection; yet it is essentially an 'organ' of a social group, and has no existence if the members are segregated.

A simple analogy may possibly help to bring out the significance of this point. A football team is made up of players individually selected for such qualities as skill, quickness, and stamina, material to their success as members of the team. The survival of the team to win the championship, however, is determined by entirely distinct criteria, namely, the tactics and ability it displays in competition with other teams, under a particular code of conventions laid down for the game. There is no difficulty in distinguishing two levels of selection here, although the analogy is otherwise very imperfect.

The hierarchy is not the only characteristic of this kind. There are genetic mechanisms, such as those that govern the optimum balance between recombination and linkage, in which the benefit is equally clearly with the group rather than the individual. Without leaving the sphere of population regulation, however, we can find a wide range of vital parameters, the optima of which must similarly be determined by intergroup selection. Among those discussed at length in the book already cited[1] are (1) the potential life-span of individuals and, coupled with it, the generation turnover rate; (2) the relative proportions of life spent in juvenile or non-sexual condition (including diapause) and in reproduction; (3) monotely (breeding only for a single season) versus polytely; (4) the basal fecundity-level, including, in any one season, the question of one brood versus more than one.

These and similar parameters, differing from one species or class to another, are interconnected. Their combined effects are being summed over the whole population at any one time, and over many generations in any given area. It is the scale of the operation in time and space that precludes an immediate experimental test of group selection. An inference that may justifiably be drawn, however, is that maladjustment sufficient to interfere persistently with the homeostatic mechanism must either cause a progressive decline in the population or, alternatively, a chronic over-exploitation and depletion of food resources; in the end either will depopulate the locality.

There still remains the central question as to how an immediate advantage to the individual can be suppressed or overridden when it conflicts with the interests of the group. What would be the effect of selection, for example, on individuals the abnormal and socially undesirable fertility of which enabled them and their hereditary successors to contribute an ever-increasing share to future generations?

Initially, groups containing individuals like this that reproduced too fast, so that the overall recruitment rate persistently tended to exceed the death-rate, must have repeatedly exterminated themselves in the manner just indicated, by overtaxing and progressively destroying their food resources. The earliest adaptations capable of protecting the group against such recurrent disasters must necessarily have been very ancient; they may even have been acquired only once in the whole of animal phylogeny, and in this respect be comparable with such basic morphological elements as the mesoderm, or, perhaps, the coelom. Once acquired, the protective adaptations could be endlessly varied and elaborated. It is inherently difficult to reconstruct the origin of systems of this kind; but genetic mechanisms exist which could give individual breeding success a low heritability, or, in other words, make it resistant to selection. This could be relatively simply achieved, for example, if the greatest success normally attached to heterozygotes for the alleles concerned, created the stable situation characteristic of genetic homeostasis.[11]

A more complex system can be discerned, as it has developed in many of the higher vertebrates where the breeding success of individuals is very closely connected with social status. This connexion must necessarily divert an enormous additional force of selection into promoting social dominance, and penalizing the less fortunate subordinates in the population that are prevented from breeding or feeding, or get squeezed out of the habitat. Yet it is self-evident that the conventional codes under which social competition is conducted are in practice not jeopardized from this cause: selection pressure, however great, does not succeed in promoting a general recourse to deadly combat or treachery between rivals, nor does it, in the course of generations, extinguish the patient compliance of subordinates with their lot.

The reason appears to be that social status depends on a summation of diverse traits, including virtually all the hereditary and environmental factors that predicate health, vigour, and survivorship in the individual. While this is favourable to the maintenance of a high-grade breeding stock, and can result in the enhancement through selection of the weapons and conventional adornments by which social dominance is secured, dominance itself is again characterized by a low heritability, as experiments have shown. In many birds and mammals, moreover, individual status, quite apart from its genetic basis, advances progressively with the individual's age. Not only are the factors that determine social and breeding success numerous and involved, therefore, but the ingredients can vary from one successful individual to the next. A substantial part of the gene pool of the population is likely to be involved and selection for social dominance or fertility at the individual level correspondingly dissipated and ineffective, except in eliminating the sub-standard fringe.

Such methods as these which protect group adaptations, including both population parameters and social structures, from short-term changes, seem capable of preventing the rise of any hereditary tendency towards anti-social self-interest among the members of a social group. Compliance with the social code can be made obligatory and automatic, and it probably is so in almost all animals that possess social homeostatic systems at all. In at least some of the mammals, on the contrary, the individual has been released from this rigid compulsion, probably because a certain amount of intelligent individual enterprise has proved advantageous to the group. In man, as we know, compliance with the social code is by no means automatic, and is reinforced by conscience and the law, both of them relatively flexible adapations.

There appears therefore to be no great difficulty in resolving the initial problem as to how intergroup selection can override the concurrent process of selection for individual advantage. Relatively simple genetic mechanisms can be evolved whereby the door is shut to one form of selection and open to the other, securing without conflict the maximum advantage from each; and since neighbouring populations differ, not only in genetic systems but in population parameters (for example, mean fecundity[12]) and in social practices (for example, local differences in migratory behaviour in birds, or in tribal conventions among primitive men), there is no lack of variation on which intergroup selection can work.

## References

[1] Wynne-Edwards, V. C., *Animal Dispersion in Relation to Social Behaviour* (Edinburgh and London, 1962).
[2] Wright, S. (1929). *Anat. Rec.*, **44**, 287.
[3] Wright, S. (1930). *Genetics*, **16**, 97.
[4] Wright, S. (1945). *Ecology*, **26**, 415.
[5] Haldane, J. B. S. (1932). *The Causes of Evolution* (London, New York and Toronto).
[6] Sturtevant, A. H. (1938). *Quart. Rev. Biol.*, **13**, 74.
[7] Williams, G. C., and Williams, D. C. (1957). *Evolution*, **11**, 32.
[8] Carr-Saunders, A. M. (1922). *The Population Problem: A Study in Human Evolution* (Oxford Univ. Press).
[9] Lack, D. (1954). *The Natural Regulation of Animal Numbers* (Oxford Univ. Press).
[10] Jenkins, D., Watson, A., and Miller, G. R. (1963). *J. Anim. Ecol.*, **32**, 317.
[11] Lerner, I. M. (1954). *Genetic Homeostatis* (Edinburgh and London).
[12] Bagenal, T. B. (1962). *J. Mar. Biol. Assoc.*, **42**, 105.

# Group selection

J. MAYNARD SMITH†

## Introduction

The purpose of this short review is to look at some recent discussions of group selection, in particular those by E. O. Wilson (1975), D. S. Wilson (1975) and M. E. Gilpin (1975). Earlier work will be referred to only briefly; the need for a review arises because the three references given either propose a blurring of the distinction between 'group' and 'kin' selection, or suggest an importance for group selection greater than it has usually been given, or both.

The first point to establish is that the argument is quantitative, not qualitative. Group selection will have evolutionary consequences; the only question is how important these consequences have been. If there are genes which, although decreasing individual fitness, make it less likely that a group (deme or species) will go extinct, then group extinction will influence evolution. It does not follow that the influence is important enough to play the role suggested for it by some biologists.

The present phase of the debate about group selection was opened by the publication of Wynne-Edwards' 'Animal Dispersion' (1962), which applied to animals a concept first proposed by Carr-Saunders (1922) to explain human population dynamics; it ascribed to group selection a major role in the evolution of population regulation. Although his thesis has had its adherents, the orthodox response from ecologists has been to argue that the patterns of behavior he described can be explained by individual selection, and from population geneticists that the mechanism he proposes is insufficient to account for the results. It is in the nature of science that once a position becomes orthodox it should be subjected to criticism; hence the papers by D. S. Wilson (1975), Gilpin (1975), Levin and Kilmer (1974), Gadgil (1975), and others. It does not follow that, because a position is orthodox, it is wrong: hence this review.

† Reprinted with permission from the author and *Quarterly Review of Biology* **51** (1976). Copyright © 1976 by Quarterly Review of Biology.

Is the argument important? In a recent review of E. O. Wilson's *Sociobiology*, C. H. Waddington referred to group selection as 'a fashionable topic for a rather foolish controversy'. Doubtless some foolish things have been said but there is an important issue at stake. If group selection has played the role suggested by Wynne-Edwards, no one can doubt its importance. But why should it be important to argue that it has not? The reason for the vehemence with which Williams (1966, 1975), Ghiselin (1974), Lack (1966) and other opponents of group selection have argued their case is, I think, their conviction that group selection assumptions, often tacit or unconscious, have been responsible for the failure to tackle important problems. So long as we fail to distinguish group and individual selection, or assume that an explanation in terms of advantage to the species is adequate to account for the evolution of some behavior pattern or genetic process, without asking what is its effect on individual fitness, we shall make little progress.

The extent of unconscious group selectionism, particularly among ecologists, has recently been documented by Ghiselin (1974). Similar views are still widespread among ethologists. It is, however, in the study of the evolution of genetic mechanisms that group selection assumptions are most pervasive. Fisher (1930) argued that sexual reproduction owed its existence to the fact that sexual reproducing populations can evolve more rapidly than asexual ones. Darlington (1939) attempted to explain a wide range of adaptations in chromosome structure and recombination frequency in terms of inter-population selection. Since that time, the interpretation of genetic mechanisms in terms of species advantage has become almost commonplace.

Of course, a group advantage cannot be ruled out *a priori* as an explanation of the evolution of ecological adaptations, of behavior or of genetic mechanisms. But the quantitative difficulties must be faced. It is plausible, for example, to suppose that, whatever its origin, sexual reproduction is maintained in higher organisms because populations which abandon sex go extinct in competition with more rapidly evolving sexual species. This plausibility depends on the assumption that the origin of new ameiotically parthenogenetic strains is a sufficiently rare event for each such origin to be balanced by the extinction of such a strain. It is less plausible to suppose that the chiasma frequency or recombination frequency within a species is determined by group selection, because there is widespread within-species genetic variation of recombination rate, and we would therefore expect individual selection acting on this variation to outweigh any long-term effects of interspecies selection.

It is not the purpose of this review to discuss the evolution of sex or recombination. The problem has been mentioned because it illustrates particularly clearly the quantitative difficulties faced by group selection explanations. In view of these difficulties, an explanation in terms of group advantage should always be explicit, and always calls for some justification in terms of the frequency of group extinction. Wynne-Edwards' great merit is that he made the assumption explicit, and in so doing forced population geneticists to make the argument quantitative.

### The distinction between group and kin selection

The frequency of a gene in a population will be influenced not only by the effects that gene has on the survival and fertility of individuals carrying it, but also by its effects on the survival and fertility of relatives of that individual. The second kind of effect has been called 'kin selection' (Maynard Smith 1964). The effect is obvious when the 'relatives' are the children or other direct descendants of the individual; it has always been appreciated that genes improving parental care will be selected. Fisher (1930) and Haldane (1932) saw clearly that the effect would work for other relatives (e.g. sibs or cousins). In the vaguer forms of 'inter-family selection' the idea goes back to Darwin's *Origin of Species.*

Wright (1945) discussed the evolution of altruistic traits in terms of 'interpopulation selection'. Despite his use of this phrase, his model, which is sketched rather than fully worked out, would fall under the heading of kin selection rather than group selection if the distinction suggested below is accepted.

Attempts to apply the idea of kin selection in detail to the evolution, first of social insects and later of other animal societies, originated with Hamilton (1963, 1964), and are the central theme of E. O. Wilson (1975).

With the almost simultaneous publication of Hamilton's papers and Wynne-Edwards' book, it seemed desirable to draw a clear distinction between the two processes of kin and group selection, and to coin a term for the former (Maynard Smith 1964). It is always difficult to draw unambiguous distinctions in biology, but it is often valuable to try. I still think that the attempt was on the right lines, although in retrospect I can see that there is one essential feature of group selection, namely group extinction, which I failed to emphasize, although it was present in my mathematical model.

The basic distinction made concerned the population structure required for the two processes. For kin selection (as described by the first sentence in this section) it is necessary that relatives live close to one another, but it is not necessary (although it may be favorable) that the population be divided into reproductively isolated groups. All that is essential for kin selection is that relatives live close to one another, so that an animal's behavior can influence the survival or fecundity of its relatives.

For group selection, the division into groups which are partially isolated from one another is an essential feature. If group selection is to be responsible for the establishment of an 'altruistic' gene, the groups must be small, or must from time to time be re-established by a few founders. This is because in a large group there is no way in which a new 'altruistic' gene can be established. If a new mutant is to be established in a large group, then it must increase the fitness of individuals carrying it, or more precisely, it must increase their 'inclusive fitness' relative to other members of the group (Hamilton 1964); if this is so, group selection is not needed to explain its spread. Small group size is not needed for the maintenance of an altruistic gene, only for its establishment. The involvement of genetic drift as an essential feature of group selection has been queried by D. S. Wilson (1975); I will return to this point later.

What I should have said in my 1964 paper, but did not, is that the extinction of some groups and the 'reproduction' of others are essential features of evolution by group selection. If groups are the units of selection, then they must have the properties of variation, multiplication, and heredity required if natural selection is to operate on them. In a finite universe, multiplication implies death. Group selection could operate for a short time on differences in group reproduction, without group extinction, but in the long run evolution by group selection requires group extinction just as evolution by individual selection requires individual death.

The relevance of group reproduction and extinction can best be illustrated by a partly imaginary example. Anubis baboons live in troops; females remain in their natal troop whereas males must move to another troop before breeding (Packer 1975). Suppose that a gene were to arise which caused females to help other females in their troop to raise their offspring (in fact, females do help one another to defend their young against males). This gene would increase in frequency in the whole population—more males would be produced by a troop whose females carried the gene, and these males would transmit the gene to other troops. Despite the existence of troops, I would regard this as an example of kin selection. The gene would not increase in frequency unless the females in a troop were related to one another. But the increase does not depend on small group size and hence on genetic drift, nor does it depend on group reproduction or extinction; indeed, it would work in much the same way if there were no groups, provided that females bred close to where they were born, and that males dispersed before breeding.

There will doubtless be cases in which the distinction is difficult to draw. E. O. Wilson (1975) has argued that 'pure kin and pure interdemic selection are the two poles at the end of a gradient of selection on ever enlarging nested sets of related individuals'. In similar vein, D. S. Wilson (1975) writes 'the traditional concepts of group and individual selection are seen as two extremes of a continuum'. The disagreement between us, if there is one, is one of semantics and the strategy of research rather than of fact. In the history of ideas as in taxonomy, there are lumpers and splitters; I am a splitter. I think we would do best to draw as clear a distinction as we can between different processes. I welcome E. O. Wilson's distinction between inter-demic and inter-species selection, although both are forms of group selection.

Why do I think it desirable to sharpen rather than to blur the distinction between different modes of selection? Ultimately, the importance or otherwise of these different modes can only be decided by comparing different species, and asking whether particular traits, such as altruistic, prudent, or selfish behaviour, are associated with particular population structures. For this to be a meaningful enterprise, we must be clear about what are the relevant features of population structure. In particular, we must be clear whether our theory asserts that the evolution of a trait requires the existence of groups, or merely that relatives should be neighbors.

To sum up, 'group' selection should be confined to processes that require the existence of partially isolated groups which can reproduce and which go extinct; the origin of new altruistic traits requires that the groups be small or be founded by a few individuals. Kin selection can operate whenever relatives live close to one another, and hence can influence one another's chances of survival and reproduction; they may or may not live in groups.

## Models of group selection

Maynard Smith (1964) proposed a model, the 'Haystack' model, whereby an 'altruistic' gene, which is eliminated by its 'selfish' allele from mixed groups, can increase in frequency by group selection. In this model all groups necessarily become extinct since they depend on a transitory resource (a 'haystack'), but groups containing only the altuistic gene produce more potential founders of new groups. In most other models of group selection, extinction has been made stochastic rather than necessary, with the probability of extinction being lower in groups with a higher frequency of the altruistic allele. Thus, in my model, groups differed in 'fecundity,' whereas in most subsequent models they differ in 'viability'.

More general models for the spread of altruistic alleles brought about by differential group extinction were proposed by Levins (1970) and by Boorman and Levitt (1973). Although different in detail, these models have in common that they confirm the logical possibility of group selection, but show that the population structure in time and space required for its operation is of a kind which may be rather infrequent in practice. Levin and Kilmer (1974) reached similar conclusions on the basis of a computer-simulated model. Genetically effective deme sizes of less than 25 and usually closer to 10 were required, and the rate of gene flow by migration could not be greater than 5 per cent per generation.

Gilpin (1975) presents a very thoroughly worked out model. His proposed mechanism is clearly a case of group selection as defined above. It is of particular interest because the altruistic trait he considers, 'prudence' on the part of predators leading them not to over-exploit their food supply, is precisely the one for which Wynne-Edwards proposed group selection. The new features which Gilpin has introduced into the argument relate to the dynamics of predator–prey systems. For a wide class of models describing the interaction between a predator and prey (the model of Rosenzweig and MacArthur, 1963, is a familiar example), a small change in the properties of the predator can cause a large change in system behavior. Suppose we start from a state of stable coexistence of predator and prey, and suppose individual selection to be acting to improve the hunting ability of the predators. The equilibrium will change gradually, with the predator numbers increasing and the prey decreasing, until a critical point is reached. If hunting ability continues to increase beyond this point, the equilibrium becomes unstable and the system passes into a stable oscillation (a limit cycle). With further improvement in hunting ability, the amplitude of the oscillation rises until, with finite populations, chance extinction of one or both species would ensue.

Gilpin imagines an environment composed of patches. A predator population may contain only genotypes $aa$, in which case stable coexistence is possible, or only $AA$, which leads to rapid extinction, or some mixture of $aa$, $Aa$ and $AA$. A mixed population evolves by individual selection to the fixation of allele $A$, and hence to extinction. In his model, Gilpin allows for migration between patches, and genetic drift within them. He shows that for a reasonably wide range of parameter values the altruistic allele $a$ can maintain itself against $A$.

The problem remains whether the range of parameter values includes cases likely to correspond to natural predator–prey systems. The very complexity of Gilpin's model makes this difficult to decide. Fortunately, however, I believe it possible to replace Gilpin's model of group selection (and many other models) by a much simpler model, and to identify a single parameter which will determine the fate of an altruistic gene. In Gilpin's model, there are at any moment three kinds of patch: 'empty' patches, E, containing no predators (although they may contain prey); 'altruistic' patches, A, containing only $aa$ predators; and 'selfish' patches, S, containing at least some $AA$ or $Aa$ predators. The types of patch, and the possible transitions between them, are shown in Fig. 1.

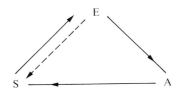

Fig. 1. A model of group selection.
For explanation see text.

The transition from E to S is shown by a broken arrow because, although it is permitted in Gilpin's model, it would not be in some other models of group selection.

The main model is formally similar to one analyzed by Maynard Smith (1974), but with a different biological interpretation. In that model, E represented empty patches; A, patches with prey individuals only; and S, patches containing both predator and prey and destined to become empty. The model is also formally similar to an unpublished model by Ivar Heuch of the University of Bergen for the evolution of the sex ratio in the butterfly *Acraea encedon*. In this species there is a 'driving' Y* chromosome, such that XY* females have only daughters. In a panmictic population this should lead to species extinction. Heuch suggests that the species survives because of its patchy distribution. Again, E represents empty patches; A represents patches with only normal XY females; S represents 'infected' patches containing at least one XY* female and destined for extinction. In effect, Heuch's model is one of balance between gametic selection for Y* and group selection for Y.

In such models, it is easy to see that the fate of the system depends on the single parameter M, which is the expected number of successful 'selfish' emigrants from an S patch during the lifetime of the patch, from the moment it becomes 'infected' to the time it becomes extinct. By a 'selfish' emigrant is meant an $AA$ or $Aa$ predator in Gilpin's model, a predator in my model, or an XY* female in Heuch's model. By 'successful' is meant that the migrant established itself and leaves descendants in a new patch (either E or A). If $M > 1$ the frequency of S patches will increase. The continued existence of the system requires $M < 1$.

In Gilpin's model, this means that during the history of a patch, from the moment when it is 'infected' by an $Aa$ or $AA$ immigrant to the time of its extinction, there must on the average be less than one successful $Aa$ or $AA$ migrant to another patch. It is thus fairly easy to understand the features of his model which make for the maintenance of prudent $aa$ genotypes. They are small population size per patch, little migration, and rapid extinction of a patch once infected. Rapid extinction arises from the nature of the predator–prey dynamics (a plausible assumption which is Gilpin's main innovation), and from the assumption that a single gene substitution in a small population is sufficient to tip the balance from stability to extinction.

It is hard to say how often the condition $M < 1$ will hold in nature. But it may be easier to decide whether group selection is a plausible mechanism in particular cases by concentrating on this inequality than by considering all the details and parameter values of the full model.

Finally, before leaving this topic, it is helpful to note the analogy between the equilibrium condition, $M = 1$, and the well-known principle of individual selection, that the mutational load equals the mutation rate. The latter principle states that each individual harmful mutation must be balanced by a death; the former that each socially harmful infection must be balanced by an extinction.

A quite different approach to the problem of group selection is taken by D. S. Wilson (1975). He considers a population which, although breeding at random, is divided for some part of the life cycle into 'trait groups,' within which altruistic or selfish inter-actions can take place. For a pair of alleles $A$ and $a$, such that $a$ determines an altruistic behavior which decreases the individual's chance of survival but increases the survival chances of all other members of the group, he shows that $a$ can increase in frequency, but only if the between-trait-group genetic variance is greater than random—i.e. than would be expected if the members of trait groups were a random sample of the whole population.

This is an interesting result, but seems to me to refer to kin selection and not to group selection. Indeed, there is no reason other than mathematical convenience why he should consider trait 'groups'. His argument would work equally well if there were no discontinuities in spatial distribution provided that there was a genetic similarity between neighbors. The obvious reason why the members of a trait group might resemble one another genet-ically is that they are relatives. Wilson's result is then fully in accord with Hamilton's (1964) assertion that an altruistic gene will increase in frequency only if $k > 1/r$, where $r$ is the coefficient of relationship between donor and recipient, and $k$ the ratio of the gain of fitness of the recipient to the loss of fitness of the donor. Clearly, if $r = 0$ altruism will not increase in frequency.

Wilson argues from his model that genetic drift is not needed for group selection; since I do not think he is discussing group selection, I do not find his argument convincing. In fact, it is not clear why Wilson does not regard his model as one of kin selection. For example, he writes 'consider a situation in which larval insects are deposited into the trait-groups by adult females. The larvae upon hatching intermix within the trait-group, and so do not fall under the traditional concept of kin selection.' This is precisely the type of situation I intended to fall under the term kin selec-tion. If, for example, animals behave with an equal degree of altruism to all their 'neighbors', or to all fellow members of their 'trait-group', and if *on average* animals are related to their neigh-bors, then I would regard this as an example of kin selection. It is not a necessary feature of kin selection that an animal should distinguish different degrees of relationship among its neighbors, and behave with greater altruism to the more closely related.

Apart from relatedness, there are other possible reasons why members of a trait-group might resemble one another genetically. If individuals of like genotype tend to associate together, either because they are attracted to one another, or because they are attracted by common features of the environment, or because they are the survivors of a common selective force, then altruism can be selected for. Before invoking this mechanism, however, it is important to remember that if an altruistic allele $a$ is to replace a selfish allele $A$, then the members of a trait-group must resemble one another *at that locus*. If they are not related in the normal sense, this would require that the altruistic locus have pleiotropic effects determining association. Thus, these other reasons for genetic similarity between neighbors seem likely to be unimportant compared with identity by descent.

## Conclusions

It is useful to distinguish as sharply as possible the processes of 'kin' and 'group' selection. The term group selection should be confined to cases in which the group (deme or species) is the unit of selection. This requires that groups be able to 'reproduce,' by splitting or by sending out propagules, and that groups should go extinct. The origin of an altruistic trait (but not its maintenance) requires that the groups be small, or that new groups be established by one or a few founders. Kin selection requires only that relatives should live close to one another. The division of the population into groups, either permanently or for part of the life cycle, may favor the operation of kin selection but is not a necessary feature.

Group selection can maintain 'altruistic' alleles—i.e. alleles which reduce individual fitness but increase the fitness of groups carrying them. The conditions under which this can happen are stringent, so that the main debate concerns whether the process has had evolutionarily important consequences. The main function of models is to indicate the circumstances in which group selection can be important. For one large class of models, it can easily be seen that the condition for the maintenance of an altruistic allele by group selection is $M < 1$, where $M$ is the expected number of successful migrants carrying the selfish allele produced by a group during the whole period from the moment when it was first 'infected' by a selfish immigrant to the time of its extinction.

It follows that the features favoring group selection are small group size, low migration rates, and rapid extinction of groups infected with the selfish allele. The ultimate test of the group selection hypothesis will be whether populations having these characteristics tend to show 'self-sacrificing' or 'prudent' behaviour more commonly than those which do not.

## References

Boorman, S. A. and Levitt, P. R. (1973). Group selection at the boundary of a stable population. *Theoret. Pop. Biol.,* 4, 85–128.

Carr-Saunders, A. M. (1922). *The Population Problem: A Study in Human Evolution.* Clarendon Press, Oxford.

Darlington, C. D. (1939). *The Evolution of Genetic Systems.* University Press, Cambridge.

Fisher, R. A. (1930). *The Genetical Theory of Natural Selection.* Clarendon Press, Oxford.

Gadgil, M. (1975). Evolution of social behavior through interpopulation selection. *Proc. Nat. Acad. Sci. U.S.A.,* 72, 1199–1201.

Ghiselin, M. T. (1974). *The Economy of Nature and the Evolution of Sex.* University of California Press, Berkeley.

Gilpin, M. E. (1975). *Group Selection in Predator-Prey Communities.* Princeton University Press.

Haldane, J. B. S. (1932). *The Causes of Evolution.* Longmans, Green, London.

Hamilton, W. D. (1963). The evolution of altruistic behavior. *Am. Natur.,* 97, 354–56.

–– (1964). The genetical theory of social behaviour, I, II. *J. Theoret. Biol.,* 7, 1–52.

Lack, D. (1966). *Population Studies of Birds.* University Press, Oxford.

Levin, B. R. and Kilmer, W. L. (1974). Interdemic selection and the evolution of altruism: a computer simulation study. *Evolution,* 28, 527–545.

Levins, R. (1970). Extinction. In M. Gerstenhaber (ed.), *Some Mathematical Problems in Biology,* p. 77–107. The American Mathematical Society, Providence.

Maynard Smith, J. (1964). Group selection and kin selection. *Nature, Lond.* 201, 1145–1147.

–– (1974). *Models in Ecology.* University Press, Cambridge.

Packer, C. (1975). Male transfer in olive baboons. *Nature, Lond.* 255, 219–220.

Rosenzweig, M. L., and MacArthur, R. H. (1963). Graphical representation and stability conditions of predator-prey interactions. *Am. Natur.,* 97, 209–223.

Williams, G. C. (1966). *Adaptation and Natural Selection: A Critique of Some Current Evolutionary Thought.* University Press, Princeton.

–– (1975). *Sex and Evolution.* University Press, Princeton.

Wilson, D. S. (1975). A theory of group selection. *Proc. Nat. Acad. Sci. U.S.A.,* 72, 143–146.

Wilson, E. O. (1975). *Sociobiology.* Harvard University Press, Cambridge.

Wright, S. (1945). *Tempo and Mode in Evolution:* A critical review. *Ecology,* 26 415–419.

Wynne-Edwards, V. C. (1962). *Animal Dispersion in Relation to Social Behaviour.* Oliver & Boyd, Edinburgh.

# The evolution of altruistic behavior

W. D. HAMILTON†

It is generally accepted that the behavior characteristic of a species is just as much the product of evolution as the morphology. Yet the kinds of behavior which can be adequately explained by the classical mathematical theory of natural selection are limited. In particular this theory cannot account for any case where an animal behaves in such a way as to promote the advantages of other members of the species not its direct descendants at the expense of its own. The explanation usually given for such cases and for all others where selfish behavior seems moderated by concern for the interests of a group is that they are evolved by natural selection favoring the most stable and co-operative groups. But in view of the inevitable slowness of any evolution based on group selection compared to the simultaneous trends that can occur by selection of the classical kind, based on individual advantage, this explanation must be treated with reserve so long as it remains unsupported by mathematical models. Fisher in the second edition of *The Genetical Theory of Natural Selection* (1958) rejects almost all explanations based on 'the benefit of the species' (e.g. p. 49). Sewall Wright (1948) in a summary of population genetics shows explicitly that a general advantage conferred on a group cannot alter the course of intragroup selection. This point is very adverse to the following model of Haldane (1932, p. 208) which seemed to offer a possibility for the evolution of altruism. Haldane supposed an increment to a group fitness (and therefore to group rate of increase) proportional to its content of altruistic members and showed that there could be an initial numerical increase of a gene for altruism provided the starting gene frequency was high enough and the individual disadvantage low enough compared to the group advantage conferred. He concluded that genetical altruism could show some advance in populations split into 'tribes' small enough for a single mutant to approximate the critical frequency. He did not,

† Reprinted with permission from the author and *American Naturalist* **97** (1963).

however, sufficiently emphasize that ultimately the gene number must begin to do what the gene frequency tends to do, *ex hypothesi*, from the very first; namely, to decrease to zero. The only escape from this conclusion (as Haldane hints) would be some kind of periodic reassortment of the tribes such that by chance or otherwise the altruists became re-concentrated in some of them.

There is, however, an extension of the classical theory, generalizing that which serves to cover parental care and still have the generation as the time-unit of progress, which does allow to a limited degree the evolution of kinds of altruism which are not connected with parental care.

As a simple but admittedly crude model we may imagine a pair of genes g and G such that G tends to cause some kind of altruistic behavior while g is null. Despite the principle of 'survival of the fittest' the ultimate criterion which determines whether G will spread is not whether the behavior is to the benefit of the behaver but whether it is to the benefit of the gene G; and this will be the case if the average net result of the behavior is to add to the gene-pool a handful of genes containing G in higher concentration than does the gene-pool itself. With altruism this will happen only if the affected individual is a relative of the altruist, therefore having an increased chance of carrying the gene, and if the advantage conferred is large enough compared to the personal disadvantage to offset the regression, or 'dilution', of the altruist's genotype in the relative in question. The appropriate regression coefficient must be very near to Sewall Wright's Coefficient of Relationship r provided selection is slow. If the gain to a relative of degree r is k times the loss to the altruist, the criterion for positive selection of the causative gene is

$$k > \frac{1}{r}.$$

Thus a gene causing altruistic behavior towards brothers and sisters will be selected only if the behavior and the circumstances are generally such that the gain is more than twice the loss; for half-brothers it must be more than four times the loss; and so on. To put the matter more vividly, an animal acting on this principle would sacrifice its life it is could thereby save more than two brothers, but not for less. Some similar illustrations were given by Haldane (1955).

It follows that altruistic behavior which benefits neighbors irrespective of relationship (such as the warning cries of birds) will only arise when (a) the risk or disadvantage involved is very slight, and (b) the average neighbor is not too distantly related.

An altruistic action which adds to the genotype-reproduction (inclusive of the reproduction of identical genes procured in the relative) by one per cent is not so strongly selected as a one per cent advantage in personal reproduction would be; for it involves also an addition of unrelated genes which are in the ratio of the existing gene-pool—an addition which must be larger the more distant the relationship.

A multi-factorial model of inheritance, which is doubtless more realistic, does not invalidate the above criterion, and provided fitness is reckoned in terms of 'inclusive' genotype-reproduction, and the dilution due to unrelated genes is allowed for, the classical treatment of dominance and epistasis can be followed closely.

Fisher in 1930 (1958, p. 177 *et seq.*) offered an explanation of the evolution of aposematic colouring based on the advantage to siblings of the self-sacrifice, by its conspicuousness, of a distasteful larva, and his discussion contains what is probably the earliest precise statement concerning a particular case of the principle presented above: 'The selective potency of the avoidance of brothers will of course be only half as great as if the individual itself were protected; against this is to be set the fact that it applies to the whole of a possibly numerous brood.' It would appear that he did not credit the possibility that selection could operate through the advantage conferred on more distant relatives, even though these must in fact tend to be still more numerous in rough inverse proportion to the coefficient of relationship.

This discussion by Fisher is one of the few exceptions to his general insistence on individual advantage as the basis of natural selection; the other notable exception from the present point of view is his discussion of putative forces of selection in primitive human societies (p. 261 *et seq.*).

## References

Fisher, R. A. (1958). *The genetical theory of natural selection.* 2nd ed. Dover Publ. Inc., New York.
Haldane, J. B. S. (1932). *The causes of evolution.* Longmans Green, London.
—— (1955). Population genetics. *New Biology* 18, 34—51.
Wright, S. (1948). Genetics of populations. *Encyclopaedia Britannica* (1961 printing) 10, 111—112.

# Section 2

# Reproductive strategies

# Section 2

# Reproductive strategies

## Introduction

We have argued in the first section that natural selection produces individuals which are adapted to maximize the spread of their own genes—sometimes through relatives but more commonly through their own offspring. This suggests that variation in breeding strategies is likely to be functional, and that differences in life history variables (e.g. litter or clutch size, egg or foetal weight, breeding frequency, growth rate, parental care, dispersal, and lifespan) should be related to differences in ecology. The idea dates from Darwin, but this area of research has developed rapidly since MacArthur and Wilson (1967) distinguished between selection favouring high fecundity and rapid development ($r$ selection) and selection favouring lower fecundity and slower development through increased competitive ability of the offspring ($K$ selection). $r$ selection is likely to be strongest where competition with conspecifics has little effect on the survival of offspring (for example, in strongly fluctuating environments where density-independent factors produce frequent changes in population density), $K$ selection where competition is intense (as is likely in areas of high environmental stability). Pianka's 1970 review (see p. 45) is important because it extends predictions concerning the ecological, physiological, and morphological correlates of $r$ and $K$ selection.

Such differences in selection are most commonly used to interpret the adaptive significance of major differences in life history variables between vertebrates and invertebrates. However, as Pianka notes, all species can be placed somewhere on an $r$–$K$ continuum and the theory may also be useful for interpreting adaptive differences between related species (E. O. Wilson 1975). Though Lack did not use this evolutionary distinction when he wrote his important review of avian breeding systems (Lack 1968, see p. 238), many species differences which he described appear to fit predictions on this basis. And among the mammals, the rodents include groups with relatively high survival rate and low reproductive capacity (e.g. the Scuiridae, Zapodidae, and the Heteromyidae) and species with relatively low survival rates and high reproductive capacity (the Microtinae and Muridae) (French, Stoddart, and Bobek 1975). The theory may also prove useful in interpreting differences in social behaviour (see E. O. Wilson 1975). However, it is already clear that associations between breeding biology and habitat stability will not be simple since, within environments, niche differences will constrain reproductive parameters in different ways.

A variety of other papers have considered functional aspects of species differences in relation to life history variables. Many of these are reviewed by Wilson and Bossert (1971) and by E. O. Wilson (1975). They include body size (Schoener 1969; Bourlière 1975), lifespan (Medawar 1957; Lewontin 1965; Hamilton 1966; Murphy 1968; Gadgil and Bossert 1970), and reproductive rate (Lewontin 1965; Holgate 1967; Williams 1966a, b; Gadgil and Bossert 1970; Fagen 1972; Emlen 1970).

There is one fundamental aspect of reproductive biology whose function we have not considered: sexuality. Many species of plants and invertebrates reproduce asexually, either through growth (fission or vegetative propogation) or specialized structures (spores, eggs, seeds). Because of its widespread distribution, sexuality must have evolved at some early stage in evolution though its function is not obvious: an asexually reproducing organism passes on all its genes to each offspring, while a sexual reproducer only passes on about 50% (except in some rare cases).

There are three theories concerning the evolutionary advantages of sexuality (see Maynard Smith 1975, p. 185). The first was proposed by Fisher (1930) and relies on group selection. The essence of Fisher's argument is that the sexual process permits faster genetic change in a population: through recombination, it allows favourable mutants to aggregate, whereas, in parthenogenetically reproducing populations, there is no way in which they can become combined in a single individual. A model which formalizes this theory (Crow and Kimura 1965) has been criticized by Maynard Smith (1968). The second theory has been propounded

by G. C. Williams (1966a, 1975). Williams argues that environmental heretogeneity selects for individuals producing genetically
variable offspring. Consider a forest in which a mature tree dies
and leaves enough room for a single replacement. Surrounding
trees will provide germinating seedlings which will compete for
the space. By reproducing sexually, a tree provides many alternative
genotypes, some of which will be more efficient competitors than
others. Williams likens the situation to a lottery where, sexual
reproducers have many tickets, each with a different number,
while asexual reproducers have several tickets all with the same
number. In this case, the 50% dilution of the parents genotype
associated with sexuality may be overridden by the increased
chance of offspring survival, though this will only occur if the
environment is variable on a generation-to-generation basis
(Maynard Smith 1968). The third theory, suggested by Price and
M. Williams and described by Maynard Smith (1975, p. 188) considers a sexually reproducing population in which a parthenogenetic
mutant arises. Initially, the mutant will spread because carriers will
not dilute their genotype but this advantage may disappear if
asexual lineages are unable to track continuing environmental
change.

   If sexuality is adaptive, then further questions are worth asking.
First, why is the sex ratio at birth approximately equal in most
species? Second, why do the sexes differ? The answer to the first
question was given by Fisher (1930), who argued that if one sex
is under-represented in a population (at less than 50%), individuals
of that sex will, on average, leave more offspring than members of
the other sex because all offspring must have a mother and a
father. Consequently, genes predisposing parents to produce an
excess of the rarer sex will increase in frequency until an equilibrium is reached when both sexes are equally common. In fact,
Fisher showed that parents should not necessarily produce equal
*numbers* of offspring of each sex, but should *invest equally* in each
sex (see also MacArthur 1965; Hamilton 1967; Leigh 1970). However, in some cases selection will not favour a 1:1 sex ratio for
other reasons. For example, if related males are likely to compete
with each other, selection should produce a sex ratio in favour of
females (Hamilton 1967). And in some cases, physiological controls
which permit the sex ratio to be varied according to the probable
breeding success of males and females have evolved (Wilson 1975,
p. 317; Trivers and Willard 1973).

   Understanding the adaptive significance of sex differences is of central importance in sociobiology and the rest of this section is devoted to the topic. The basis of current theory in this area was supplied by Darwin in *The descent of Man, and selection in relation to sex* (1871) who argued that sex differences could arise either through competition between members of the same sex ('intra-sexual selection') or through preferences by members of one sex for choosing mates of a particular kind ('epigamic selection'; Huxley 1938).

Sex differences produced by intrasexual selection are ultimately attributable to variation in reproductive potential between the sexes. The paper by Trivers (1972) reproduced here deals with this topic at length and we shall only describe his conclusions cursorily. Trivers introduces the concept of 'investment' (see p. 55), arguing that parents should distribute their available resources to that number of offspring or relatives which will maximize their own inclusive fitness. For example, it is presumably advantageous for *r*-selected species to invest little in each of a large number of off-spring , and for *K*-selected ones to invest heavily in a small number. Investments are measured in terms of the extent to which they reduce the individual's future ability to invest.

The central point of Trivers' paper is that the sexes invest differently. In most vertebrates, females invest more in individual progeny than males. For example, in mammals the male invests a small amount of time and energy during courtship and copulation, whereas the female invests much more heavily during pregnancy and lactation. The number of offspring that a female can produce (either per unit time or during her life) is therefore smaller than the number that can be produced by a male. Consequently, males can increase their reproductive success by mating with several females whereas females usually cannot increase theirs by mating with several males. As a result, competition between males for access to females should be stronger than competition between females for access to males, and traits favouring success in breeding competi-tion (e.g. weapon development, fighting skills, body size, and harem or territorial defence) should be more developed in males. This will not always be the case. In monogamous animals, the number of offspring that a male can produce (unless he mates out-side the pair bond and successfully 'cheats' another male into rearing his offspring) will be the same as the number his mate can produce and (assuming that the sex ratio is approximately equal) sex differences in breeding competition should be minimal; differ-ences in competitive equipment also should be small in so far as they are the product of intrasexual competition. As Trivers points out, this tends to be the case and recent studies of grouse (Wiley 1974) and primates (Clutton-Brock and Harvey 1977) have demonstrated significant relationships between sexual dimorphism in body weight and the average number of females mated by each breeding male. In polyandrous species, such as the jacana (*Jacana spinosa*) where females lay several clutches of eggs which are subsequently incubated by several different males, we might expect breeding competition to be stronger among females than among males and the usual sex differences to be reversed. This is generally the case in such species and females tend to be larger and more strongly territorial than males (Jenni 1974). However, not all species in which females are larger than males are polyandrous (Lack 1968; Ralls 1976) and there are probably several potential causes for this trait.

From the previous argument, it should be clear that polygynous breeding will usually be to a male's advantage but may be to the disadvantage of the female, whose offspring may receive less paternal investment than if their mother had bred monogamously (see Orians 1969). The distribution of polygyny in birds and mammals is discussed in the paper by Orians reproduced on p. 115. He argues that, in those species where polygynous breeding occurs, it must also be to the *female's* advantage to breed polygynously (despite the apparent reduction in paternal investment), otherwise she would select a monogamous mate. This may occur in species where successful males can defend resources necessary for reproduction (e.g. nest sites, food supplies), forcing subordinates to breed in areas of low resource availability. In such a situation, it might be to the female's advantage to breed polygynously with a successful mate rather than monogamously with a subordinate. (It is worth noticing that, by this manoeuvre, males may still lower the *average* reproductive success of females: Wilson 1971). A similar argument is used by Clutton-Brock and Harvey (1977) to explain the distribution of monogamy in primates: they suggest that monogamy develops in those species where selection for territoriality is strong and where the male cannot defend resources adequate for more than one female and her offspring.

As Orians points out, polyandry should be less common than polygyny. There are no known cases of polyandry in mammals, while in birds it is rare and irregularly distributed (Jenni 1974). It is difficult to find any single explanation of its distribution, though it is often associated with an imbalanced sex ratio in favour of males (e.g. Maynard Smith and Ridpath 1972; Jenni 1974). Why the latter should evolve is obscure.

The relative roles of the male and female in caring for the young or eggs differ widely among vertebrates. In mammals, the mother's contribution to rearing is generally greater than the father's while, in birds, either sex may incubate and/or care for the young though maternal care is more common. Among fish, the pattern is reversed and paternal care is at least as common as maternal (Breder and Rosen 1966). To provide a functional explanation of variation in parental care, it is necessary to understand the costs and benefits to each sex of deserting its mate after egg laying or parturition. Trivers (1972) argues that, at any point during rearing, the sex which has invested less may be tempted to desert, thus forcing the partner to complete the rearing alone while it mates again. He goes on to suggest (see p. 64) that it may be adaptive for males to desert females since the female's initially heavy investment is likely to commit her to further investment (though this is less likely to occur late in the rearing process when the male's investment has increased). One possible strategy that the female can adopt is to require heavy male investment prior to mating (e.g. by playing 'coy' and extending the process of courtship), thus increasing the risks of subsequent desertion to the male.

The first argument has been successfully contested by Dawkins and Carlisle (1976) on the grounds that it is not the amount which an individual has already invested but the *future* investment necessary to bring the offspring to maturity which should determine the parent's willingness to indulge in additional investment. However, predictions made on this basis are similar to those made by Trivers, who has proved (on this point) to be largely right for the wrong reasons (though see p. 98). A more important objection (see Dawkins and Carlisle 1976) is that in order to predict which sex will desert we should consider the partner's probable reaction to desertion. Maynard Smith's (1977) paper (see p. 98) argues that one parent should desert if a single parent can raise more than half as many offspring as two. Where this is the case, the young will normally be reared by the male if the female has invested so much in eggs that she cannot afford to invest further or if there is an excess of males in the population; they will be reared by the female if males have a better chance of remating than females or if there is an excess of females. However, in many situations, either male or female desertion may be favoured and it is not possible to predict firmly which sex will rear the young.

Epigamic selection has received less attention than intrasexual selection, though it is usual to attribute the evolution of behavioural and morphological traits associated with courtship displays to this form of selection (Darwin 1871; Selander 1972). The importance of selection for mate quality will depend on the breeding system and is likely to differ between the sexes. In monogamous species, both males and females should carefully select their partners while, in polygynous and promiscuous species, females are more likely to do so than males. Trivers (1972, see p. 83) reviews the criteria on which females might judge their prospective mates.

An alternative explanation, based on female choice, of many sex differences has been proposed by Zahavi (1975). His argument, which is sometimes called the 'handicap principle', is that females should mate selectively with males of the highest genetic 'quality'. Identification of 'quality' may initially depend on some functional trait, such as large muscles (Dawkins 1976). However, this is likely to lead to selection for males which 'cheat' by developing supranormal muscles, out of proportion to their actual strength, and thus to counter-acting selection pressures on females against using this trait as a measure of quality. The one trait which cannot be faked, Zahavi argues, is a handicap such as a long tail or striking colouration. Males which have survived to maturity despite possessing the trait must be of high 'quality' and the offspring of females which mate with them should show high fitness: the females because they will inherit 'high quality' genes, the males because they will carry the handicap and be selected by the next generation of females. The theory faces the problem that handicaps must be sex-limited, otherwise they would also be expressed in female offspring. It is unsatisfactory in several respects: there are many traits (especially behavioural ones) which are inevitably related to physical prowess and which would not form a permanent handicap; no explanation of the degree of sex differences is provided; and there is no evidence that females mate selectively with males showing particularly well-developed handicaps. Lastly, although attempts have been made (Davis and O'Donald 1976; Maynard Smith 1976), it has not yet proved possible to produce a formal model for the evolution of such handicaps.

One final point should be made concerning the evolution of sex differences. Although we have assumed that they are the product of sexual selection, they may also develop if selection favours divergence of male and female feeding niches. There is evidence that dimorphism in several bird species has probably evolved for this reason (Selander 1966, 1972). In such cases we can predict that sex differences should be associated with variation in feeding behaviour and should be more marked in feeding apparatus (e.g. bill shape, tooth form) than in traits associated with fighting ability (see Selander 1972).

# On r– and K–selection

E. R. PIANKA†

Dobzhansky (1950) proposed that natural selection in the tropics operates in a fundamentally different way than it does in temperate zones. He argued that much of the mortality in the temperate zones is *relatively* independent of the genotype (and phenotype) of the organism concerned, and has little to do with the size of the population. Traditional examples of mass winter kills of fish and sparrows are extremes of this sort. Dobzhansky reasoned that in the relatively constant tropics, most mortality is more directed, generally favoring those individuals with better competitive abilities. Thus, in the temperate zones selection often favors high fecundity and rapid development, whereas in the tropics lower fecundity and slower development could act to increase competitive ability. By putting more energy into each offspring and producing fewer total offspring, overall individual fitness is increased. The small clutch sizes characteristic of many tropical birds are consistent with Dobzhansky's hypothesis. Dobzhansky's ideas were framed in terms too specific to reach the general ecological audience and have gone more or less unnoticed until fairly recently.

MacArthur and Wilson (1967) coined the terms '$K$-selection' and '$r$-selection' for these two kinds of selection, which are clearly not restricted to the tropics and the temperate zones ($K$ refers to carrying capacity and $r$ to the maximal intrinsic rate of natural increase [$r_{max}$]). To the extent that these terms invoke the much overused logistic equation, they are perhaps unfortunate. However, it is clear that there are two opposing kinds of selection, which usually have to be compromised. Certainly, no organism is completely '$r$-selected' or completely '$K$-selected', but all must reach some compromise between the two extremes. Fisher (1930) stated the problem as follows: 'It would be instructive to know not only

† Reprinted with permission from the author and *American Naturalist,* **104** (1970). Copyright © 1970 by the University of Chicago Press. All rights reserved.

by what physiological mechanism a just apportionment is made between the nutriment devoted to the gonads and that devoted to the rest of the parental organism, but also what circumstances in the life-history and environment would render profitable the diversion of a greater or lesser share of the available resources towards reproduction.' Fisher's early statement is one of the clearest on the idea of the budgeting of time, matter, and energy into these components (see also Williams 1966; Gadgil and Bossert 1970). Presumably, natural selection will usually act to maximize the amounts of matter and energy gathered per unit time; the problem is to understand how this matter and energy are partitioned among somatic and reproductive tissues and activities.

We can visualize an $r$–$K$ continuum, and a particular organism's position along it. The $r$-endpoint represents the quantitative extreme—a perfect ecologic vacuum, with no density effects and no competition. Under this situation, the optimal strategy is to put all possible matter and energy into reproduction, with the smallest practicable amount into each individual offspring, and to produce as many total progeny as possible. Hence $r$-selection leads to high productivity. The $K$-endpoint represents the qualitative extreme—density effects are maximal and the environment is saturated with organisms. Competition is keen and the optimal strategy is to channel all available matter and energy into maintenance and the production of a few extremely fit offspring. Replacement is the keynote here. $K$-selection leads to increasing efficiency of utilization of environmental resources. Table 1 summarizes some of the correlates of the $r$- and $K$-selected extremes. However, even in a perfect ecologic vacuum, as soon as the first organism replicates itself, there is the possibility of some competition, and natural selection should favour compromising a little more toward the $K$-endpoint. Hence, as an ecologic vacuum is filled, selection will shift a population from the $r$- toward the $K$-endpoint (MacArthur and Wilson 1967).

One whole class of terrestrial organisms (vertebrates) seems to be relatively $K$-selected, while another large group (most insects, and perhaps terrestrial invertebrates in general) apparently is relatively $r$-selected. There are, of course, a few exceptions among both the insects (e.g. 17-year cicada) and the vertebrates (some amphibians). Nevertheless, many of the correlates of the two kinds of selection listed in Table 1 are characteristic of these two natural groups of terrestrial organisms. Presumably perennial and annual plants differ in a similar way. Aquatic organisms do not appear to

Table 1. Some of the correlates of $r$- and $K$-selection

|  | $r$-Selection | $K$-Selection |
| --- | --- | --- |
| Climate | Variable and/or unpredictable: uncertain | Fairly constant and/or predictable: more certain |
| Mortality | Often catastrophic, nondirected, density-independent | More directed, density-dependent |
| Survivorship | Often Type III (Deevey 1947) | Usually Type I and II (Deevey 1947) |
| Population size | Variable in time, nonequilibrium; usually well below carrying capacity of environment; unsaturated communities or portions thereof; ecologic vacuums; recolonization each year | Fairly constant in time, equilibrium; at or near carrying capacity of the environment; saturated communities; no recolonization necessary |
| Intra- and interspecific competition | Variable, often lax | Usually keen |
| Relative abundance | Often does not fit MacArthur's broken stick model (King 1964) | Frequently fits the MacArthur model (King 1964) |
| Selection favours | 1. Rapid development 2. High $r_{max}$ 3. Early reproduction 4. Small body size 5. Semelparity: single reproduction | 1. Slower development, greater competitive ability 2. Lower resource thresholds 3. Delayed reproduction 4. Larger body size 5. Iteroparity: repeated reproductions |
| Length of life | Short, usually less than 1 year | Longer, usually more than 1 year |
| Leads to | Productivity | Efficiency |

obey this generalization; fish, in particular, span the range of the $r$–$K$ continuum.

While the existence of this dichotomy can (and doubtless will) be challenged, there are a number of reasons to believe it is real. For instance, Fig. 1 shows the distribution of body lengths for a wide variety of terrestrial insects and vertebrates from eastern North America. Body length is far from the most desirable measurement to demonstrate the polarity, but the strong inverse correlation of $r_{max}$ with generation time and body size (below) suggests that, when frequency distributions for the former two parameters become available, a similar bimodality will emerge.

Some interesting and important generalizations have been made concerning the relationships between body size and $r_{max}$ and generation time. Bonner (1965) plotted the logarithm of body length

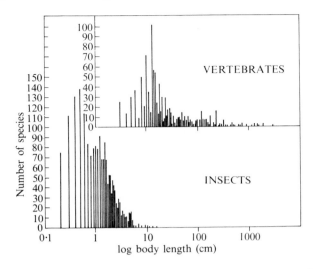

Fig. 1. Frequency distributions of body lengths for many species of terrestrial insects and vertebrates from eastern North America (data taken from numerous field guides and taxonomic accounts).

against the logarithm of generation time for a wide variety of organisms and demonstrated a strong, nearly linear, positive correlation. Smith (1954) demonstrated a similar, but inverse, correlation on a log-log plot of $r_{max}$ versus generation time. Smith pointed out that $r_{max}$ measures the rate at which an organism can fill an ecologic vacuum (at zero density); it is therefore one of the better indices of an organism's position on the $r$–$K$ continuum. He also noted that $r_{max}$ was inversely related to body size (i.e. that larger organisms are usually more $K$-selected than smaller ones). Now, $r_{max}$ is inversely related to generation time $T$ by the following formula: $r_{max} = \log_e R_0/T$, where $R_0$ is the net reproductive rate. From this equation it can be seen that variations in $R_0$ alter $r_{max}$ only slightly compared to changes in $T$. Hence a hyperbola is expected. When Smith's data and that of Howe (1953) are plotted on arithmetic axes (Fig. 2), the data do fit a strong hyperbola, the arms of which appear to represent the two 'natural' groups of organisms referred to earlier. Cole (1954) presents a similar plot showing the reduction in $r_{max}$ associated with delayed reproduction (or increased generation time).

Over a long period of time, the average $r$ of any stable population must equal zero. Smith (1954) pointed out that organisms like *Escherichia coli* with very high $r_{max}$ values are much further from

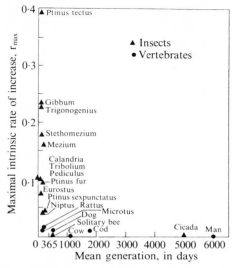

Fig. 2. Arithmetic plot of $r_{max}$ against generation time, using data of Howe (1953) and Smith (1954).

realizing their full 'biotic potential' than organisms with low $r_{max}$ values, such as man. He suggested that $r_{max}$ represents a rate of increase necessary for the population to persist in the face of its inevitable 'environmental resistance'. Furthermore, he argued that the resistance of the environment must exactly balance the biotic potential for an actual *r* equal to zero. Hence $r_{max}$ also measures environmental resistance. Increased body size no doubt reduces environmental resistance in many ways; an obvious one is that a larger organism has fewer potential predators. Larger organisms are also better buffered from changes in their physical environment. But the dividends of reduced environmental resistance attained by increased body size are offset by loss of biotic potential. Nonetheless, the fossil record does show frequent evolutionary trends toward larger size (Newell 1949).

There are three major cycles in nature—daily, lunar, and annual. Most organisms live longer than a day. Lunar cycles are probably of relatively little importance to the majority of terrestrial organisms. Thus it is unlikely that either daily or lunar rhythms underlie the apparent bimodality of these organisms. However, the annual cycle is another case. To survive, any resident organism with a generation time greater than a year must be adapted to cope with a full range of physical and biotic conditions which prevail at a given locality. An organism which lives less than a year encounters only a portion of the total annual range of conditions. The latter usually survive the harshest periods by forming resting eggs or pupae. Their population sizes vary with the particular local climate conditions and the length of time during which $r$ is positive (for descriptions and discussions of such organisms, see Andrewartha and Birch 1954). Because longer-lived larger organisms are better buffered from environmental vicissitudes, their population sizes do not vary as much as those of smaller, shorter-lived organisms. Furthermore, presumably their competitive relationships are also more predictable and constant.

The attainment of a generation time exceeding a year may well be a threshold event in the evolutionary history of a population. When perenniality is reached, there are substantially fewer environmental 'surprises' and a rather drastic shift from $r$- to $K$-selection. I conclude that there may well be a natural bimodality in environmental resistance.

There are, of course, other possible reasons for the apparent bimodality of relatively $r$- and relatively $K$-selected organisms in nature. It could be simply historic accident that body sizes and generation times of insects and vertebrates are largely non-overlapping. It is also quite possible that an either/or strategy is usually superior to some compromise. This might be a simple consequence of the hyperbolic inverse relationship between $r_{max}$ and $T$.

## References

Andrewartha, H. G., and Birch, L. C. (1954). *The distribution and abundance of animals.* University of Chicago Press.

Bonner, J. T. (1965). *Size and cycle: an essay on the structure of biology.* Princeton University Press.

Cole, L. C. (1954). The population consequences of life history phenomena. *Quart. Rev. Biol.* 29, 103–37.

Deevey, E. S. (1947). Life tables for natural populations of animals. *Quart. Rev. Biol.* 22, 283–314.

Dobzhansky, T. (1950). Evolution in the tropics. *Amer. Sci.* 38, 209–21.

Fisher, R. A. (1930). *The genetical theory of natural selection.* 2nd revised ed. Dover, New York, 1958.

Gadgil, M., and Bossert, W. H. (1970). Life historical consequences of natural selection. *Amer. Natur.* 104, 1–24.

Howe, R. W. (1953). Studies on beetles of the family Ptinidae. VIII. The intrinsic rate of increase of some ptinid beetles. *Ann. Appl. Biol.* 40, 121–34.

King, C. E. (1964). Relative abundance of species and MacArthur's model. *Ecology* 45, 716–27.

MacArthur, R. H., and Wilson, E. O. (1967). *The theory of island biogeography.* Princeton University Press.

Newell, N. D. (1949). Phyletic size increase—an important trend illustrated by fossil invertebrates. *Evolution* 3, 103—24.

Smith, F. E. (1954). Quantitative aspects of population growth, 277—294. In E. Boell (ed.), *Dynamics of growth processes.* Princeton University Press.

Williams, G. C. (1966). *Adaptation and natural selection.* Princeton University Press.

# Parental investment and sexual selection

ROBERT L. TRIVERS†

## Introduction

Charles Darwin's (1871) treatment of the topic of sexual selection was sometimes confused because he lacked a general framework within which to relate the variables he perceived to be important: sex-linked inheritance, sex ratio at conception, differential mortality, parental care, and the form of the breeding system (monogamy, polygyny, polyandry, or promiscuity). This confusion permitted others to attempt to show that Darwin's terminology was imprecise, that he misinterpreted the function of some structures, and that the influence of sexual selection was greatly overrated. Huxley (1938), for example, dismisses the importance of female choice without evidence or theoretical argument, and he doubts the prevalence of adaptations in males that decrease their chances of surviving but are selected because they lead to high reproductive success. Some important advances, however, have been achieved since Darwin's work. The genetics of sex has now been clarified, and Fisher (1958) has produced a model to explain sex ratios at conception, a model recently extended to include special mechanisms that operate under inbreeding (Hamilton 1967). Data from the laboratory and the field have confirmed that females are capable of very subtle choices (for example, Petit and Ehrman 1969), and Bateman (1948) has suggested a general basis for female choice and male–male competition, and he has produced precise data on one species to support his argument.

This paper presents a general framework within which to consider sexual selection. In it I attempt to define and interrelate the key variables. No attempt is made to review the large, scattered literature relevant to sexual selection. Instead, arguments are presented on how one might *expect* natural selection to act on the sexes, and some data are presented to support these arguments.

## Variance in reproductive success

Darwin defined sexual selection as (1) competition within one sex for members of the opposite sex and (2) differential choice by members of one sex for members of the opposite sex, and he pointed out that this usually meant males competing with each other for females and females choosing some males rather than others. To study these phenomena one needs accurate data on differential reproductive success analysed by sex. Accurate data on female reproductive success are available for many species, but similar data on males are very difficult to gather, even in those species that tend towards monogamy. The human species illustrates this point. In any society it is relatively easy to assign accurately the children to their biological mothers, but an element of uncertainty attaches to the assignment of children to their biological fathers. For example, Henry Harpending (personal communication) has gathered biochemical data on the Kalahari Bushmen showing that about two per cent of the children in that society do not belong to the father to whom they are commonly attributed. Data on the human species are, of course, much more detailed than similar data on other species.

To gather precise data on both sexes Bateman (1948) studied a single species, *Drosophila melanogaster,* under laboratory conditions. By using a chromosomally marked individual in competition with individuals bearing different markers, and by searching for the markers in the offspring, he was able to measure the reproductive success of each individual, whether female or male. His method consisted of introducing five adult males to five adult female virgins, so that each female had a choice of five males and each male competed with four other males.

Data from numerous competition experiments with *Drosophila* revealed three important sexual differences: (1) Male reproductive success varied much more widely than female reproductive success. Only four per cent of the females failed to produce any surviving offspring, while 21 per cent of the males so failed. Some males, on the other hand, were phenomenally successful, producing nearly three times as many offspring as the most successful female. (2) Female reproductive success did not appear to be limited by ability to attact males. The four per cent who failed to copulate were apparently courted as vigorously as those who did copulate. On the other hand, male reproductive success was severely limited by ability to attract or arouse females. The 21 per cent who failed to reproduce showed no disinterest in trying to copulate, only an inability to be accepted. (3) A female's reproductive success did not increase much, if any, after the first copulation and not at all after the second; most females were uninterested in copulating more than once or twice. As shown by genetic markers in the offspring, males showed an almost linear increase in reproductive success with increased copulations. (A corollary of this finding is that males tended not to mate with the same female twice.) Although these results were obtained in the laboratory, they may apply with even greater force to the wild, where males are not limited to five females and where females have a wider range of males from which to choose.

Bateman argued that his results could be explained by reference to the energy investment of each sex in their sex cells. Since male *Drosophila* invest very little metabolic energy in the production of a given sex cell, whereas females invest considerable energy, a male's reproductive success is not limited by his ability to produce sex cells but by his ability to fertilize eggs with these cells. A female's reproductive success is not limited by her ability to have her eggs fertilized but by her ability to produce eggs. Since in almost all animal and plant species the male produces sex cells that are tiny by comparison to the female's sex cells, Bateman (1948) argued that his results should apply very widely, that is, to 'all but a few very primitive organisms, and those in which monogamy combined with a sex ratio of unity eliminated all intra-sexual selection'.

Good field data on reproductive success are difficult to find, but what data exist, in conjunction with the assumption that male reproductive success varies as a function of the number of copulations,[†] support the contention that in all species, except those mentioned below in which male parental care may be a limiting resource for females, male reproductive success varies more than female reproductive success. This is supported, for example, by data from dragonflies (Jacobs 1955), baboons (DeVore 1965), common frogs (Savage 1961), prairie chickens (Robel 1966), sage grouse (Scott 1942), black grouse (Koivisto 1965), elephant seals (LeBoeuf and Peterson 1969), dung flies (Parker 1970a), and some

[†]Selection should favor males producing such an abundance of sperm that they fertilize all a female's available eggs with a single copulation. Furthermore, to decrease competition among offspring, natural selection may favor females who prefer single paternity for each batch of eggs (see Hamilton 1964). The tendency for females to copulate only once or twice per batch of eggs is supported by data for many species (see, for example, Bateman 1948, Savage 1961, Burns 1968 but see also Parker 1970b).

anoline lizards (Rand 1967 and Trivers, in preparation, discussed below). Circumstantial evidence exists for other lizards (for example, Blair 1960; Harris 1964) and for many mammals (see Eisenberg 1965). In monogamous species, male reproductive success would be expected to vary as female reproductive success, but there is always the possibility of adultery and differential female mortality (discussed below) and these factors should increase the variance of male reproductive success without significantly altering that of the female.

### Relative parental investment

Bateman's argument can be stated in a more precise and general form such that the breeding system (for example, monogamy) as well as the adult sex ratio become functions of a single variable controlling sexual selection. I first define parental investment as *any investment by the parent in an individual offspring that increases the offspring's chance of surviving (and hence reproductive success) at the cost of the parent's ability to invest in other off-spring.* So defined, parental investment includes the metabolic investment in the primary sex cells but refers to any investment (such as feeding or guarding the young) that benefits the young. It does not include effort expended in finding a member of the opposite sex or in subduing members of one's own sex in order to mate with a member of the opposite sex, since such effort (except in special cases) does not affect the survival chances of the resulting offspring and is therefore not *parental* investment.

Each offspring can be viewed as an investment independent of other offspring, increasing investment in one offspring tending to decrease investment in others. I measure the size of a parental investment by reference to its negative effect on the parent's ability to invest in other offspring: a large parental investment is one that strongly decreases the parent's ability to produce other offspring. There is no necessary correlation between the size of parental investment in an offspring and its benefit for the young. Indeed, one can show that during a breeding season the benefit from a given parental investment must decrease at some point or else species would not tend to produce any fixed number of offspring per season. Decrease in reproductive success resulting from the negative effect of parental investment on *nonparental* forms of reproductive effort (such as sexual competition for mates) is excluded from the measurement of parental investment.

In effect, then, I am here considering reproductive success as if the only relevant variable were parental investment.

For a given reproductive season one can define the total parental investment of an individual as the sum of its investments in each of its offspring produced during that season, and one assumes that natural selection has favored the total parental investment that leads to maximum net reproductive success. Dividing the total parental investment by the number of individuals produced by the parent gives the typical parental investment by an individual per offspring. Bateman's argument can now be reformulated as follows. Since the total number of offspring produced by one sex of a sexually reproducing species must equal the total number produced by the other (and assuming the sexes differ in no other way than in their typical parental investment per offspring)† then the sex whose typical parental investment is greater than that of the opposite sex will become a limiting resource for that sex. Individuals of the sex investing less will compete among themselves to breed with members of the sex investing more, since an individual of the former can increase its reproductive success by investing successively in the offspring of several members of the limiting sex. By assuming a simple relationship between degree of parental investment and number of offspring produced, the argument can be presented graphically (Fig. 1). The potential for sexual competition in the sex investing less can be measured by calculating the ratio of the number of offspring that sex optimally produces (as a function of parental investment alone, assuming the opposite sex's investment fixed at its optimal value) to the number of offspring the limiting sex optimally produces (L/M in Fig. 1).

*What governs the operation of sexual selection is the relative parental investment of the sexes in their offspring.* Competition for females usually characterizes males because males usually invest almost nothing in their offspring. Where male parental investment per offspring is comparable to female investment one would expect male and female reproductive success to vary in similar ways and for female choice to be no more discriminating than male choice (except as noted below). Where male parental investment strongly exceeds that of the female (regardless of which sex invests more in

† In particular, I assume an approximately 50/50 sex ratio at conception (Fisher 1958) and no differential mortality by sex, because I later derive differential mortality as a function of reproductive strategies determined by sexual selection. (Differential maturation, which affects the adult sex ratio, can also be treated as a function of sexual selection.) For most species the disparity in parental investment between the sexes is so great that the assumptions here can be greatly relaxed.

Per reproductive episode.

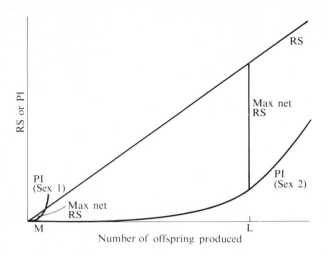

Fig. 1. Reproductive success (RS) and decrease in future reproductive success resulting from parental investment (PI) are graphed as functions of the number of offspring produced by individuals of the two sexes. At M and L the net reproductive success reaches a maximum for sex 1 and sex 2 respectively. Sex 2 is limited by sex 1 (see text). The shape of the PI curves need not be specified exactly.

the sex cells) one would expect females to compete among themselves for males and for males to be selective about whom they accept as a mate.

Note that it may not be possible for an individual of one sex to invest in only part of the offspring of an individual of the opposite sex. When a male invests less per typical offspring than does a female but more than one-half what she invests (or vice versa) then selection may not favor male competition to pair with more than one female, if the offspring of the second female cannot be parcelled out to more than one male. If the net reproductive success for a male investing in the offspring of one female is larger than that gained from investing in the offspring of two females, then the male will be selected to invest in the offspring of only one female. This argument is graphed in Fig. 2 and may be important to understanding differential mortality in monogamous birds, as discussed below.

Fisher's (1958) sex ratio model compares the parental expenditure (undefined) in male offspring with that in female offspring and suggests energy and time as measures of expenditure. Restate-

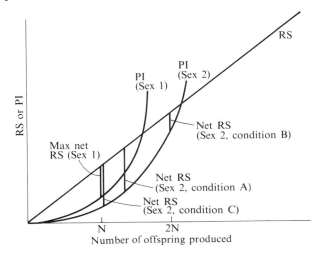

Fig. 2. RS and PI as functions of the number of offspring produced for two sexes. Sex 2 invests per typical offspring more than half of what sex 2 invests. Condition A: maximum net RS for a member of sex 1 assuming he can invest in any number of offspring between N and 2N. Condition B: net RS assuming member of sex 2 invests in N offspring. If member of sex 2 must invest in an integral multiple of N offspring, natural selection favors condition C.

ments of Fisher's model (for example, Kolman 1960; Willson and Pianka 1963; T. Emlen 1968; Verner 1965; Leigh 1970) employ either the undefined term, parental expenditure, or the term energy investment. In either case the key concept is imprecise and the relevant one is parental investment, as defined above. Energy investment may often be a good approximation of parental investment, but it is clearly sometimes a poor one. An individual defending its brood from a predator may expend very little energy in the process but suffer a high chance of mortality; such behavior should be measured as a large investment, not a small one as suggested by the energy involved.

## Parental investment patterns

Species can be classified according to the relative parental investment of the sexes in their young. In the vast majority of species, the male's only contribution to the survival of his offspring is his sex cells. In these species, female contribution clearly exceeds male and by a large ratio.

A male may invest in his offspring in several ways. He may provide his mate with food as in baloon flies (Kessel 1955) and some other insects (Engelmann 1970), some spiders, and some birds (for example, Calder 1967; Royama 1966; Stokes and Williams, 1971). He may find and defend a good place for the female to feed, lay eggs or raise young, as in many birds. He may build a nest to receive the eggs, as in some fish (for example, Morris 1952). He may help the female lay the eggs, as in some parasitic birds (Lack 1968). The male may also defend the female. He may brood the eggs, as in some birds, fish, frogs, and sala-manders. He may help feed the young, protect them, provide opportunities for learning, and so on, as in wolves and many monogamous birds. Finally, he may provide an indirect group benefit to the young (such as protection), as in many primates. All of these forms of male parental investment tend to decrease the disparity in investment between male and female resulting from the initial disparity in size of sex cells.

To test the importance of relative parental investment in con-trolling sexual selection one should search for species showing greater male than female parental investment (see Williams 1966, pp. 185–6). The best candidates include the Phalaropidae and the polyandrous bird species reviewed by Lack (1968). In these species, a female's parental investment ends when she lays her eggs; the male alone broods the eggs and cares for the young after hatching. No one has attempted to assess relative parental invest-ment in these species, but they are striking in showing very high male parental investment correlating with strong sex role reversal: females tend to be more brightly colored, more aggressive and larger than the males, and tend to court them and fight over them. In the phalaropes there is no evidence that the females lay multiple broods (Höhn 1967; Johns 1969), but in some polyandrous species females apparently go from male to male laying successive broods (for example, Beebe 1925; see also Orians 1969). In these species the female may be limited by her ability to induce males to care for her broods, and female reproductive success may vary more than male. Likewise, high male parental investment in pipefish and seahorses (syngnathidae) correlates with female courtship and bright coloration (Fiedler 1954), and female reproductive success may be limited by male parental investment. Field data for other groups are so scanty that is not possible to say whether there are any instances of sex role reversal among them, but available data for some dendrobatid frogs suggest at least the possibility. In these species, the male carries one or more young on his back for an unknown length of time (for example, Eaton 1941). Females tend to be more brightly colored than males (rare in frogs) and in at least one species, *Dendrobates aurata,* several females have been seen pursuing, and possibly courting, single males (Dunn 1941). In this species the male carries only one young on his back, until the tadpole is quite large, but females have been found with as many as six large eggs inside, and it is possible that females compete with each other for the backs of males. There are other frog families that show male parental care, but even less is known of their social behavior.

In most monogamous birds male and female parental investment is probably comparable. For some species there is evidence that the male invests somewhat less than the female. Kluijver (1933, cited in Coulson 1960) has shown that the male starling (*Sturnus vulgaris*) incubates the eggs less and feeds the young less often than the female, and similar data are available for the other passerines (Verner and Willson 1969). The fact that in many species males are facultative polygynists (von Haartman 1969) suggests that even when monogamous the males invest less in the young than their females. Because sex role reversal, correlating with evidence of greater male than female parental investment, is so rare in birds and because of certain theoretical considerations discussed below, I tentatively classify most monogamous bird species as showing somewhat greater female than male investment in the young.

A more precise classification of animals, and particularly of similar species, would be useful for the formulation and testing of more subtle hypotheses. Groups of birds would be ideal to classify in this way, because slight differences in relative parental investment may produce large differences in social behavior, sexual dimorphism and mortality rates by sex. It would be interesting to compare human societies that differ in relative parental investment and in the details of the form of the parental investment, but the specification of parental investment is complicated by the fact that humans often invest in kin other than their children. A wealthy man supporting brothers and sisters (and their children) can be viewed functionally as a polygynist if the contributions to his fitness made by kin are devalued appropriately by their degree of relationship to him (see Hamilton 1964). There is good evidence that premarital sexual permissiveness affecting females in human societies relates to the form of parental investment in a way that

would, under normal conditions, tend to maximize female reproductive success (Goethals 1971).

## The evolution of investment patterns

The parental investment pattern that today governs the operation
of sexual selection apparently resulted from an evolutionarily very
early differentiation into relatively immobile sex cells (eggs)
fertilized by mobile ones (spermatozoa). An undifferentiated
system of sex cells seems highly unstable: competition to fertilize
other sex cells should rapidly favor mobility in some sex cells,
which in turn sets up selection pressures for immobility in the
others. In any case, once the differentiation took place, sexual
selection acting on spermatozoa favored mobility at the expense
of investment (in the form of cytoplasm). This meant that as long
as the spermatozoa of different males competed directly to fertilize
eggs (as in oysters) natural selection favoring increased parental
investment could act only on the female. Once females were able
to control which male fertilized their eggs, female choice or mortality selection on the young could act to favor some new form
of male investment in addition to spermatozoa. But there exist
strong selection pressures against this. Since the female already
invests more than the male, breeding failure for lack of an additional
investment selects more strongly against her than against the male.
In that sense, her initial very great investment commits her to
additional investment more than the male's initial slight investment
commits him. Furthermore, male–male competition will tend to
operate against male parental investment, in that any male investment in one female's young should decrease the male's chances of
inseminating other females. Sexual selection, then, is both controlled by the parental investment pattern and a force that tends to
mold that pattern.

The conditions under which selection favors male parental
investment have not been specified for any group of animals.
Except for the case of polygyny in birds, the role of female choice
has not been explored; instead, it is commonly assumed that,
whenever two individuals can raise more individuals together than
one alone could, natural selection will favor male parental investment (Lack 1968, p. 149), an assumption that overlooks the effects
of both male–male competition and female choice.

*Initial parental investment*

An important consequence of the early evolutionary differentiation of the sex cells and subsequent sperm competition is that male sex cells remain tiny compared to female sex cells, even when selection has favored a total male parental investment that equals or exceeds the female investment. The male's initial parental investment, that is, his investment at the moment of fertilization, is much smaller than the female's, even if later, through parental care, he invests as much or more. Parental investment in the young can be viewed as a sequence of discrete investments by each sex. The relative investment may change as a function of time and each sex may be more or less free to terminate its investment at any time. In the human species, for example, a copulation costing the male virtually nothing may trigger a nine-month investment by the female that is not trivial, followed, if she wishes, by a fifteen-year investment in the offspring that is considerable. Although the male may often contribute parental care during this period, he need not necessarily do so. After a nine-month pregnancy, a female is more or less free to terminate her investment at any moment but doing so wastes her investment up until then. Given the initial imbalance in investment the male may maximize his chances of leaving surviving offspring by copulating and abandoning many females, some of whom, alone or with the aid of others, will raise his offspring. In species where there has been strong selection for male parental care, it is more likely that a mixed strategy will be the optimal male course—to help a single female raise young, while not passing up opportunities to mate with other females whom he will not aid.

In many birds, males defend a territory which the female also uses for feeding prior to egg laying, but the cost of this investment by the male is difficult to evaluate. In some species, as outlined above, the male may provision the female before she has produced the young, but this provisioning is usually small compared to the cost of the eggs. In any case, the cost of the copulation itself is always trivial to the male, and in theory the male need not invest anything else in order to copulate. If there is any chance the female can raise the young, either alone or with the help of others, it would be to the male's advantage to copulate with her. By this reasoning one would expect males of monogamous species to retain some psychological traits consistent with promiscuous habits. A male would be selected to differentiate between a female he will only impregnate and a female with whom he will also raise young. Toward the former he should be more eager for sex and less discriminating in choice of sex partner than the female toward him, but toward the latter he should be about as discriminating as she toward him.

If males within a relatively monogamous species are, in fact, adapted to pursue a mixed strategy, the optimal is likely to differ for different males. I know of no attempt to document this possibility in humans, but psychology might well benefit from attempting to view human sexual plasticity as an adaptation to permit the individual to choose the mixed strategy best suited to local conditions and his own attributes. Elder (1969) shows that steady dating and sexual activity (coitus and petting) in adolescent human females correlate inversely with a tendency to marry up the socioeconomic scale as adults. Since females physically attractive as adolescents tend to marry up, it is possible that females adjust their reproductive strategies in adolescence to their own assets.

### Desertion and cuckoldry

There are a number of interesting consequences of the fact that the male and female of a monogamous couple invest parental care in their offspring at different rates. These can be studied by graphing and comparing the cumulative investment of each parent in their offspring, and this is done for two individuals of a hypothetical bird species in Fig. 3. I have graphed no parental investment

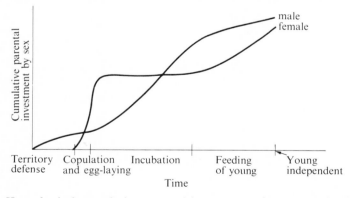

Fig. 3. Hypothetical cumulative parental investment of a male and a female bird in their offspring as a function of time. *Territory defense*: male defends area for feeding and nest building. *Copulation and egg-laying*: Female commits her eggs to male who commits his defended nest to the female. *Incubation*: Male incubates eggs while female does nothing relevant to offspring. *Feeding of young*: Each parent feeds young but female does so at a more rapid rate.

by the female in her young before copulation, even though she may be producing the eggs before then, because it is not until the act of copulation that she commits the eggs to a given male's genes. In effect, then, I have graphed the parental investment of each individual in the other individual's offspring. After copulation, this is the same as graphing investment in their own offspring, assuming, as I do here, that the male and female copulate with each other and each other only.

To discuss the problems that confront paired individuals ostensibly cooperating in a joint parental effort, I choose the language of strategy and decision, as if each individual contemplated in strategic terms the decisions it ought to make at each instant in order to maximize its reproductive success. This language is chosen purely for convenience to explore the adaptations one might expect natural selection to favor.

At any point in time the individual whose cumulative investment is exceeded by his partner's is theoretically tempted to desert, especially if the disparity is large. This temptation occurs because the deserter loses less than his partner if no offspring are raised and the partner would therefore be more strongly selected to stay with the young. Any success of the partner will, of course, benefit the deserter. In Fig. 3, for example, desertion by the male right after copulation will cost him very little, if no offspring are raised, while the chances of the female raising some young alone may be great enough to make the desertion worthwhile. Other factors are important in determining the adaptiveness of abandonment, factors such as the opportunities outside the pair for breeding and the expected shape of the deserter's investment curve if he does not desert. If the male's investment curve does not rise much after copulation, then the female's chances of raising the young alone will be greater and the time wasted by the male investing moderately in his offspring may be better spent starting a new brood.

What are the possible responses of the deserted individual? If the male is deserted before copulation, he has no choice but to attempt to start the process over again with a new female; whatever he has invested in that female is lost. If either partner is deserted after copulation, it has three choices. (1) It can desert the eggs (or eat them) and attempt to breed again with another mate, losing thereby all (or part of) the initial investment. (2) It can attempt to raise the young on its own, at the risk of overexertion and failure. Or, (3) it can attempt to induce another partner to help it raise the young. The third alternative, if successful, is the most adaptive for it, but this requires deceiving another organism into doing something contrary to its own interests, and adaptations should evolve to guard individuals from such tasks. It is difficult to see how a male could be successful in deceiving a new female, but if a female acts quickly, she might fool a male. As time goes on (for example, once the eggs are laid), it is unlikely that a male could easily be fooled. The female could thus be programmed to try the third strategy first, and if it failed, to revert to the first or second. The male deserter gains most if the female succeeds in the third strategy, nothing if she chooses the first strategy, and possibly an intermediate value if she chooses the second strategy.

If neither partner deserts at the beginning, then as time goes on, each invests more and more in the young. This trend has several consequences. On the one hand, the partner of a deserter is more capable of finishing the task alone and natural selection should favor its being more predisposed to try, because it has more to lose. On the other hand, the deserter has more to lose if the partner fails and less to gain if the partner succeeds. The balance between these opposing factors should depend on the exact form of the cumulative investment curves as well as the opportunities for further breeding outside the pair.

There is another effect with time of the increasing investment by both parents in the offspring. As the investments increase, natural selection may favor *either* partner deserting even if one has invested more in the young than the other. This is because the desertion may put the deserted partner in a cruel bind: he has invested so much that he loses considerably if he also deserts the young, even though, which should make no difference to him, the partner would lose even more. The possibility of such binds can be illustrated by an analogous situation described by Rowley (1965). Two neighboring pairs of wrens happened to fledge their young simultaneously and could not tell their young apart, so both pairs fed all six young indiscriminately, until one pair 'deserted' to raise another brood, leaving their neighbors to feed all six young, which they did, even though this meant they were, in effect, being taken advantage of.

Birds should show adaptations to avoid being deserted. Females, in particular, should be able to guard against males who will only copulate and not invest subsequent parental effort. An instance of such an adaptation may be found in the red-necked phalarope, *Phalaropus lobatus.* In phalaropes the male incubates the eggs alone and alone cares for the young after hatching (Höhn 1967; Johns

1969), so that a graph of cumulative parental investment would show an initial large female investment which then remains the same through time, whereas the initial male investment is nil and increases steadily, probably to surpass the female investment. Only the female is vulnerable to being deserted and this right after copulation, since any later desertion by the male costs him his investment in incubation, the young being almost certain to perish. Tinbergen (1935) observed a female vigorously courting a male and then flying away as soon as he responded to the courtship by attempting to copulate. This coy performance was repeated numerous times for several days. Tinbergen attributed it to the 'waxing and waning of an instinct', but the behavior may have been a test of the male's willingness to brood the female's eggs. The male under observation was, in fact, already brooding eggs and was courted when he left the eggs to feed on a nearby pond. In order to view a complete egg-laying sequence, Tinbergen destroyed the clutch the male was brooding. Within a half day the female permitted the male sexual access, and he subsequently brooded her eggs. The important point is that the female could apparently tell the difference between a free and an encumbered male, and she withheld sex from the latter. Courtship alternating with flight may be the test that reveals the male's true attachments: the test can show, for example, whether he is free to follow the female.

It is likely that many adaptations exist in monogamous species to guard against desertion, but despite evidence that desertion can be common (Rowley 1965) no one had attempted to analyze courtship with this danger in mind. Von Haartman (1969) has reviewed some evidence for adaptations of females to avoid being mated to a polygynous male, and being so mated is sometimes exactly equivalent to being deserted by the male (von Haartman 1951).

External fertilization requires a synchrony of behavior such that the male can usually be certain he is not attempting to fertilize previously fertilized eggs. With the evolution of internal fertilization the male cannot be so certain. For many species (for example, most mammals), the distinction is not important because the male loses so little by attempting to fertilize previously fertilized eggs. Where male parental care is involved, however, the male runs the risk of being cuckolded, of raising another male's offspring. For Fig. 1 it was assumed that the pair copulated with each other and each other only, but the male can usually not be sure that such is the case and what is graphed in such a situation is the male's investment in the *female's* offspring. Adaptations should evolve to help guarantee that the female's offspring are also his own, but these can partly be countered by the evolution of more sophisticated cuckolds.

One way a male can protect himself is to ensure that other males keep their distance. That some territorial aggression of monogamous male birds is devoted to protecting the sanctity of the pair bond seems certain, and human male aggression toward real or suspected adulterers is often extreme. Lee (1969), for example, has shown that, when the cause is known, the major cause of fatal Bushman fights is adultery or suspected adultery. In fact, limited data on other hunter-gathering groups (including Eskimos and Australian aborigines) indicate that, while fighting is relatively rare (in that organized intergroup aggression is infrequent), the 'murder rate' may be relatively high. On examination, the murderer and his victim are usually a husband and his wife's real or suspected lover. In pigeons (*Columba livia*) a new male arriving alone at a nocturnal roosting place in the fall is attacked day after day by one or more resident males. As soon as the same male appears with a mate, the two are treated much more casually (Trivers, unpublished data), suggesting that an unpaired male is more threatening than a paired one.

I have argued above that a female deserted immediately after copulation may be adapted to try to induce another male to help raise her young. This factor implies adaptations on the part of the male to avoid such a fate. A simple method is to avoid mating with a female on first encounter, sequester her instead and mate with her only after a passage of time that reasonably excludes her prior impregnation by another male. Certainly males guard their females from other males, and there is a striking difference between the lack of preliminaries in promiscuous birds (Scott 1942; Kruijt and Hogan 1967) and the sometimes long lag between pair bonding and copulation in monogamous birds (Nevo 1956), a lag which usually seems to serve other functions as well.

Biologists have interpreted courtship in a limited way. Courtship is seen as allowing the individual to choose the correct species and sex, to overcome antagonistic urges and to arouse one's partner (Bastock 1967). The above analysis suggests that courtship could also be interpreted in terms of the need to guard oneself from the several possibilities of maltreatment at the hands of one's mate.

### Differential mortality and the sex ratio

Of special interest in understanding the effects of sexual selection are accurate data on differential mortality of the sexes, especially of immature individuals. Such data are, however, among the most difficult to gather, and the published data, although important, are scanty (for example, Emlen 1940; Hays 1947; Chapman, Casida, and Cote 1938; Robinette *et al.* 1957; Coulson 1960; Potts 1969; Darley 1971; Myers and Krebs 1971). As a substitute one can make use of data on sex ratios within given age classes or for all age classes taken together. By assuming that the sex ratio at conception (or, less precisely, at birth) is almost exactly 50/50, significant deviations from this ratio for any age class or for all taken together should imply differential mortality. Where data exist for the sex ratio at birth and where the sex ratio for the entire local population is unbalanced, the sex ratio at birth is usually about 50/50 (see above references, Selander 1965; Lack 1954). Furthermore, Fisher (1958) has shown, and others refined (Leigh 1970), that parents should invest roughly equal energy in each sex. Since parents usually invest roughly equal energy in each individual of each sex, natural selection, in the absence of unusual circumstances (see Hamilton 1967), should favor approximately a 50/50 sex ratio at conception.

It is difficult to determine accurately the sex ratio for any species. The most serious source of bias is that males and females often make themselves differentially available to the observer. For example, in small mammals sexual selection seems to have favored male attributes, such as high mobility, that tend to result in their differential capture (Beer, Frenzel, and MacLeod 1958; Myers and Krebs 1971). If one views one's capture techniques as randomly sampling the existing population, one will conclude that males are more numerous. If one views one's capture techniques as randomly sampling the effects of mortality on the population, then one will conclude that males are more prone to mortality (they are captured more often) and therefore are less numerous. Neither assumption is likely to be true, but authors routinely choose the former. Furthermore, it is often not appreciated what a large sample is required in order to show significant deviations from a 50/50 ratio. A sample of 400 animals showing a 44/56 sex ratio, for example, does not deviate significantly from a 50/50 ratio. (Nor, although this is almost never pointed out, does it differ significantly from a 38/62 ratio.)

Mayr (1939) has pointed out that there are numerous deviations from a 50/50 sex ratio in birds and I believe it is likely that, if data were sufficiently precise, most species of vertebrates would show a significant deviation from a 50/50 sex ratio. Males and females differ in numerous characteristics relevant to their different reproductive strategies and these characters are unlikely to have equivalent effects on survival. Since it is not advantageous for the adults of each sex to have available the same number of adults of the opposite sex, there will be no automatic selective agent for keeping deviations from a 50/50 ratio small.

A review of the useful literature on sex ratios suggests that (except for birds) when the sex ratio is unbalanced it is usually unbalanced by there being more females than males. Put another way, males apparently have a tendency to suffer higher mortality rates than females. This is true for those dragonflies for which there are data (Corbet, Longfield, and Moore 1960), for the house fly (Rockstein 1959), for most fish (Beverton and Holt 1959), for several lizards (Tinkle 1967; Harris 1964; Hirth 1963; Blair 1960; Trivers, discussed below), and for many mammals (Bouliere and Verschuren 1960; Cowan 1950; Eisenberg 1965; Robinette *et al.* 1957; Beer, Frenzel, and MacLeod 1958; Stephens 1952; Tyndale-Biscoe and Smith 1969; Myers and Krebs 1971; Wood 1970). Hamilton (1948) and Lack (1954) have reviewed studies on other animals suggesting a similar trend. Mayr (1939) points out that where the sex ratio can be shown to be unbalanced in monogamous birds there are usually fewer females, but in polygynous or promiscuous birds there are fewer males. Data since his paper confirm this finding. This result is particularly interesting since in all other groups in which males tend to be less numerous monogamy is rare or nonexistent.

*The chromosomal hypothesis*

There is a tendency among biologists studying social behavior to regard the adult sex ratio as an independent variable to which the species reacts with appropriate adaptations. Lack (1968) often interprets social behavior as an adaptation in part to an un-balanced (or balanced) sex ratio, and Verner (1964) has sum-marized other instances of this tendency. The only mechanism that will generate differential mortality independent of sexual differences clearly related to parental investment and sexual section is the chromosomal mechanism, applied especially to humans and the other mammals: the unguarded X chromosome of the male is presumed to predispose him to higher mortality. This mechanism is inadequate as an explanation of differential mortality for three reasons.

1. The distribution of differential mortality by sex is not predicted by a knowledge of the distribution of sex determining mechanisms. Both sexes of fish are usually homogametic, yet males suffer higher mortality. Female birds are heterogametic but suffer higher mortality only in monogamous species. Homogametic male meal moths are outsurvived by their heterogametic female counterparts under laboratory conditions (Hamilton and Johansson 1965).

2. Theoretical predictions of the degree of differential mortality expected by males due to their unguarded X chromosome are far lower than those observed in such mammals as dogs, cattle and humans (Ludwig and Boost 1951). It is possible to imagine natural selection favoring the heterogametic sex determining mechanism if the associated differential mortality is slight and balanced by some advantage in differentiation or in the homogametic sex, but a large mortality associated with heterogamy should be counteracted by a tendency toward both sexes becoming homogametic.

3. Careful data for humans demonstrate that castrate males (who remain of course heterogametic) strongly outsurvive a control group of males similar in all other respects and the earlier in life the castration, the greater the increase in survival (Hamilton and Mestler 1969). The same is true of domestic cats (Hamilton, Hamilton and Mestler 1969), but not of a species (meal moths) for which there is no evidence that the gonads are implicated in sexual differentiation (Hamilton and Johansson 1965).

## An adaptive model of differential mortality

To interpret the meaning of balanced or unbalanced sex ratios one needs a comprehensive framework within which to view life historical phenomena. Gadgil and Bossert (1970) have presented a model for the adaptive interpretation of differences between species' life histories; for example, in the age of first breeding and in the growth and survival curves. Although they did not apply this model to sexual differences in these parameters, their model is precisely suited for such differences. One can, in effect, treat the sexes as if they were different species, the opposite sex being a resource relevant to producing maximum surviving offspring. Put this way, female 'species' usually differ from male species in that females compete among themselves for such resources as food but not for members of the opposite sex, whereas males ultimately compete only for members of the opposite sex, all other forms of competition being important only insofar as they affect this ultimate competition.

To analyze differential mortality by sex one needs to correlate different reproductive strategies with mortality, that is, one must show how a given reproductive strategy entails a given risk of mortality. One can do this by graphing reproductive success (RS) for the first breeding season as a function of reproductive effort expended during that season, and by graphing the diminution in future reproductive success (D) in units of first breeding season reproductive success. (Gadgil and Bossert show that the reproductive value of a given effort declines with age, hence the need to convert future reproductive success to comparable units.) For simplicity I assume that the diminution, D, results entirely from mortality between the first and second breeding seasons. The diminution could result from mortality in a later year (induced by reproductive effort in the first breeding season) which would not change the form of the analysis, or it could result from decreased ability to breed in the second (or still later) breeding season, which sometimes occurs but which is probably minor compared to the diminution due to mortality, and which does not change the analysis as long as one assumes that males and females do not differ appreciably in the extent to which they suffer this form of diminution.

Natural selection favors an individual expending in the first breeding season the reproductive effort (RE) that results in a maximum net reproductive success (RS—D). The value of D at this RE gives the degree of expected mortality between the first and second breeding seasons (see Figs. 4 and 5). Differences between the sexes in D will give the expected differential mortality. The same analysis can be applied to the $n$th breeding season to predict mortality between it and the $n$th + 1 breeding season. Likewise, by a trivial modification, the analysis can be used to generate differences in juvenile mortality: let D represent the diminution in chances of surviving to the first breeding season as a function of RE at first breeding. Seen this way, one is measuring the cost in survival of developing during the juvenile period attributes relevant to adult reproductive success.

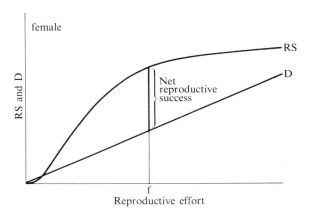

Fig. 4. Female reproductive success during the first breeding season (RS) and diminution of future reproductive success (D) as functions of reproductive effort during first breeding. D is measured in units of first breeding (see text). At f the net reproductive success reaches a maximum. Species is one in which there is very little male parental investment.

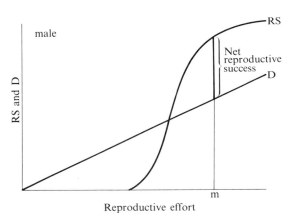

Fig. 5. Same as Fig. 4 except that it is drawn for the male instead of the female. At m the net reproductive success reaches a maximum.

*Species with little or no male parental investment*

In Fig. 4, I have graphed RS and D as functions of reproductive effort in the first breeding season for females of a hypothetical species in which males invest very little parental care. The RS function is given a sigmoidal shape for the following reasons. I assume that at low values of RE, RS increases only very gradually because some investment is necessary just to initiate reproduction (for example, enlarging the reproductive organs). RS then increases more rapidly as a function of RE but without achieving a very steep slope. RS finally levels off at high values of RE because of increased inefficiencies there (for example, inefficiencies in foraging; see Schoener 1971). I have graphed the value, f, at which net reproductive success for the female reaches a maximum. Technically, due to competition, the shape of the RS function for any given female will depend partly on the reproductive effort devoted by other females; the graph therefore assumes that other females tend to invest near the optimal value, f, but an important feature of a female's RS is that it is *not* strongly dependent on the RE devoted by other females: the curve would not greatly differ if all other females invested much more or less. I have graphed D as a linear function of RE. So doing amounts to a definition of reproductive effort, that is, a given increment in reproductive effort during the first breeding season can be detected as a proportionately increased chance of dying between the first and the second breeding seasons. Note that reproductive effort for the female is essentially synonymous with parental investment.

Male RS differs from female RS in two important ways, both of which stem from sexual selection. (1) A male's RS is highly dependent on the RE of other males. When other males invest heavily, an individual male will usually not outcompete them unless he invests as much or more. A considerable investment that is slightly below that of other males may result in zero RS. (2) A male's RS is potentially very high, much higher than that of a conspecific female, but only if he outcompetes other males. There should exist some factor or set of factors (such as size, aggressiveness, mobility) that correlates with high male RS. The effect of competition between males for females is selection for increased male RE, and this selection will continue until greater male than female RE is selected as long as the higher associated D is offset by the potentially very high RS. This argument is graphed in Fig. 5, where the steep slope of RS reflects the high interaction between one male's RS and the RE of the other males. Note that the argument here depends on the existence of a set of factors correlated with high male reproductive success. If these factors exist, natural selection will predispose the male to higher mortality rates than the female. Where a male can achieve very high RS in a breeding season (as in landbreeding seals, Bartholemew 1970), differential mortality will be correspondingly high.

*Species with appreciable male parental investment*

The analysis here applies to species in which males invest less parental care than, but probably more than one-half, what females invest. I assume that most monogamous birds are so characterized, and I have listed reasons and some data above supporting this assumption. The reasons can be summarized by saying that because of their initial large investment, females appear to be caught in a situation in which they are unable to force greater parental investment out of the males and would be strongly selected against if they unilaterally reduced their own parental investment.

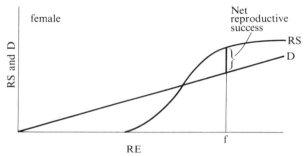

Fig. 6. Female reproductive success and diminution in future reproductive success as functions of reproductive effort (RE) assuming male reproductive effort of $m_1$. Species is a hypothetical monogamous bird in which males invest somewhat less than females in parental care (see Figs 7 and 8).

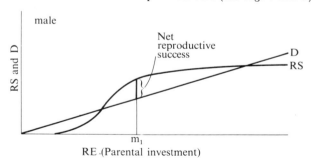

Fig. 7. Male reproductive success and diminution in future reproductive success as functions of reproductive effort, assuming female reproductive effort of f. Species is same as in Fig. 6. Reproductive effort of male is invested as parental care in one female's offspring. Net reproductive success is a maximum at $m_1$.

Functions relating RS to parental investment are graphed for females and males in Figs. 6 and 7, assuming for each sex that the opposite sex shows the parental investment that results for it in a maximum net reproductive success. The female curve is given a sigmoidal shape for the reasons that apply to Fig. 4; in birds the female's initial investment in the eggs will go for nothing if more is not invested in brooding the eggs and feeding the young, while beyond a certain high RE further increments do not greatly affect RS. Assuming the female invests the value, f, male RS will vary as a function of male parental investment in a way similar to female RS, except the function will be displaced to the left (Fig. 7) and some RS will be lost due to the effects of the cuckoldry graphed in Fig. 8.

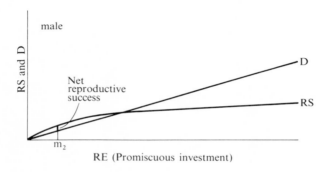

Fig. 8. Male reproductive success and diminution of future reproductive success as a function of reproductive effort solely devoted to promiscuous behavior. Net reproductive succes at $m_2$ is a maximum. Same species as in Figs 6 and 7.

Because males invest in parental care more than one-half what females invest and because the offspring of a given female tend to be inseminated by a single male, selection does not favor males competing with each other to invest in the offspring of more than one female. Rather, sexual selection only operates on the male to inseminate females whose offspring he will not raise, especially if another male will raise them instead. Since selection presumably does not strongly favor female adultery and may oppose it (if, for example, detection leads to desertion by the mate), the opportunities for cuckoldry are limited: high investment in promiscuous activity will bring only limited RS. This argument is graphed in

Fig. 8. The predicted differential mortality by sex can be had by comparing D (f) with D (m$_1$ + m$_2$).

It may seem ironic, but in moving from a promiscuous to a monogamous life, that is, in moving toward *greater* parental investment in his young, the male tends to *increase* his chances of surviving relative to the female. This tendency occurs because the increased parental investment disproportionately decreases the male's RE invested in male–male competition to inseminate females.

Note that in both cases above differential mortality tends to be self-limiting. By altering the ratio of possible sexual partners to sexual competitors differential mortality sets up forces that tend to keep the differential mortality low. In species showing little male parental investment differential male mortality increases the average number of females available for those males who survive. Other things being equal, this increase tends to make it more difficult for the most successful males to maintain their relative advantage. In monogamous birds differential female mortality induces competition among males to secure at least one mate, thereby tending to increase male mortality. Such competition presumably also increases the variance in male reproductive success above the sexual differential expected from cuckoldry.

*Species with greater male than female parental investment*

Since the above arguments were made with reference to relative parental investment and not sex, they apply to species in which males invest more parental effort than females, except that there is never apt to be a female advantage to cuckolding other females, and this advantage is always alive with males. Where females invest more than one-half what males invest, one would predict differential female mortality. Where females invest less than one-half what males invest, one would predict competition, and a resulting differential female mortality.

**Male–male competition**

Competition between males does not necessarily end with the release of sperm. Even in species with internal fertilization, competition between sperm of different males can be an important component of male–male competition (see the excellent review by Parker 1970*b*). In rare cases, competition between males may continue after eggs are fertilized. For example an adult male langur (*Presbytis entellus*) who ousts the adult male of a group may systematically kill the infants of that group (presumably fathered by the ousted male) thereby bringing most of the adult females quickly into estrus again (Sugiyama 1967). While clearly disadvantageous for the killed infants and their mothers, such behavior, benefiting the new male, may be an extreme product of sexual selection. Female mice spontaneously abort during the first four days of pregnancy when exposed to the smell of a strange male (Bruce 1960, reviewed in Sadleir in 1967), a situation subject to several interpretations including one based on male–male competition.

Sperm competition may have important effects on competition between males prior to release of sperm. In those insects in which later-arriving sperm take precedence in fertilizing eggs, selection favors mating with a female just prior to release of eggs, thereby increasing competition at ovulation sites and intensifying selection by a postovulatory guarding phase by the male (see Parker 1970*bc d*; Jacobs 1955). I here concentrate on male–male competition prior to the release of sperm in species showing very little male parental investment.

The form of male–male competition should be strongly influenced by the distribution in space and time of the ultimate resource affecting male reproductive success, namely, conspecific breeding females. The distribution can be described in terms of three parameters: the extent to which females are clumped or dispersed in space, the extent to which they are clumped or dispersed in time, and the extent to which their exact position in space and time is predictable. I here treat females as if they are a passive resource for which males compete, but female choice may strongly influence the form of male–male competition, as, for example, when it favors males clumping together on display grounds (for example, S. Emlen 1968) which females then search out (see below under 'Female choice')'.

*Distribution in space*

Cervids differ in the extent to which females are clumped in space or randomly dispersed (deVos, Broky and Geist 1967) as do antelopes (Eisenberg 1965), and these differences correlate in a predictable way with differences in male attributes. Generally male–male aggression will be the more severe the greater the number of females two males are fighting over at any given moment. Searching behavior should be more important in highly dispersed species especially if the dispersal is combined with unpredictability.

*Distribution in time*

Clumped in time refers to highly seasonal breeders in which many females become sexually available for a short period at the same moment (for example, explosive breeding frogs; Bragg 1965; Rivero and Estevez 1969), while highly dispersed breeders (in time) are species (such as chimpanzees; Van Lawick-Goodall 1968) in which females breed more or less randomly throughout the year. One effect of extreme clumping is that it becomes more difficult for any one male to be extremely successful: while he is copulating with one female, hundreds of other females are simultaneously being inseminated. Dispersal in time, at least when combined with clumping in space, as in many primates, permits each male to compete for each newly available female and the same small number of males tend repeatedly to inseminate the receptive females (DeVore 1965).

*Predictability*

One reason males in some dragonflies (Jacobs 1955) may compete with each other for female oviposition sites is that those are highly predictable places at which to find receptive females. Indeed, males display several behaviors, such as testing the water with the tips of their abdomen, that apparently aid them in predicting especially good oviposition sites, and such sites can permit very high male reproductive success (Jacobs 1955). In the cicada killer wasp (*Sphecius spheciosus*) males establish mating territories around colony emergency holes, presumably because this is the most predictable place at which to find receptive females (Lin 1963).

The three parameters outlined interact strongly, of course, as when very strong clumping in time may strongly reduce the predicted effects of strong clumping in space. A much more detailed classification of species with non-obvious predictions would be welcome. In the absence of such models I present a partial list of factors that should affect male reproductive success and that may correlate with high male mortality.

*Size*

There are very few data showing the relationship between male size and reproductive success but abundant data showing the relationship between male dominance and reproductive success: for example, in elephant seals (LeBoeuf and Peterson 1969), black grouse (Koivisto 1965; Scott 1942), baboons (DeVore 1965), and rainbow lizards (Harris 1964). Since dominance is largely established through aggression and larger size is usually helpful in aggressive encounters, it is likely that these data partly reveal the relationship between size and reproductive success. (It is also likely that they reflect the relationship between experience and reproductive success.)

Circumstantial evidence for the importance of size in aggressive encounters can be found in the distribution of sexual size dimorphism and aggressive tendencies among tetrapods. In birds and mammals males are generally larger than females and much more aggressive. Where females are known to be more aggressive (that is, birds showing reversal in sex roles) they are also larger. In frogs and salamanders females are usually larger than males, and aggressive behavior has only very rarely been recorded. In snakes, females are usually larger than males (Kopstein 1941) and aggression is almost unreported. Aggression has frequently been observed between sexually active crocodiles and males tend to be larger (Allen Greer, personal communication). In lizards males are often larger than females, and aggression is common in some families (Carpenter 1967). Male aggressiveness is also common, however, in some species in which females are larger, for example, *Sceloporus*, (Blair 1960). There is a trivial reason for the lack of evidence of aggressiveness in most amphibians and reptiles: the species are difficult to observe and few behavioral data of any sort have been recorded. It is possible, however, that this correlation between human ignorance and species in which females are larger is not accidental. Humans tend to be more knowledgeable about those species that are also active diurnally and strongly dependent on vision, for example, birds and large mammals. It may be that male aggressiveness is more strongly selected in visually oriented animals because vision provides long-range information on the behavior of competitors. The male can, for example, easily observe another male beginning to copulate and can often quickly attempt to intervene (for example, baboons, DeVore 1965 and sage grouse, Scott 1942).

Mammals and birds also tend towards low, fixed clutch sizes and this may favor relatively smaller females, since large female size may be relatively unimportant in reproductive success. In many fish, lizards, and salamanders female reproductive success as measured by clutch size is known to correlate strongly within species with size (Tinkle, Wilbur and Tilley 1970; Tilley 1968).

Measuring reproductive success by frequency of copulation,

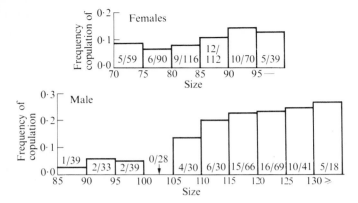

Fig. 9. Reproductive success in male and female *A. garmani* as a function of size. Reproductive success is measured by the number of copulations observed per number of individuals (male or female) in each nonoverlapping 5 mm size category. Data combined from five separate visits to study area between summer 1969 and summer 1971.

I have analyzed male and female reproductive success as a function of size in *Anolis garmani* (Fig. 9). Both sexes show a significant positive correlation between size and reproductive success, but the trend in males is significantly stronger than the trend in females ($p < 0.01$). Consistent with this tendency, males grow faster at all sizes than females (Fig. 10) and reach an adult weight two and

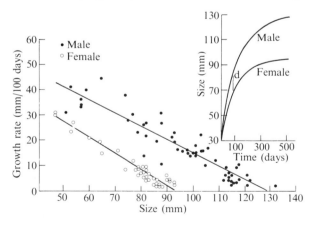

Fig. 10. Male and female growth rates in *A. garmani* as a function of initial size based on summer 1970 recaptures of animals marked 3 to 4 months before. A line has been fitted to each set of data; d indicates how much larger a male is when a similar aged female reaches sexual maturity.

one-half times that of adult females. The sex ratio of all animals is unbalanced in favor of females, which would seem to indicate differential mortality, but the factors that might produce the difference are not known. Males are highly aggressive and territorial, and large males defend correspondingly large territories with many resident females. No data are available on size and success in aggressive encounters, but in the closely related (and behaviorally very similar) *A. lineatopus,* 85 per cent of 182 disputes observed in the field were won by the larger animal (Rand 1967). Females lay only one egg at a time, but it is likely that larger adult females lay eggs slightly more often than smaller ones, and this may partly be due to advantages in feeding through size-dependent aggressiveness, since larger females wander significantly more widely than smaller adult ones. An alternate interpretation (based on ecological competition between the sexes) has been proposed for sexual dimorphism in size among animals (Selander 1966), and the interpretation may apply to *Anolis* (Schoener 1967).

*Metabolic rate*

Certainly more is involved in differential male mortality than size, even in species in which males grow to a larger size than females. Although data show convincingly that nutritional factors strongly affect human male survival *in utero,* a sexual difference in size among humans is not detected until the twenty-fourth week after conception whereas differences in mortality appear as soon as the twelfth week. Sellers *et al.* (1950) have shown that male rats excrete four times the protein females do; the difference is removed by castration. Since males suffer more from protein-deficient diets than females (they gain less weight and survive less well) the sex-linked proteinuria, apparently unrelated to size, may be a factor in causing lower male survival in wild rats (Schein 1950). (The connection between proteinuria and male reproductive success is obscure.) Again, although human male survival is more adversely affected by poor nutritional conditions than female survival, Hamilton (1948) presents evidence that the higher metabolic rate of the male is an important factor increasing his vulnerability to many diseases which strike males more heavily than females. Likewise, Taber and Dasmann (1954) argue that greater male mortality in the deer, *Odocoileus hemionus,* results from a higher metabolic rate. Higher metabolic rate could relate to both aggressiveness and searching behavior.

*Experience*

If reproductive success increases more rapidly in one sex than the other as a function of age alone (for example, through age-dependent experience), then one would expect a postponement of sexual maturity in that sex and a greater chance of surviving through a unit of time than in the opposite sex. Thus, the adult sex ratio might be biased in favor of the earlier maturing sex but the sex ratio for all ages taken together should be biased in favor of the later maturing sex. Of course, if reproductive success for one sex increases strongly as a function of experience and experience only partly correlates with age, then the sex may be willing to suffer increased mortality if this mortality is sufficiently offset by increases in experience. Selander (1965) has suggested that the tendency of immature male blackbirds to exhibit some mature characteristics may be adaptive in that it increases the male's experience, although it also presumably increases his risk of mortality.

*Mobility*

Data from mammals (reviewed by Eisenberg 1965 and Brown 1966) and from some salamanders (Madison and Shoop 1970) and numerous lizards (Tinkle 1967; Blair 1960) suggest that males often occupy larger home ranges and wander more widely than females *even when males are smaller* (Blair 1965). Parker (1970a) has quantified the importance of mobility and search behavior in dung flies. If females are a dispersed resource, then male mobility may be crucial in exposing the male to a large number of available females. Again, males may be willing to incur greater mortality if this is sufficiently offset by increases in reproductive success. This factor should only affect the male during the breeding season (Kikkawa 1964) unless factors relevant to mobility (such as speed, agility or knowledge of the environment) need to be developed prior to the reproductive season. Lindburg (1969) has shown that macaque males, but not females, change troops more frequently during the reproductive season than otherwise and that this mobility increases male reproductive success as measured by frequency of copulation, suggesting that at least in this species, greater mobility can be confined to the reproductive season (see also Miller 1958). On the other hand, Taber and Dasmann (1954) present evidence that as early as six months of age male deer wander more widely from their mothers than females—a difference whose function, of course, is not known. Similar very early differences in mobility have been demonstrated for a lizard (Blair 1960) and for several primates, including man (Jensen, Bobbitt and Gordon 1968).

## Female choice

Although Darwin (1971) thought female choice an important evolutionary force, most writers since him have relegated it to a trivial role (Huxley 1938; Lack 1968; but see Fisher 1958, and Orians 1969). With notable exceptions the study of female choice has limited itself to showing that females are selected to decide whether a potential partner is of the right species, of the right sex and sexually mature. While the adaptive value of such choices is obvious, the adaptive value of subtler discriminations among broadly appropriate males is much more difficult to visualize or document. One needs both theoretical arguments for the adaptive value of such female choice and detailed data on how females choose. Neither of these criteria is met by those who casually ascribe to female (or male) choice the evolution of such traits as the relative hairlessness of both human sexes (Hershkovitz 1966) or the large size of human female breasts (Morris 1967). I review here theoretical considerations of how females might be expected to choose among the available males, along with some data on how females do choose.

### Selection for otherwise neutral or disfunctional male attributes

The effects of female choice will depend on the way females choose. If some females exercise a preference for one type of male (genotype) while others mate at random, then other things being equal, selection will rapidly favor the preferred male type and the females with the preference (O'Donald 1962). If each female has a specific image of the male with whom she prefers to mate and if there is a decreasing probability of a female mating with a male as a function of his increasing deviation from her preferred image, then it is trivial to show that selection will favor distributions of female preferences and male attributes that coincide. Female choice can generate continuous male change only if females choose by a relative rather than an absolute criterion. That is, if there is a tendency for females to sample the male distribution and to prefer one extreme (for example, the more brightly colored males), then selection will move the male distribution toward the favored extreme. After a one generation lag, the distribution of female preferences will also move toward a greater percentage of females with extreme desires, because the granddaughters of females preferring the favored extreme will be more numerous than the granddaughters of females favoring other male attributes. Until countervailing selection intervenes, this female preference will, as first pointed out by Fisher (1958), move both male attributes and female preferences with increasing rapidity in the same direction. The female preference is capable of overcoming some countervailing selection on the male's ability to survive to reproduce, if the increased reproductive success of the favored males when mature offsets their chances of surviving to reproduce.

There are at least two conditions under which one might expect females to have been selected to prefer the extreme male of a sample. When two species, recently speciated, come together, selection rapidly favors females who can discriminate the two species of males. This selection may favor females who prefer the appropriate extreme of an available sample, since such a mechanism would minimize mating mistakes. The natural selection of females with such a mechanism of choice would then initiate sexual selection in the same direction, which in the absence of countervailing selection would move the two male phenotypes further apart than necessary to avoid mating error.

Natural selection will always favor female ability to discriminate male sexual competence, and the safest way to do this is to take the extreme of a sample, which would lead to runaway selection for male display. This case is discussed in more detail below.

### Selection for otherwise functional male attributes

As in other aspects of sexual selection, the degree of male investment in the offspring is important and should affect the criteria of female choice. Where the male invests little or nothing beyond his sex cells, the female has only to decide which male offers the ideal genetic material for her offspring, assuming that male is willing and capable of offering it. This question can be broken down to that of which genes will promote the survival of her offspring and which will lead to reproductive success, assuming the offspring survive to adulthood. Implicit in these questions may be the relation between her genes and those of her mate: do they complement each other?

Where the male invests parental care, female choice may still involve the above question of the male's genetic contribution but should also involve, perhaps primarily involve, questions of the male's willingness and ability to be a good parent. Will he invest in the offspring? If willing, does he have the ability to contribute much? Again, natural selection may favor female attentiveness to complementarity: do the male's parental abilities complement her own? Can the two parents work together smoothly? Where males invest considerable parental care, most of the same considerations that apply for female choice also apply to male choice. The alternate criteria for female choice are summarized in Table 1.

Table 1. Theoretical criteria for female choice of males

---

I.  All species, but especially those showing little or no male parental investment

    A.  Ability to fertilize eggs
        (1)  correct species
        (2)  correct sex
        (3)  mature
        (4)  sexually competent

    B.  Quality of genes
        (1)  ability of genes to survive
        (2)  reproductive ability of genes
        (3)  complementarity of genes

II. Only those species showing male parental investment

    C.  Quality of parental care
        (1)  willingness of male to invest
        (2)  ability of male to invest
        (3)  complementarity of parental attributes

---

## Sexual competence

Even in males selected for rapid repeated copulations the ability to do so is not unlimited. After three or four successive ejaculations, for example, the concentration of spermotozoa is very low in some male chickens (Parker, McKenzie and Kempster 1940), yet males may copulate as often as 30 times in an hour (Guhl 1951). Likewise, sperm is completely depleted in male *Drosophila melanogaster* after the fifth consecutive mating on the same day (Demerec and Kaufmann 1941; Kaufmann and Demerec 1942). Duration of copulation is cut in half by the third copulation of a male dung fly on the same day and duration of copulation probably correlates with sperm transferred (Parker 1970a). In some species females may be able to judge whether additional sperm are needed (for example, house flies; Riemann, Moen and Thorson 1967) or whether a copulation is at least behaviorally successful (for example, sea lions; Peterson and Bartholomew 1967), but in many species females may guarantee reproductive success by mating with those males who are most vigorous in courtship, since this vigor may correlate with an adequate supply of sperm and a willingness to transfer it.

When the male is completely depleted, there is no advantage in his copulating but selection against the male doing so should be much weaker than selection against the female who accepts him. At intermediate sperm levels, the male may gain something from copulation, but the female should again be selected to avoid him. Since there is little advantage to the male in concealing low reproductive powers, a correlation between vigor of courtship and sperm level would not be surprising. Females would then be selected to be aroused by vigorous courtship. If secondary structures used in display, such as bright feathers, heighten the appearance of vigorousness, then selection may rapidly accentuate such structures. Ironically, the male who has been sexually most successful may not be ideal to mate with if this success has temporarily depleted his sperm supply. Males should not only be selected to recover rapidly from copulations but to give convincing evidence that they have recovered. It is not absurd to suppose that in some highly promiscuous species the most attractive males may be those who, having already been observed to mate with several females, are still capable of vigorous display toward a female in the process of choosing.

*Good genes*

Maynard Smith (1956) has presented evidence that, given a choice, female *Drosophila subobscura* discriminate against inbred males of that species and that this behavior is adaptive: females who do not so discriminate leave about ¼ as many viable offspring as those who do. Females may choose on the basis of courtship behavior: inbred males are apparently unable to perform a step of the typical courtship as rapidly as outbred males. The work is particularly interesting in revealing that details of courtship behavior may reveal a genetic trait, such as being inbred, but it suffers from an artificiality. If inbred males produce mostly inviable offspring, then, even in the absence of female discrimination, one would expect very few, if any, inbred males to be available in the adult population. Only because such males were artificially selected were there large numbers to expose to females in choice experiments. Had that selection continued one generation further, females who chose inbred males would have been the successful females.

Maynard Smith's study highlights the problem of analyzing the potential for survival of one's partner's genes: one knows of the adult males one meets that they have survived to adulthood; by what criterion does one decide who has survived better? If the female can judge age, then all other things being equal, she should choose older males, as they have demonstrated their capacity for long survival. All other things may not be equal, however, if old age correlates with lowered reproductive success, as it does in some ungulates (Fraser 1968) through reduced ability to impregnate. If the female can judge the physical condition of males she encounters, then she can discriminate against undernourished or sickly individuals, since they will be unlikely to survive long, but discrimination against such individuals may occur for other reasons, such as the presumed lowered ability of such males to impregnate successfully due to the weakened condition.

In some very restricted ways it may be possible to second-guess the future action of natural selection. For example, stabilizing selection has been demonstrated to be a common form of natural selection (see Mayr 1963) and under this form of selection females may be selected to exercise their own discrimination against extreme types, thereby augmenting the effects of any stabilizing selection that has occurred prior to reproduction. Mason (1969) has demonstrated that females of the California Oak Moth discriminate against males extreme in some traits, but no one has shown independent stabilizing selection for the same traits. Discrimination against extreme types may run counter to selection for diversity; the possible role of female choice in increasing or decreasing diversity is discussed below as a form of complementarity.

Reproductive success, independent of ability to survive is easier for the female to gauge because she can directly observe differences in reproductive success before she chooses. A striking feature of data on lek behavior of birds is the tendency for females to choose males who, through competition with other males, have already increased their likelihood of mating. Female choice then greatly augments the effects of male–male competition. On the lek grounds there is an obvious reason why this may be adaptive. By mating with the most dominant male a female can usually mate more quickly, and hence more safely, than if she chooses a less dominant individual whose attempts at mating often result in

interference from more dominant males. Scott (1942) has shown
that many matings with less dominant individuals occur precisely
when the more dominant individuals are unable, either because of
sexual exhaustion or a long waiting line, to quickly service the
female. Likewise, Robel (1970) has shown that a dominant
female prevents less dominant individuals from mating until she
has mated, presumably to shorten her stay and to copulate while
the dominant male still can. A second reason why choosing a mate
with more dominant males may be adaptive is that the female
allies her genes with those of a male who, by his ability to dominate other males, has demonstrated his reproductive capacity. It is
a common observation in cervids that females placidly await the
outcome of male strife to go with the victor. DeVore (1965) has
quantified the importance of dominance in male baboon sexual
success, emphasizing the high frequency of interference by other
males in copulation and the tendency for female choice, when it is
apparent, to be exercised in favor of dominant males. That previous
success may increase the skill with which males court females is
suggested by work on the black grouse (Kruijt, Bossema and
de Voss, in press), and females may prefer males skillful at courting
in part because their skill correlates with previous success.

In many species the ability of the male to find receptive females
quickly may be more important than any ability to dominate other
males. If this is so, then female choice may be considerably simpli-
fied: the first male to reach her establishes thereby a *prima facie*
case for his reproductive abilities. In dung flies, in which females
must mate quickly while the dung is fresh, male courtship behavior is virtually nonexistent (Parker 1970*a*). The male who first
leaps on top of a newly arrived female copulates with her. This
lack of female choice may also result from the *prima facie* case the
first male establishes for his sound reproductive abilities. Such a
mechanism of choice may of course conflict with other criteria
requiring a sampling of the male population, but in some species
this sampling could be carried out prior to becoming sexually
receptive.

There are good data supporting the importance of complementarity of genes to female choice. Assortative mating in the
wild has been demonstrated for several bird species (Cooch and
Beardmore 1959; O'Donald 1959) and disassortative mating for a
bird species and a moth species (Lowther 1961; Sheppard 1952).
Petit and Ehrman (1969) have demonstrated the tendency in
several *Drosophila* species for females to prefer mating with the

rare type in choice experiments, a tendency which in the wild leads to a form of complementarity, since the female is presumably usually of the common type. These studies can all be explained plausibly in terms of selection for greater or lesser genetic diversity, the female choosing a male whose genes complement her own, producing an 'optimal' diversity in the offspring.

## Good parent

Where male parental care is involved, females certainly sometimes choose males on the basis of their ability to contribute parental care. Orians (1969), for example, has recently reviewed arguments and data suggesting that polygyny evolves in birds when becoming the second mate of an already mated male provides a female with greater male parental contribution than becoming the first mate of an unmated male would. This will be so, for example, if the already mated male defends a territory considerably superior to the un-mated male's. Variability in territory quality certainly occurs in most territorial species, even in those in which territories are not used for feeding. Tinbergen (1967), for example, has documented the tendency for central territories in the black-headed gull to be less vulnerable to predation. If females compete among themselves for males with good territories, or if males exercise choice as well, then female choice for parental abilities will again tend to augment intra-male competition for the relevant resources (such as territor-ies). The most obvious form of this selection is the inability of a nonterritory holding male to attract a female.

Female choice may play a role in selecting for increased male parental investment. In the roadrunner, for example, food caught by a male seems to act on him as an aphrodisiac: he runs to a female and courts her with the food, suggesting that the female would not usually mate without such a gift (Calder 1967). Male parental care invested after copulation is presumably not a result of female choice after copulation, since she no longer has anything to bargain with. In most birds, however, males defend territories which initially attract the females (Lack 1940). Since males without suitable territories are unable to attract a mate, female choice may play a role in maintaining male territorial behavior. Once a male has invested in a territory in order to attract a mate his options after copulating with her may be severely limited. Driving the female out of his territory would almost certainly result in the loss of his investment up until then. He could establish another terri-tory, and in some species some males do this (von Haartman 1951), but in many species this may be difficult, leaving him with the option of aiding, more or less, the female he has already mated. Female choice, then, exercised *before* copulation, may indirectly force the male to increase his parental investment *after* copulation.

There is no reason to suppose that males do not compete with each other to pair with those females whose breeding potential appears to be high. Darwin (1871) argued that females within a species breeding early for nongenetic reasons (such as being in excellent physical condition) would produce more offspring than later breeders. Sexual selection, he argued, would favor males competing with each other to pair with such females. Fisher (1968) has nicely summarized this argument, but Lack (1968, p. 157) dismisses it as being 'not very cogent', since 'the date of breeding in birds has been evolved primarily in relation to two different factors, namely the food supply for the young and the capacity of the female to form eggs'. These facts are, of course, fully consistent with Darwin's argument, since Darwin is merely supposing a developmental plasticity that allows females to breed earlier if they are capable of forming the eggs, and data presented elsewhere in Lack (1968) support the argument that females breeding earlier for nongenetic reasons (such as age or duration of pair bond) are more successful than those breeding later (see also, for example, Fisher 1969, and Coulson 1966). Goforth and Baskett (1971) have recently shown that dominant males in a penned Mourning Dove population preferentially pair with dominant females; such pairs breed earlier and produce more surviving young than less dominant pairs. It would be interesting to have detailed data from other species on the extent to which males do compete for females with higher breeding potential. Males are certainly often initially aggressive to females intruding in their territories, and this aggressiveness may act as a sieve, admitting only those females whose high motivation correlates with early egg laying and high reproductive potential. There is good evidence that American women tend to marry up the socioeconomic scale, and physical attractiveness during adolescence facilitiates such movement (Elder 1969). Until recently such a bias in female choice presumably correlated with increased reproductive success, but the value, if any, of female beauty for male reproductive success is obscure.

The importance of choice by both female and male for a mate who will not desert nor participate in sex outside the pair bond has been emphasized in an earlier section ('Desertion and cuckoldry'). The importance of complementarity is documented in a study by Coulson (1966).

*Criteria other than male characters*

In many species male–male competition combined with the import-ance of some resource in theory unrelated to males, such as oviposition sites, may mitigate against female choice for male characters. In the dragonfly *Parthemis tenera* males compete with each other to control territories containing good oviposition sites, probably because such sites are a predictable place at which to find receptive females and because sperm competition in insects usually favors the last male to· copulate prior to oviposition (Parker 1970*b*). It is clear that the females choose the oviposition site and not the male (Jacobs 1955), and male courtship is geared to advertise good oviposition sites. A male maintaining a territory containing a good oviposition site is *not* thereby contributing parental investment unless that maintenance benefits the resulting young.

Female choice for oviposition sites may be an especially import-ant determinant of male competition in those species, such as frogs and salamanders, showing external fertilization. Such female choice almost certainly predisposed these species to the evolution of male parental investment. Female choice for good oviposition sites would tend to favor any male investment in improving the site, and if attached to the site to attract other females the male would have the option of caring more or less for those eggs already laid. A similar argument was advanced above for birds. Internal fertilization and development mitigate against evolution of male parental care in mammals, since female choice can then usually only operate to favor male courtship feeding, which in herbivores would be nearly valueless. Female choice may also favor males who mate away from oviposition sites if so doing reduced the probability of predation.

Where females are clumped in space the effects of male com-petition may render female choice almost impossible. In a monkey troop a female preference for a less dominant male may never lead to sexual congress if the pair are quickly broken up and attacked by more dominant males. Apparent female acquiescence in the results of male–male competition may reflect this factor as much as the plausible female preference for the male victor outlined above.

**Summary**

The relative parental investment of the sexes in their young is the key variable controlling the operation of sexual selection. Where one sex invests considerably more than the other, members of the latter will compete among themselves to mate with members of the former. Where investment is equal, sexual selection should operate similarly on the two sexes. The pattern of relative parental investment in species today seems strongly influenced by the earlier evolutionary differention into mobile sex cells fertilizing immobile ones, and sexual selection acts to mold the pattern of relative parental investment. The time sequence of parental investment analyzed by sex is an important parameter affecting species in which both sexes invest considerable parental care: the individual initially investing more (usually the female) is vulnerable to desertion. On the other hand, in species with internal fertilization and strong male parental investment, the male is always vulnerable to cuckoldry. Each vulnerability has led to the evolution of adaptations to decrease the vulnerability and to counter-adaptations.

Females usually suffer higher mortality rates than males in monogamous birds, but in nonmonogamous birds and all other groups, males usually suffer higher rates. The chromosomal hypothesis is unable to account for the data. Instead, an adaptive interpretation can be advanced based on the relative parental investment of the sexes. In species with little or no male parental investment, selection usually favors male adaptations that lead to high reproductive success in one or more breeding seasons at the cost of increased mortality. Male competition in such species can only be analyzed in detail when the distribution of females in space and time is properly described. Data from field studies suggest that in some species, size, mobility, experience and metabolic rate are important to male reproductive success.

Female choice can augment or oppose mortality selection. Female choice can only lead to runaway change in male morphology when females choose by a relative rather than absolute standard, and it is probably sometimes adaptive for females to so choose. The relative parental investment of the sexes affects the criteria of female choice (and of male choice). Throughout, I emphasize that sexual selection favors different male and female reproductive strategies and that even when ostensibly cooperating in a joint task male and female interests are rarely identical.

# References

Bartholomew, G. A. (1970). A model of the evolution of pinniped polygyny. *Evolution* 24, 546—59.

Bastock, M. (1967). *Courtship: An ethological study*. Chicago: Aldine.

Bateman, A. J. (1948). Intrasexual selection in Drosophila. *Heredity* 2, 349—68.

Beebe, W. (1925). The variegated Tinamou *Crypturus variegatus variegatus* (Gmelin). *Zoologica* 6, 195—227.

Beer, J. R., Frenzel, L. D. and MacLeod, C. F. (1958). Sex ratios of some Minnesota rodents. *Am. Midl. Natur.* 59, 518—24.

Beverton, J. M., and Holt, S. J. (1959). A review of the lifespan and mortality rates of fish in nature and their relation to growth and other physiological characteristics. In *The lifespan of animals* (ed. G. Wolstenhome and M. O'Connor), pp. 142—177. Churchill, London.

Blair, W. F. (1960). *The Rusty Lizard*. University of Texas, Austin.

Bouliere, Z. F., and Verschuren, J. (1960). *Introduction a l'ecologie des ongules du Parc National Albert*. Bruxelles: Institut des Parcs Nationaux du Congo Belge.

Bragg, A. N. (1965). *Gnomes of the night*. University of Pennslyvania Press, Philadelphia.

Brown, L. E. (1966). Home range and movement of small mammals. *Symposium of the Zoological Society of London* 18, 111—42.

Bruce, H. (1960). A block of pregnancy in the mouse caused by the proximity of strange males. *Reprod. Fert.* 1, 96—103.

Burns, J. M. (1968). Mating frequency in natural populations of skippers and butterflies as determined by spermatophore counts. *Proc. Natn. Acad. Sci. U.S.A.* 61, 852—9,

Calder, W. A. (1967). Breeding behavior of the Roadrunner, *Geococcyx californianus*. *Auk* 84, 597—8.

Carpenter, C. (1967). Aggression and social structure in Iguanid Lizards. In *Lizard ecology* (ed. W. Milstead). University of Missouri, Columbia.

Chapman, A. B., Casida, L. E. and Cote, A. (1938). Sex ratios of fetal calves. *Proc. Amer. Soc. anim. Prod.* 1938, pp. 303—304.

Cooch, F. G., and Beardmore, M. A. (1959). Assortative mating and reciprocal difference in the Blue-Snow Goose complex. *Nature, Lond.* 183, 1833—4.

Corbet, P., Longfield, C., and Moore, W. (1960). *Dragonflies*. Collins, London.

Coulson, J. C. (1960). A study of the mortality of the starling based on ringing recoveries. *J. anim. Ecol.* 29, 251—71.

—— (1966). The influence of the pair-bond and age on the breeding biology of the kittiwake gull *Rissa tridactyla. J. anim. Ecol.* 35, 269—79.

Cowan, I. M. (1950). Some vital statistics of big game on overstocked mountain range. *Transactions of North American Wildlife Conference* 15, 581—8.

Darley, J. (1971). Sex ratio and mortality in the brown-headed cowbird. *Auk* 88, 560—6,

Darwin, C. (1871). *The descent of man, and selection in relation to sex*. John Murray, London.

Demerec, M., and Kaufmann, B. P. (1941). Time required for *Drosophila* males to exhaust the supply of mature sperm. *Amer. Natur.* 75, 366—79.

DeVore, I. (1965). Male dominance and mating behavior in baboons. In *Sex and behavior* (ed. Frank Beach). John Wiley and Sons, New York.

deVos, A., Broky, P., and Geist, V. (1967). A review of social behavior of the North American Cervids during the reproductive period. *Amer. Midl. Natur.* 77, 390—417.

Dunn, E. R. (1941). Notes on *Dendrobates auratus. Copeia* 1941, pp. 88—93.

Eaton, T. H. (1941). Notes on the life history of *Dendrobates auratus. Copeia* 1941, pp. 93—5.

Eisenberg, J. F. (1965). The social organizations of mammals. *Handbuch der Zoologie* 10 (7), 1—92.

Elder, G. (1969). Appearance and eduction in marriage mobility. *Amer. Sociolog. Rev.* 34, 519—33.

94    Reproductive strategies

Emlen, J. M. (1968). A note on natural selection and the sex-ratio. *Amer. Natur.* 102, 94–5.
Emlen, J. T. (1940). Sex and age ratios in the survial of the California Quail. *J. Wildl. Mgmt.* 4, 91–9.
Emlen, S. T. (1968). Terrioriality in the bullfrog. *Rana catesbeiana. Copeia* 1968, pp. 240–3.
Englemann, F. (1970). *The physiology of insect reproduction.* Pergamon Press, Oxford.
Fiedler, K. (1954). Vergleichende Verhaltensstudien an Seenadeln, Schlangennadeln und Seepferdchen (Syngnathidae). *Zeitsch. Tierpsych.* 11, 358–416.
Fisher, H. (1969). Eggs and egg-laying in the Laysan Albatross, *Diomedea immutabilis. Condor* 71, 102–12.
Fisher, R. A. (1958). *The genetical theory of natural selection.* Dover Publications, New York.
Fraser, A. F. (1968). *Reproductive behavior in Ungulates.* Academic Press, London and New York.
Gadgil, M., and Bossert, W. H. (1970). Life historical consequences of natural selection. *Amer. Natur.* 104, 1–24.
Goethals, G. W. (1971). Factors affecting permissive and nonpermissive rules regarding premarital sex. In *Studies in the sociology of sex: a book of readings,* (ed. J. M. Henslin). Appleton-Century-Croft, New York.
Goforth, W., and Baskett, T. (1971). Social organization of penned Mourning Doves. *Auk* 88, 528–42.
Guhl. A. M. (1951). Measurable differences in mating behaviour of cocks. *Poultry Sci.* 30, 687.
Haartman, L. von. (1951). Successive polygamy. *Behavior* 3, 256–74.
—— (1969). Nest-site and evolution of polygamy in European Passerine birds. *Ornis Fennica* 46, 1–12.
Hamilton, J. B. (1948). The role of testicular secretions as indicated by the effects of castration in man and by studies of pathological conditions and the short lifespan associated with maleness. *Recent Progress in Hormone Research* 3, 257–322.
—— and Johansson, M. (1965). Influence of sex chromosomes and castration upon lifespan: studies of meal moths, a species in which sex chromosomes are homogenous in males and heterogenous in females. *Anatomical Record* 24, 565–78.
—— and Mestler, G. E. (1969). Mortality and survival: comparison of eunuchs with intact men and women in a mentally retarded population. *J. Gerontol.* 24, 395–411.
——, Hamilton, R. S. and Mestler, G. E. (1969). Duration of life and causes of death in domestic cats: influence of sex, gonadectomy and inbreeding. *J. Gerontol.* 24, 427–37.
Hamilton, W. D. (1964). The genetical evolution of social behavior. *J. Theoret. Biol.* 7, 1–52.
—— (1967). Extraordinary sex ratios. *Science* 156, 477–88.
Harris, V. A. (1964). *The life of the Rainbow Lizard.* Hutchinson Tropical Monographs, London.
Hays, F. A. (1947). Mortality studies in Rhode Island Reds II. *Massachusetts Agricultural Experiment Station Bulletin* 442, 1–8.
Hershkovitz, P. (1966). Letter to *Science* 153, 362.
Hirth, H. F. (1963). The ecology of two lizards on a tropical beach. *Ecological Monographs* 33, 83–112.
Höhn, E. O. (1967). Observations on the breeding biology of Wilson's Phalarope (*Steganopus tricolour*) in Central Alberta. *Auk* 84, 220–44.
Huxley, J. S. (1938). The present standing of the theory of sexual selection. In *Evolution* (ed. G. DeBeer). Oxford University Press, New York.
Jacobs, M. (1955). Studies in territorialism and sexual selection in dragonflies. *Ecology* 36, 566–86.

Jensen, G. D., Bobbitt, R. A., and Gordon, B. N. (1968). Sex differences in the development of independence of infant monkeys. *Behavior* 30, 1–14.

Johns, J. E. (1969). Field studies of Wilson's Phalarope. *Auk* 86, 660–70.

Kaufmann, B. P., and Demerec, M. (1942). Utilization of sperm by the female *Drosophila melanogaster*. *Amer. Natur.* 76, 445–69.

Kessel, E. L. (1955). The mating activities of baloon flies. *Systematic Zoology* 4, 97–104.

Kikkawa, J. (1964). Movement, activity and distribution of small rodents *Clethrionomys glareolus* and *Apodemus sylvaticus* in woodland. *J. anim. Ecol.* 33, 259–99.

Kluijver, H. N. (1933). Bijrage tot de biologie en de ecologie van den spreeuw (*Sturnus vulgaris* L.) gedurende zijn voortplantingstijd. *Versl. Plantenziektenkundigen dienst, Wageningen,* 69, 1–145.

Koivisto, I. (1965). Behaviour of the black grouse during the spring display. *Finnish Game Research* 26, 1–60.

Kolman, W. (1960). The mechanism of natural selection for the sex ratio. *Amer. Natur.* 94, 373–7.

Kopstein, F. (1941). Über Sexualdimorphismus bei Malaiischen Schlangen. *Temminckia,* 6, 109–85.

Kruijt, J. P., Bossema, I. and deVos, G. J. (in press). Factors underlying choice of mate in Black Grouse. *15th Congr. Intern. Ornith.,* The Hague, 1970.

Kruijt, J. P., and Hogan, J. A. (1967). Social behavior on the lek in Black Grouse, *Lyrurus tetrix tetrix* (L.) *Ardea* 55, 203–39.

Lack, D. (1940). Pair-formation in birds. *Condor* 42, 269–86.

—— (1954). *The natural regulation of animal numbers.* Oxford University Press, New York.

—— (1968). *Ecological adaptations for breeding in birds.* Methuen, London.

LeBoeuf, B. J., and Peterson R. S. (1969). Social status and mating activity in elephant seals. *Science* 163, 91–3.

Lee, R. (1969). King Bushman violence. Paper presented at meeting of American Anthropoligcal Association, November, 1969.

Leigh, E. G. (1970). Sex ratio and differential mortality between the sexes. *Amer. Natur.* 104, 205–10.

Lin, N. (1963). Territorial behavior in the Cicada killer wasp *Sphecius spheciosus* (Drury) (Hymenoptera: Sphecidae.) I. *Behaviour* 20, 115–33.

Lindburg, D. G. (1969). Rhesus monkeys: mating season mobility of adult males. *Science* 166, 1176–8.

Lowther, J. K. (1961). Polymorphism in the white-throated sparrow, *Zonotrichia albicollis* (Gmelin). *Can. J. Zool.* 39, 281–92.

Ludwig, W., and Boost, C. (1951). Über Beziehungen zwischen Elternalter, Wurfgrösse und Geschlechtsverhältnis bei Hunden. *Zeitschrift fur indukt. Abstammungs und Vererbungslehre* 83, 383–91.

Madison, D. M., and Shoop, C. R. (1970). Homing behavior, orientation and home range of salamanders tagged with tantalum-182. *Science* 168, 1484–7.

Mason, L. G. (1969). Mating selection in the California Oak Moth (Lepidoptera, Droptidae). *Evolution* 23, 55–8.

Maynard Smith, J. (1956). Fertility, mating behaviour and sexual section in *Drosophila subobscura*. *J. Genet.* 54, 261–79.

Mayr, E. (1939). The sex ratio in wild birds. *Amer. Natur.* 73, 156–79.

—— (1963). *Animal species and evolution.* Harvard University Press.

Miller, R. S. (1958). A study of a wood mouse population in Wytham Woods, Berkshire. *J. Mammalogy* 39, 477–93.

Morris, D. (1952). Homosexuality in the Ten-spined Stickleback (*Pygosteus pungitius*). *Behaviour* 4, 233–61.

—— (1967). *The naked ape.* McGraw Hill, New York.

Myers, J., and Krebs, C. (1971). Sex ratios in open and closed vole populations: demographic implications. *Amer. Natur.* 105, 325–44.

Nevo, R. W. (1956). A behavior study of the red-winged blackbird. 1. Mating and nesting activities. *Wilson Bull.* 68, 5—37.

O'Donald, P. (1959). Possibility of assortative mating in the Arctic Skua. *Nature, Lond.* 183, 1210.

—— (1962). The theory of sexual selection. *Heredity* 17, 541—52.

Orians, G. H. (1969). On the evolution of mating systems in birds and mammals. *Amer. Natur.* 103, 589—604.

Parker, G. A. (1970*a*). The reproductive behaviour and the nature of sexual selection in *Scatophaga stercoraria* L. (Diptera: Scatophagidae) 2. The fertilization rate and the spatial and temporal relationships of each sex around the site of mating and oviposition. *J. anim. Ecol.* 39, 205—28.

—— (1970*b*). Sperm competition and its evolutionary consequences in the insects. *Biol. Revs.* 45, 525—68.

—— (1970*c*). The reproductive behavior and the nature of sexual selection in *Scatophaga stercoraria* L. (Diptera: Scatophagidae) VI. The adaptive significance of emigration from the oviposition site during the phase of genital contact. *J. anim. Ecol.* 40, 215—33.

—— (1970*d*). The reproductive behaviour and the nature of sexual selection in *Scatophaga stercoraria* L. (Diptera: Scatophagidae). VI. The adaptive significance of evolution of the passive phase. *Evolution* 24, 774—88.

Parker, J. E., McKenzie, F. F., and Kempster, H. L. (1940). Observations on the sexual behavior of New Hampsire males. *Poultry Sci.* 19, 191—7.

Peterson, R. S., and Bartholomew, G. A. (1967). *The natural history and behavior of the California Sea Lion.* Special Publications #1, American Society of Mammalogists.

Petit, C., and Ehrman, L. (1969). Sexual selection in *Drosophila.* In *Evolutionary biology,* vol. 5 (eds. T. Dobzhansky, M. K. Hecht, W. C. Steere). Appleton-Century-Crofts, New York.

Potts, C. R. (1969). The influence of eruptive movements, age, population size and other factors on the survival of the Shag (*Phalacrocorax aristotelis* L.). *J. anim. Ecol.* 38, 53—102.

Rand, A. S. (1967). Ecology and social organization in the Iguanid lizard *Anolis lineatopus. Proc. U.S. Nat. Mus.* 122, 1—79.

Riemann, J. G., Moen, D. J., and Thorson, B. J. (1967). Female monogamy and its control in house flies. *Insect Physiology* 13, 407—18.

Rivero, J. A., and Estevez, A. E. (1969). Observations on the agonistic and breeding behavior of *Leptodotylus pentaductylus* and other amphibian species in Venezuela. *Breviora No.* 321, 1—14.

Robel, R. J. (1966). Booming territory size and mating success of the Greater Prairie Chicken (*Tympanuchus cupido pinnatus*). *Anim. Behav.* 14, 328—31.

—— (1970). Possible role of behavior in regulating greater prairie chicken populations. *J. Wildl. Mgmt.* 34, 306—12.

Robinette, W. L., Gashwiler, J. S., Low, J. B., and Jones, D. A. (1957). Differential mortality by sex and age among mule deer. *J. Wildl. Mgmt.* 21, 1—16.

Rockstein, M. (1959). The biology of ageing insects. In *The lifespan of animals,* (eds. G. Wolstenhome and M. O'Connor), pp. 247—64. J. A. Churchill, London.

Rowley, I. (1965). The life history of the Superb Blue Wren *Malarus cyaneus. Emu* 64, 251—97.

Royama, T. (1966). A re-interpretation of courtship feeding. *Bird Study* 13, 116—29.

Sadleir, R. (1967). *The ecology of reproduction in wild and domestic mammals.* Methuen, London.

Savage, R. M. (1961). *The ecology and life history of the common frog.* Sir Isaac Pitman and Sons, London.

Schein, M. (1950). The relation of sex ratio to physiological age in the wild brown rat. *Amer. Natur.* 84, 489—96.

Schoener, T. W. (1967). The ecological significance of sexual dimorphism in size in the lizard *Anolis conspersus. Science* 155, 474—7.

—— (1971). Theory of feeding strategies. *Ann. Rev. Ecol. Syst.* 2, 369–404.

Scott, J. W. (1942). Mating behavior of the Sage Grouse. *Auk* 59, 477–98.

Selander, R. K. (1965). On mating systems and sexual selection. *Amer. Natur.* 99, 129–41.

—— (1966). Sexual dimorphism and differential niche utilization in birds. *Condor* 68, 113–51.

Sellers, A., Goodman, H., Marmorston, J., and Smith, M. (1950). Sex differences in proteinuria in the rat. *Amer. J. Physiol.* 163, 662–7.

Sheppard, P. M. (1952). A note on the non-random mating in the moth *Panaxia dominula*. (L.) *Heredity* 6, 239–41.

Stephens, M. N. (1952). Seasonal observations on the Wild Rabbit (*Oryctolagus cuniculus cuniculus* L.) in West Wales. *Proc. Zool. Soc., London* 122, 417–34.

Stokes, A., and Williams, H. (1971). Courtship feeding in gallinaceous birds. *Auk* 88, 543–59.

Sugiyama, Y. (1967). Social organization of Hanuman langurs. In *Social communication among primates* (ed. S. Altmann). University of Chicago Press.

Taber, R. D. and Dasmann, R. F. (1954). A sex difference in mortality in young Columbian Black-tailed Deer. *J. Wildl. Mgmt.* 18, 309–15.

Tilley, S. (1968). Size-fecundity relationships and their evolutionary implications in five Desmognathine salamanders. *Evolution* 22, 806–16.

Tinbergen, N. (1935). Field observations of East Greenland birds. 1. The behavior of the Red-necked Phalarope (*Phalaropus lobatus* L.) in Spring. *Ardea* 24, 1–42.

—— (1967). Adaptive features of the Black-headed Gull *Larus ridibundus* L. *Proceedings of the International Ornithological Congress* 14, 43–59.

Tinkle, D. W. (1967). The life and demography of the Side-blotched Lizard, *Uta stansburiana. Miscellaneous Publications of the Museum of Zoology, University of Michigan* 132, 1–182.

——, Wilbur, H., and Tilley, S. (1970). Evolutionary strategies in lizard reproduction. *Evolution* 24, 55–74.

Tyndale-Biscoe, C. H. and Smith, R. F. C. (1969). Studies on the marsupial glider, *Schoinobates volans* (Kerr). 2. Population structure and regulatory mechanisms. *J. anim. Ecol.* 38, 637–50.

Van Lawick-Goodall, J. (1968). The behavior of free-living chimpanzees in the Gombe Stream Reserve. *Animal Behaviour Monographs* 1, 161–311.

Verner, J. (1964). Evolution of polygamy in the long-billed marsh wren. *Evolution* 18, 252–61.

—— (1965). Selection for sex ratio. *Amer. Natur.* 99, 419–21.

—— and Willson, M. (1969). Mating systems, sexual dimorphism, and the role of male North American passerine birds in the nesting cycle. *Ornithological Monographs* 9, 1–76.

Williams, G. C. (1966). *Adaptation and natural selection.* Princeton University Press.

Willson, M., and Pianka, E. (1963). Sexual selection, sex ratio, and mating system. *Amer. Natur.* 97, 405–6.

Wood, D. H. (1970). An ecological study of *Antechinus stuartii* (Marsupialia) in a Southeast Queensland rain forest. *Aust. J. Zool.* 18, 185–207.

# Parental investment: a prospective analysis*

J. MAYNARD SMITH†

## Introduction

Trivers (1972), in a stimulating paper, analysed the evolution of sexual strategies in terms of 'parental investment', defined as 'any investment by the parent in an individual offspring that increases the offspring's chance of surviving (and hence reproductive success) at the cost of the parent's ability to invest in other offspring'. One of his central ideas is that selection will tend to favour desertion by whichever parent has invested least in the offspring up to that moment, and to favour continued investment by the other parent.

Dawkins and Carlisle (1976), while accepting most of Trivers's conclusions, have criticized his approach, on the grounds that selection will favour that strategy which maximizes the number of offspring which will be produced by an individual in the future, regardless of what that individual has done in the past. There can be no question that Dawkins and Carlisle are correct in saying that it is future expectations and not past expenditure which determine the selective value of behaviour. Trivers could defend himself by pointing out that he *defined* parental investment in terms of loss of future offspring, so that his argument is in effect prospective, even when it seems to have a retrospective form.

Formally, Trivers would be justified in such a defence. However, it is easy to be misled by his formulation, and it therefore seems better to do as Dawkins and Carlisle (1976) propose, and discuss the evolution of sexual strategies explicitly in terms of future consequences.

The most likely reason for error when analysing strategies from an investment point of view is that one will forget that a male (or female) which deserts cannot produce more offspring unless it can find another mate. Consider the following example. A species has a 1:1 sex ratio and an accurately synchronized breeding season.

* The Niko Tinbergen Lecture, 1976.
† Reprinted by permission of the author and *Animal Behaviour* 25 (1977).

Females expend considerable energy on maturing eggs before mating occurs, whereas males expend very little. A female will only mate with a male who has courted for some definite time, say 1 week. An uncritical application of Trivers's argument might lead one to conclude that selection would favour male desertion, whereas females should care for the young. But clearly a male gains nothing by desertion unless he can find a second female who will mate with him, which, in the artificial case I propose, he would not be able to do. In Trivers's sense, by courting the female for 1 week, the male has invested very heavily, because he has thereby lost all opportunity of mating any other female. But if one thinks of investment in terms of energy expenditure or risk of mortality, one would overlook this, and draw the wrong conclusion.

In this paper, I propose some simple mathematical models of sexual strategies. Such models have an obvious air of unreality when compared to the qualitative and verbal models discussed by Trivers. They have the corresponding advantage of forcing one to make one's assumption clearer. The purpose of mathematical formulation in this case is almost entirely to clarify the assumptions made; mathematical manipulation is minimal.

In developing these models, I have taken up two other suggestions made by Dawkins and Carlisle, one general and the other specific. The general point is that the appropriate formulation for such models is in terms of evolutionarily stable strategies or ESSs (Maynard Smith and Price 1973); we are seeking for a pair of strategies, say $I_m$ for males and $I_f$ for females, such that it would not pay a male to diverge from strategy $I_m$ so long as females adopt $I_f$, and would not pay females to diverge from $I_f$ so long as males adopt $I_m$. It will be important to remember that, for a given biological situation there may be more than one pair of strategies which are evolutionarily stable; if so, a species could evolve to one of two stable states, depending on its initial state.

The specific point raised by Dawkins and Carlisle, and discussed also by Trivers, is the contrast between the situation in mammals and birds, in which, if either sex contributes more to the care of the young, it is usually the female, and the situation in fish, in which it is at least equally often the male (Breder and Rosen 1966). It will be convenient to refer to the case in which the male deserts early and the female cares for the young as the 'duck' strategy, and the opposite as the 'stickleback' strategy. At first sight, the 'stickleback' appears to contradict Trivers's argument about investment. I shall offer an explanation of it which seems to be in accord with the spirit of Trivers's approach. My explanation differs from that offered by Dawkins and Carlise; however, I would not have looked at the problem in this way if they had not suggested it to me.

There follow three simple models. Model 1 assumes discrete breeding seasons. The success of a pair varies only with post-copulatory investment. This corresponds to the situation typical of birds (Lack 1968), in which the number of young successfully raised depends on the capacity of the parents to guard and feed the young, but is not limited by the number of eggs a female can lay: it would not help her to lay any more eggs. Model 2 also assumes discrete breeding seasons. The success of a pair varies both with post-copulatory investment, and also with the pre-copulatory investment made by the female; a female who invests heavily in eggs cannot guard as well as one who lays fewer eggs. Model 3 assumes continuous breeding.

## Model 1. Success limited mainly by parental care

Suppose that there are discrete breeding seasons, and that the number of surviving young per pair is $V_2$ if both parents care for the young and $V_1$ if only one parent does so; if both parents desert, no young survive. The female must make some pre-copulatory investment of time and energy before she can lay eggs, but, above a necessary minimum, differences in breeding success do not depend on differences in the extent of this investment.

A male who deserts his mate after she has laid her eggs (or at least, when it is too late for her to mate with a second male) has a chance $p$ of mating a second female. His fitness is then $V_1 + pV_2$, assuming for simplicity that he does not desert the second female. Therefore, desertion will be favoured if $V_1 + pV_2 > V_2$, or $p > (V_2 - V_1)/V_2$. That is, desertion is favoured if there is a good chance of finding a second female, and if one parent is almost as good as two in caring for the young.

A female whose mate deserts has fitness $V_1$, compared to $V_2$ if her mate stays. Is there a strategy that the female could adopt which makes it more likely that the male will stay? There is such an ESS. Consider a population of females which mate with a male only after the male has been paired with them for an appreciable time. Then, if the sex ratio is close to unity and if there is a fair degree of synchrony in pairing, a male which deserts has little chance of finding a second mate. That is, $p$ is small. Then, if $V_2$ is substantially greater than $V_1$, it would not pay a male to desert.

In order to ensure faithfulness, it is not necessary that the male should be compelled to invest anything other than time. However, once an ESS has evolved in which the male has nothing to gain by desertion after copulation, he would increase his fitness by investing before copulation, for example in nest-building.

What of the alternative ESS, in which females desert after laying their eggs, leaving the eggs to be cared for by the male, and seek a second male? Given the basic assumptions of the model, this 'stickleback' ESS is as likely as the 'duck' ESS already discussed. It is interesting that Kluijver (1951) reports the 'stickleback' strategy being adopted by individual female great tits. A possible reason why the 'stickleback' ESS is rarer, at least in birds, is as follows. Even in the absence of any delaying tactics on the part of the male, there is an inevitable time delay between the moment when a female copulates and the moment when she had laid her eggs and is ready to mate again. She is therefore unlikely to find an unmated male ready to care for her second batch of eggs. It is, however, worth noting that a distortion of the adult sex ratio could favour the establishment of one or other ESS, favouring desertion by the rarer sex.

## Model 2. Success limited by egg-laying and by parental care

Suppose now that the success of a pair during the breeding season depends not only on post-copulatory care, but also on the magnitude of investment by the female before egg-laying or parturition. It is assumed that the investment in eggs is substantial, so that a female who lays more eggs is less able to guard them, and vice-versa.

Let $P_0$, $P_1$ and $P_2$ be the probabilities of survival of eggs which are unguarded, guarded by one parent and guarded by two parents, respectively; $P_2 \geqslant P_1 \geqslant P_0$.

In each sex there are two possible strategies, $G$ and $D$, 'guard' and 'desert'. A male who deserts has a chance $p$ of mating again. A female who deserts lays $W$ eggs and one who guards lays $w$ eggs;

$$W \geqslant w$$

The pay-off matrix is then

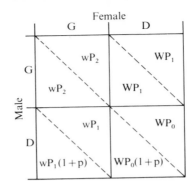

There are four possible ESSs:

ESS 1. $D♀$ and $D♂$
   requires $WP_0 > wP_1$, or the female will guard, and $P_0(1 + p) > P_1$, or the male will guard.

ESS 2. $D♀$ and $G♂$, the 'stickleback' ESS.
   requires $WP_1 > wP_2$, or the female will guard, and $P_1 > P_0(1 + p)$, or the male will desert.

ESS 3. $G♀$ and $D♂$, the 'duck' ESS.
   requires $wP_1 > WP_0$, or the female will desert, and $P_1(1 + p) > P_2$, or the male will guard.

ESS 4. $G♀$ and $G♂$.
   requires $wP_2 > WP_1$, or the female will desert, and $P_2 > P_1(1 + p)$, or the male will desert.

For given values of $W$, $w$, $P_0$, $P_1$, $P_2$ and $p$, ESS 1 and ESS 4 can be alternative possibilities, as can ESS 2 and ESS 3.
   The 'stickleback' ESS is favoured if

(i)  $W > w$; that is, if the female does not guard, she can lay more eggs, and

(ii) $P_1 \gg P_0$; $P_2$ not much greater than $P_1$; that is, one parent guarding is better than none, but two parents are not much better than one.

However, the 'duck' ESS is often an alternative possibility, especially if one parent is an adequate guard and if a male who deserts has a good chance of mating again.

## Model 3. 'Continuous' breeding

Consider a population in which there are no breeding seasons, and in which each individual breeds many times. 'Fitness' can then be measured by the *rate* at which an individual produces offspring which survive to enter the adult breeding population. Let these rates be $R_m$ and $R_f$, respectively, for males and females.

If the breeding sex ratio is $S$ males to one female, then, since the total number of offspring born to males in the population equals the number born to females,

$$S \cdot R_m = R_f \tag{1}$$

Let $T_m, T_f$ = times spent by males and females respectively caring for the offspring,

$M, F$ = time before a male or female which has bred is ready to breed again (i.e. required pre-copulatory investment).

$S_m, S_f$ = time spent searching for a new mate, by males and females respectively,

and $V$ = number of surviving offspring per mating, a function of $T_f$ and $T_m$.

The time spent by a female between starting one family and starting the next is $T_f + F + S_f$.

Therefore and
$$\left. \begin{array}{l} R_f = V/T_f + F + S_f \\ R_m = V/T_m + M + S_m \end{array} \right\} \tag{2}$$

From (1)

$$T_m + M + S_m = (T_f + F + S_f)S \tag{3}$$

In the following analysis, it is assumed that $T_m$ and $T_f$ are evolutionary variables, tending to an ESS, $T_m{}^*$, $T_f{}^*$. $M$ and $F$ are constant, and $S_m$ and $S_f$ are not under genetic control, but must satisfy (3). It is an implication of (3) that, with $S \simeq 1$, if males desert early ($T_m \ll T_f$) or if males invest less before copulation ($M \ll F$), then they must spend more time searching for a mate ($S_m \gg S_f$).

Consider first the case in which the number of surviving offspring, $V$, increases linearly with time invested in caring, up to a total time per parent of $\overline{T}$, as shown in Fig. 1. (The more plausible case of non-linear dependence is dealt with in the appendix: it is more difficult to follow but leads to similar conclusions.)

Suppose first that $\beta > a$; that is, two parents are not twice as

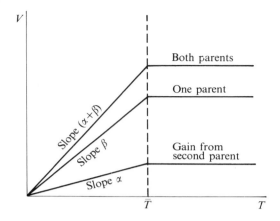

Fig. 1. Number of surviving offspring, $V$, as a function of the time $T$ that the parents spend guarding them.

good as one. In this case, it is possible that

$$a < R_m, R_f < \beta \qquad (4)$$

If so, there are two possible ESSs, the 'stickleback' and the 'duck' strategy. It will always pay one parent to desert, but it will then pay the other parent to stay on.

Given that two ESSs often exist, it is hard to predict which will actually evolve, since this depends on the initial condition.

Consider first the initial condition of no parental care. If there is internal fertilization, as for example in reptiles, female guarding is more likely to evolve, because only the female may be present when the eggs are laid. With external fertilization and pair formation, as in the anura, either ESS could evolve, but male guarding is more probable for the reasons discussed under model 2; in effect, the male has more to give. With communal external fertilization, as in shoaling fish, no parental guarding is likely to evolve, if only because no parent can identify its own offspring.

If the initial condition is for both parents to guard, either ESS is possible, but male desertion is more likely for the following reason. With 1:1 sex ratio, and $T_f^* = T_m^* = T$ say, we have $T + M + S_m = T + F + S_f$. That is, since $M < F$, only the males have a substantial searching time. The mean value of $S_m$ is fixed, but selection will favour males whose $S_m$ is less than the average. This could reduce the efficiency of the males in guarding, and so tip evolution in the direction of the 'duck' strategy. If $\beta < a$, implying that one parent is less than half as good as two, then as soon as it pays one parent to desert, it will pay the other to do so. The only possible ESSs are for both parents to care for the young, or for neither to do so.

## Discussion

It is a feature of all three models that two alternative ESSs are often possible; the 'duck' strategy in which the male deserts and only the female cares for the young, and the 'stickleback' strategy in which only the male cares for the young. These two possibilities are likely if one parent is almost as effective as two in caring for the young, and if the prospects of a deserting parent mating again are good. It is important to remember that, although in species without parental care it pays a male to mate as often as possible and a female to mate only once, in species with parental care a female may have as much to gain as a male by mating twice.

In those species in which the ESS is for one parent to desert, it will tend to be the male which cares for the young if the female has invested so much in eggs that she cannot effectively do so, or if there is an excess of males; it will tend to be the female which cares for the young if the timing of mating is such that a deserting male has a better chance of re-mating than a deserting female, or if there is an excess of females. As will become apparent below, however, it is often far from obvious why evolution has taken one path rather than the other.

In contrast, if two parents can raise twice as many offspring as one, or if the chance that a deserting parent will re-mate is small, then monogamy with both parents caring for the young is the likely ESS.

These conclusions overlap with those drawn by Orians (1969). His essential argument is that polygyny will always be advantageous to males, and hence its presence or absence will depend on the advantages and disadvantages to females. The main factors making for polygyny will therefore be (i) a variable habitat (e.g. marshes, early successional stages), so that it may pay a female to mate bigamously with a male in an optimal territory than monogamously with a male in a marginal one, and (ii) a way of life in which the male can contribute little to the survival of the young. In my models I have not allowed for the variability of the habitat; instead, I have attempted to allow for the second factor, and at the same time, following Trivers (1972), to take into account differences in the extent and timing of parental investment.

I will now review very briefly the strategies observed in different vertebrate groups, and indicate some features which are comprehensible in terms of the preceding analysis and some which are unexplained.

In mammals, the dominant factor is that only females feed the young on milk. In herbivorous and fish-eating mammals, as opposed to land carnivores, the males do not feed their young. In some species the males protect their young from predation, but the main selective force on males seems to have been to increase the number of females mated, for example by holding a harem or a territory containing several females. Monogamy is found in some carnivores and primates. In primates it is always associated with territoriality; Clutton-Brock and Harvey (1977) have argued that monogamy is characteristic of primate species in which a male can hold a territory large enough to support a single female and her offspring. If a male can hold a larger territory than this, territoriality is combined with polygyny. If a male cannot hold an area sufficient for one female, territoriality is replaced by a group with a home range.

Thus in mammals female lactation makes female desertion unlikely to be an ESS. In species in which males do not feed their offspring, polygyny is common. There are, however, monogamous species, and it is odd that no such species has evolved male lactation. Model 2 suggests that, since females must make a large investment before parturition, it would be evolutionarily stable for the male to make a larger investment after parturition. It may be that a species with a typical mammalian physiology and behaviour cannot easily evolve the 'stickleback' strategy, even though this would be stable once it had evolved.

Birds differ from mammals in that the absence of lactation makes desertion by either sex a possible ESS. They differ from lower vertebrates in that the number of offspring successfully produced in a breeding season is not usually limited by the number of eggs a female can lay, since the egg number is restricted to the number of offspring the parents can feed, or, if the young are not fed, to the number of eggs the parent can incubate (Lack 1968). Thus model 1 is the relevant model. If the parents feed the young, either in the nest as in passerine and raptorial birds, or after leaving the nest as, for example, in grebes (*Podiceps*) and divers (*Gavia*), the typical pattern is for both members of a monogamous pair to feed the young. This is consistent with the model; if $V_2 \gg V_2$ it does not pay either parent to desert.

If the parents do not feed the young it is common for the male to desert after copulation. In ducks (*Anas*) and most game birds, only the female incubates and cares for the young. In both groups the clutch size is large; as argued under model 1, this makes male desertion more likely to evolve than female desertion, because a deserting female would be too late to find a second mate. In some species, particularly game birds showing lek behaviour, selection on the males is clearly acting to increase the number of females mated rather than the productivity of a single mating.

The evolution of breeding systems in grouse (Tetraonidae) has been reviewed by Wiley (1974). Some species (*Tetrastes, Lagopus*) form more or less permanent pair bonds, although only in the red grouse, *Lagopus lagopus,* does the male contribute significantly to the care of the young. It is not easy to see the ecological or other reasons why some species form pair bonds, whereas others are promiscuous. Among the latter, lek formation (i.e. males congregating to display) is characteristic of open-country species, whereas forest species usually display from widely dispersed sites; Wiley suggests that this is because of the greater vulnerability to predation of the latter. A striking point made by Wiley is that the polygnous species not only show greater sexual dimorphism in size, but are also absolutely larger in both sexes. A similar correlation between large size and sexual dimorphism exists in primates (Clutton-Brock and Harvey 1977). If large size helps a male to mate with more females, sexual dimorphism is to be expected (for a recent discussion, see Selander 1972), but the association with larger absolute size is less easy to explain. I believe that the most plausible explanation is that the primary selection is sexual selection for greater size in males, and that the increased size of females is an unselected and possibly deleterious side effect.

There is an interesting contrast between the ducks (*Anas*) on the one hand, and geese (*Anser*) and swans (*Cygnus*) on the other. Geese and swans are usually monogamous, and although only the female incubates, the male remains on guard near the nest and helps protect the young. The explanation may be that ducks protect their young by nest concealment and by injury-feigning, whereas geese and swans often breed in habitats where nest concealment is difficult, and are large enough to attack most potential predators. Two geese are better than one, but a duck may do better unaided by a conspicuous drake.

Kear (1970) argues that biparental care is primitive in waterfowl, and that male desertion has arisen in those genera (*Anas, Aythya, Oxyura*) in which the ducklings have the most perfect adaptation to water, and are therefore better insulated and less in need of brooding. One end point of this evolutionary path is the stifftail, *Heteronetta,* which is an obligate brood parasite, the duckling receiving no parental care after hatching. From model 2, it is easy to see that if $P_0$ (the chance of survival without parental care) is as great, or almost as great, as $P_1$ or $P_2$, then the ESS will be for both parents to desert (ESS 1).

The waders are interesting in including most of the rare examples of polyandry found among birds (Jenni 1974). In most waders, both parents incubate and care for the young, which must feed themselves. The American jacana, *Jacana spinosa* (Jenni and Collier 1972), breeds throughout the year. The female holds and defends a large territory, within which several males hold separate territories. The males construct the nest and incubate and care for the young. The female is 75 per cent heavier than the male. Jenni (1974) makes the interesting suggestion that an intermediate stage between typical monogamy and polyandry of this type is the habit of double-clutching found among waders (e.g. the Sanderling, *Calidris alba* and Temminck's Stint, *Calidris temminckii*). In these species, the female produces two clutches, leaving the first to be incubated and reared by the male while she incubates and rears the second. This can be interpreted as an adaptation to a short and unpredictable breeding season.

Whatever the adaptive reason for double-clutching it is clear that in such species the male must care for the first clutch. It is then a short step to the situation found in the spotted sandpiper, *Actitis macularia* (Hays 1972; Oring and Knudson 1972), in which, if there is a local excess of males, a female will produce a series of clutches cared for by different males, staying with the last male to help rearing. It is then possible to imagine a stage in which a female could obtain additional males to raise her young by driving off other females, thus leading to the situation in the jacana. It should be remembered, however, that double-clutching is found in arctic waders with a short and unpredictable breeding season, whereas the polyandrous jacana breeds continuously in a tropical environment.

*Rhea americana* affords a different and unique type of polyandry (Bruning 1973). There is a defined breeding season, at the start of which males establish a dominance hierarchy. Dominant males acquire a harem of up to 15 females, and copulate with each female every 2 to 3 days. These females lay an egg every other day for from 7 to 10 days, in a nest constructed by the male, until up to 50 eggs have been laid. The male starts incubating a few days after the first egg is laid, and, soon after, drives the females away. Each female, however, then consorts with a second male, and may consort with and lay eggs for up to seven males in series during the season.

It is easy to see how this system can be evolutionarily stable. A dominant male gains by obtaining as many eggs as he can incubate, fathered by himself, and produced early in the season at the optimal time (Bruning 1973, shows that eggs laid late in the season are less likely to survive). A female gains by ensuring that all the eggs she can lay will be incubated without delay. It does not seem that either sex could improve its fitness by a change in strategy. If this explanation is correct, then the rhea differs from most birds in that the capacity of a female to lay eggs rapidly is an important factor limiting her breeding success.

In the mallee-fowl, *Leipoa ocellata* (Frith 1962), the major role in incubation is played by the male, who constructs a complex mound of dead leaves and sand as an incubator. The female lays eggs over a period of 3 to 4 months. The total laid varies with the female and the season, but a female may lay up to 30 eggs, each 10 per cent of her own weight. During the egg-laying period the male alone constructs and tends the mound. As in the rhea, breeding success is limited by the number of eggs a female can lay, but unlike the rhea, the mallee-fowl is monogamous for life. This cannot be taken as an example of the 'stickleback' strategy, since male expenditure in tending the mound is simultaneous with female expenditure in egg-laying. It does, however, illustrate the conclusion that if the female's capacity to lay eggs is limiting (i.e. model 2 rather than model 1), then monogamy with the male tending the eggs is an ESS.

I can say little useful about the reptiles and amphibia. Parental care, other than nest-building, is absent in tortoises and very rare in lizards, although female *Eumeces* and *Ophisaurus* brood their eggs, and in one species, *E. obsoletus,* the female stays with the young for some 10 days after hatching. In the amphibia, parental care is unusual, but can take a variety of forms. It seems to be about equally frequent for males and females to care for the young; it is the male in *Phyllobates, Dendrobates, Rhinoderma* and the midwife toad *Alytes obstetricans,* and the female in *Pipa* and *Gastrotheca.* A comparative study of the anura would be particularly valuable.

In the bony fish, if only one sex cares for the young it is often the male, as in stickleback, pipefish, seahorses and many others. There is, however, an important and interesting exception to this rule in *Tilapia*, a large genus of cichlid fishes from the African great lakes (Fryer and Iles 1972). In many species, the female broods the young in her mouth. In these species, the males occupy territories within which they construct 'nests' which are in fact mating sites. A male mates with any female entering his territory. In effect, these are 'lek' species. Other species of *Tilapia* are monogamous, and jointly defend their territory and protect their young. In two species, *T. galilaea* and *T. multifasciata,* monogamy is associated with mouth-brooding, and in these both parents mouth-brood.

It will be clear from this brief review that in many groups of vertebrates uniparental care of the young is found in species in which the parent guards but does not feed the young. This is as expected, and a plausible explanation can be offered for some of the exceptions (e.g. geese). It is by no means so clear what determines which sex shall care for the young. In a sense, this is also expected, since theory predicts that in many cases the 'duck' or the 'stickleback' strategy could be stable, so that the actual state of affairs now depends on the 'initial conditions' in an ancestral species of which we can know nothing except by inference.

Some reasons can, however, be tentatively offered for present asymmetries. It is a prediction of model 2 that if breeding success is limited by female investment before the eggs are laid (or the young born), then it will be the male which will guard the young. This is confirmed by the fact that it is the males which guard in many bony fish (the exceptions in the genus *Tilapia* are lek species), and by some of the rare cases of male guarding in birds (*Rhea, Leipoa*). But it is not clear why female guarding is found among amphibia (*Pipa, Gastrotheca*), nor why monogamous mammals have not evolved male lactation.

But there are many problems still to be investigated. The cases of female guarding in the amphibia (e.g. *Pipa, Gastrotheca*) and in the bony fishes call for an explanation, even if it is a historical one. And perhaps most intriguing of all, why don't male mammals lactate?

## Appendix: Continuous breeding

Consider the case in which the number of surviving offspring, $V$, depends non-linearly on the times $T_m$ and $T_f$ for which the male

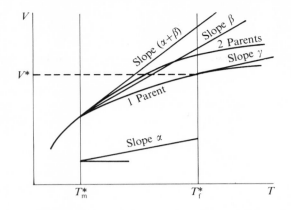

Fig. 2. Number of surviving offspring, $V$, as a function of the time $T$ that the parents spend guarding them. $T_m^*$ and $T_f^*$ are the times when typical males and females desert. The slope $\alpha$ (obtained as the difference between $(\alpha + \beta)$ and $\beta$) measures the rate at which a mutant male would increase his breeding success if he stayed after time $T_m^*$. The slope $\beta$ measures the rate at which a female increases her breeding success by staying after time $T_m^*$. The slope $\gamma$ measures the rate at which a mutant female would increase her breeding success by staying after time $T_f^*$.

and female parents guard the young. This is analysed in Fig. 2, in which it is assumed that $T_f^* > T_m^*$. The graphs of $V$ are taken as fixed; that is, values of $T_m$ and $T_f$ can evolve, but the dependence of $V$ on $T$ does not.

The slopes $a$, $\beta$ and $\gamma$ are defined in Fig. 2. The male ESS, $T_m^*$, must satisfy the condition $a = R_m^*$, because as soon as $a < R_m^*$ it will pay a male to desert. By similar reasoning, the female ESS, $T_f^*$, must satisfy the condition $\gamma = R_f^*$. With a 1:1 sex ratio, $R_m^* = R_f^*$ and hence $a = \gamma$.

If, as is biologically plausible, $d^2V/dT^2$ is negative (that is, if there is a law of diminishing returns from guarding), then $\gamma < \beta$, and hence $a < \beta$. This inequality must be satisfied if there is to be an ESS in which one parent deserts before the other. That is to say, uniparental guarding will evolve only if two parents are less than twice as good as one. If $a > \beta$ (that is, two parents are more than twice as good as one), then as soon as one parent deserts it will pay the other to do so.

If $a \ll \beta$, it is usually possible to find two ESSs, one with $T_f^* > T_m^*$ (the 'duck' strategy') and one with $T_m^* > T_f^*$ (the 'stickleback' strategy). The actual values will depend on what assumptions are made about $S_f$ and $S_m$. These searching times are subject to the constraint of equation (3). A plausible assumption would be that if $(T_m + M) < (T_f + F)S$ then $S_f = 0$, and otherwise $S_m = 0$. Once one of these values is determined, the other is given by equation (3). Once $S_m$ and $S_f$ have been determined, the actual ESS can be found by a process of successive approximation, but this is of little biological interest and therefore is not described here. What is of interest, however, is the effect of $M$ and $F$ on the

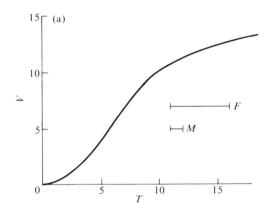

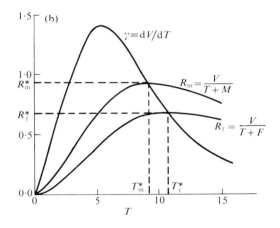

Fig. 3. Alternative ESSs with continuous breeding. (a) fixed values of $F$, $M$, and of $V$ as a function of $T$. (b) graphs of $\gamma = dV/dT$, $R_m$ and $R_f$. The possible ESSs are $T_m = 0$, $T_f = T_f{}^*$, and $T_m = T_m{}^*$, $T_f = 0$.

nature of the ESS. $M$ and $F$ are the investments, measured in time, which must be made by a male or a female respectively after the end of guarding before the individual is ready to breed again. In general we would expect $F > M$; it takes more time and energy to make eggs than sperm.

This question is analysed for the simple case of a 1:1 sex ratio, and $a = 0$; that is, one parent is as good as two, so that at the ESS one or other parent deserts immediately. Fig. 3(a) shows the assumed dependence of $V$ on $T$. Consider first the ESS with $T_\mathrm{m} = 0$. Then the average rate at which females are producing offspring is $R_\mathrm{f}* = V*/(T_\mathrm{f}* + M)$. We can find $T_\mathrm{f}*$ by solving the equation

$$\gamma = \mathrm{d}V/\mathrm{d}T = V/(T + F)$$

since as soon as $\gamma$ falls below $V/(T + F)$ it would pay the female to desert. This is done graphically in Fig. 3(b). Notice that this is mathematically equivalent to choosing that $T$ which maximizes the rate of production, $V/(T + F)$, because

$$\frac{\mathrm{d}}{\mathrm{d}T} \frac{V}{(T + F)} = \frac{1}{(T + F)} \left\{ \frac{\mathrm{d}V}{\mathrm{d}T} - \frac{V}{T + F} \right\}$$

We can find in the same way the value of $T_\mathrm{m}*$ which is an ESS when $T_\mathrm{f} = 0$. Note that the productivity of the population is higher for the 'stickleback' ESS; this conclusion does not depend on the assumption that $a = 0$. Given that two ESSs often exist, this greater productivity is irrelevant to which will evolve, unless we invoke group selection. Possible factors determining which ESS will evolve were discussed under model 3.

## References

Breder, C. M. and Rosen, D. E. (1966). *Modes of Reproduction in Fishes*. New York: Natural History Press.

Bruning, D. F. (1973). The Greater Rhea chick and egg delivery route. *Natural History*, 82, 68–75.

Clutton-Brock, T. H. and Harvey, P. H. (1977). Primate ecology and social organisation. *J. zool* 183, 1–39.

Dawkins, R. and Carlisle, T. R. (1976). Parental investment and mate desertion: a fallacy. *Nature, Lond.* 262, 131–133.

Frith, H. J. (1962). *The Mallee-Fowl.* Angus and Robertson, Sydney.

Fryer, G. and Iles, T. D. (1972). *The Cichlid Fishes of the Great Lakes of Africa.* Oliver and Boyd, Edinburgh.

Hays, H. (1972). Polyandry in the Spotted Sandpiper. *Living Bird,* 11, 43–57.

Jenni, D. A. (1974). Evolution of polyandry in birds. *Am. Zool.,* 14, 129–144.

—— and Collier, G. (1972). Polyandry in the American Jacana (*Jacana spinosa*). *Auk.* 89, 743–765.

Kear, J. (1970). The adaptive radiation of parental care in waterfowl. In: *Social Behaviour in Birds and Mammals* (ed. by J. H. Crook). Academic Press, London.

Kluijver, H. N. (1951). The population ecology of the Great Tit, *Parus m. major* L. *Ardea*, 39, 1–135.

Lack, D. (1968). *Ecological Adaptations for Breeding in Birds*. Methuen, London.

Maynard Smith J. and Price, G. R. (1973). The logic of animal conflicts. *Nature, Lond.*, 246, 15–18.

Orians, G. H. (1969). On the evolution of mating systems in birds and mammals. *Am. Nat.*, 103, 589–603.

Oring, L. W. and Knudsen, M. L. (1972). Monogamy and polyandry in the Spotted Sandpiper. *Living Bird*, 11, 59–73.

Selander, R. K. (1972). Sexual selection and dimorphism in birds. In: *Sexual Selection and the Descent of Man* (ed. by B. Campbell). Heinemann, London.

Trivers, R. L. (1972). Parental investment and sexual selection. In: *Sexual Selection and the Descent of Man* (ed. by B. Campbell). Heinemann, London.

Wiley, R. H. (1974). Evolution of social organisation and life-history patterns among grouse. *Q. Rev. Biol.*, 49, 201–227.

# On the evolution of mating systems in birds and mammals

GORDON H. ORIANS†

Mating systems and the selective factors that molded them have had an important place in the history of the theory of natural selection. Darwin (1871) himself gave considerable thought to the nature of sexual selection and its consequences for sexual dimorphism and mating patterns. He proposed two major forces in the evolution of sexual differences. First, that the fighting and display among males for the possession of females, which is especially prominent among mammals, accounted for the evolution of secondary sexual characteristics, such as horns and antlers, which are useful in battle. This aspect of sexual selection has been generally accepted. Second, Darwin suggested that the extreme development of plumage characters among males of some birds, such as pheasants and birds of paradise—features which did not seem of use in intermale combat—could be explained as being due to the cumulative effects of sexual preference exerted by the females at the time of mating. This aspect of his theory of sexual selection was challenged by a number of workers, but Fisher (1958) clearly showed that the notion of female choice is reasonable, notwithstanding the fact that direct evidence was then scarce for species other than man.

More recently, mating systems and sexual dimorphism have been assigned an important role in the theory of Wynne-Edwards (1962) as a device for regulating the reproductive output of populations. According to Wynne-Edwards, polygyny is one of a series of restrictive population adaptations arising through group selection which controls populations by reducing collective fecundity. He argues (1962, p. 515) that this restriction of breeding activity is possible because the territorial males of polygynous and promiscuous species can be fully informed about their own reproductive activity and, if the species engages in displays at communal mating grounds, of the total of matings performed by the group as

† Reprinted with permission from the author and *American Naturalist,* **103** (1969).

well. These males could be conditioned to respond by becoming sexually inactive when an appropriate number of matings had taken place. The value of polygyny and group displays would lie in the fact that the assessment of total reproductive output by the population would be much easier than with a monogamous mating system in which the individuals are spaced out through the environment. He further suggested (1962, p. 525) that a balanced sex ratio would be maintained in nonmonogamous species because it would facilitate more intensive intermale competition and thereby provide a more sensitive index of population density and total reproductive output.

This theory would best be tested by direct demonstration of the processes that are postulated to occur. For example, if females of polygynous species are unable to mate because males withhold coition after a certain number of copulations have been achieved, if low-ranking males do not attempt to solicit copulations after the quota has been reached, or if females are not receptive to their advances if they are made, then confidence in the theory would be strengthened.

Such evidence is extremely difficult to gather in the field, but there are now available data from a number of intensively studied polygynous species of birds, indicating that all females which appear in the breeding areas are able to obtain males and raise young. Jared Verner (personal communication) has not found any evidence for unmated females in the long-billed marsh wren (*Telmatodytes palustris*), and my own intensive work on red-winged blackbirds (*Agelaius phoeniceus*) and yellow-headed blackbirds (*Xanthocephalus xanthocephalus*) has failed to reveal a nonbreeding population of females in either of these species. In both species, there is a large and readily observable floating population of males. In the great-tailed grackle (*Quiscalus mexicanus*), a highly promiscuous species, adult males remain in full breeding condition and continue to attempt copulations after all females are nesting and are no longer receptive (Selander 1965). Therefore, in these species if there is some mechanism for limiting reproductive output, it must be due either to the failure of more females to present themselves at the breeding grounds or to a lowered effort per reproducing individual. At present, there is no evidence from any species that males withhold coition from receptive females.

It is the purpose of this paper to present an alternative theory of mating systems among birds and mammals which is based upon the assumption that the evolution of mate-selection behavior by

individuals of both sexes has been influenced primarily by the consequences of these choices for individual fitness. The model is
based upon a mate-selection processes that can be observed directly
in the field, and it is capable of generating a set of predictions
which can be tested against the general mating patterns of broad
groups of species for which there are no detailed observations of
the factors influencing mating behavior.

### A natural selection model of mating systems

This model is built upon the work of a number of people especially
Maynard Smith (1958), Verner (1964), Verner and Willson (1966),
Lack (1968), and Willson and Pianka (1963), to which I have
added some original ideas. Existing knowledge of the mating
patterns of birds and mammals has been summarized by Lack
(1968) and Eisenberg (1966), respectively. All theories of sexual
selection involve an element of choice, and mine is no exception.
In order for discrimination to be selected for, it is necessary that
(a) the acceptance of one mate generally precludes the acceptance
of another, and (b) the failure to accept one mate will be followed
by an opportunity to mate with other individuals with such a high
probability that the loss in reproductive output resulting from
rejection of a potential mate is, on the average, less than the average
gains that can be realized by obtaining a mate of superior fitness
(Fisher 1958, p. 144).

The first condition is met by both sexes of many species and by
females in virtually all species. Basically, a female produces gametes
with a large amount of stored energy, while a male produces
gametes with a complete set of genetic instructions but no significant amount of stored energy. Consequently, the number of
gametes that can be produced by males is potentially, and in most
cases actually, very large. On the other hand, the number of eggs
produced by females is ultimately limited by the amount of energy
that can be mobilized for their production or subsequent care. It
follows that males can be expected to increase the number of
offspring they produce by mating with more than one female, but
females should not have more offspring by successive matings with
more than one male unless one male were to provide insufficient
gametes to fertilize all the eggs, an unlikely condition (Maynard
Smith 1958, p. 146).

For the same reason, errors in mate selection are more serious
for females than for males. An interspecific mating that produced

inviable or sterile offspring might claim the entire season's gamete production for a female, while the male could have erred to the extent of no more than a few minutes and a few readily replaced gametes.

The inescapable conclusion is that mate selection will be practised whenever sensory capabilities and locomotor abilities permit it and that females will, in the vast majority of cases, exercise a stronger preference. It is a well-known fact that males of many species court rather indiscriminately and can, especially when deprived of sexual activity for some time, be induced to mate with remarkably incomplete stimulus objects. Such behavior could not have evolved if errors were strongly selected against among males. For this reason, the following model assumes that females make a choice among available males. Since polygyny must always be advantageous to males, its presence or absence must depend primarily upon the advantages or disadvantages of the females.

I also assume that the environment inhabited by a species is variable and that mean reproductive success uncomplicated by density effects is correlated with this variation in quality of the habitat. For the purposes of graphic presentation, environments are treated as though they can be ordered linearly with respect to their intrinsic quality, as measured by reproductive success, but this is not essential to the argument. A model based upon these assumptions is presented in Fig. 1.

There are two bases upon which female choice could be made. The first, already mentioned, is the genetic quality of the male, that is, the nature of the genes that will be given to the offspring from a mating with that male. Given the existence of such differences, female choice must inevitably be under strong selection, since those females mating with more. fit males will thereby produce offspring that are more fit, on the average, than females mating with less fit males. Therefore, females should evolve to be especially responsive to those morphological and/or behavioral traits of males which reflect their fitness (Fisher 1958, p. 151).

In many species, however, the role of the male in reproduction is more extensive, involving provision of food for the offspring, possession of a territory within which resources can be gathered, protection against predation or inclement weather, and so forth. In these cases, I expect selection by the females to be influenced also by the quality of the territory and the probability that the male is capable of and disposed toward taking an active role in the

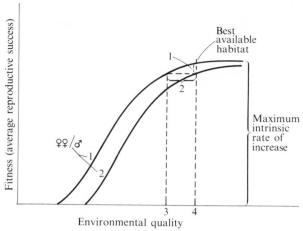

Fig. 1. Graphic model of conditions necessary for the evolution of polygynous mating patterns. Average reproductive success is assumed to be correlated with environmental differences and females are assumed to choose their mates from available males. The distance *1* is the difference in fitness between females mated monogamously and females mated bigamously in the same environment; the distance *2* is the *polygyny threshold*, which is the minimum difference in quality of habitat held by males in the same general region sufficient to make bigamous matings by females favored by natural selection.

care and defense of the offspring. The model accomodates both cases.

The model assumes that mean reproductive success is correlated with the quality of the environment in which the individuals are living. The exact shape of the function is immaterial as long as the slope is everywhere positive. Given this condition, the best strategy for a female is to mate with a male on the best quality habitat and to rear her offspring, with or without his help, in that environment. However, as more individuals settle in these optimal environments, the average reproductive success will be expected to drop for three major reasons. First, the more individuals that are exploiting a given environment, the greater the likelihood that reduction of resources will adversely influence reproductive success. Second, the higher density of individuals may attract more predators, thereby increasing the probability of nest destruction. Third, if the male does play a role in the care of the offspring and females settle at a greater density than the males, then the aid of the male will have to be shared among the females, each getting less than if the male were able to devote his full attention to the offspring of one of them. For this reason, a curve representing the average reproductive success of females mated to males having more than one mate is drawn below the curve for females involved in monogamous matings.

Whatever the relative positions of these two curves, a situation may eventually be reached at which the quality of habitat on the territories of unmated males is such that the expected reproductive success of a newly arriving female is higher if she attempts to mate with a male already with one female but on a superior-quality habitat, rather than mating with an unmated male on poorer habitat. The difference in quality of habitats occupied by mated and unmated males required to make a bigamous mating advantageous for a female may be designated the *polygyny threshold* (Verner and Willson 1966), and polygyny is expected to evolve only when this situation regularly presents itself to the females of a species.

It follows that the likelihood that polygyny will evolve is influenced by all factors which determine how broad a range of environmental conditions will be occupied by the individuals of a species. For example, if the territories of the males of a species are nearly equal in quality and there are still unpaired males available, it is unlikely that the mean reproductive success of females attempting bigamous matings would be higher. Similarly, all factors that influence the amount of difference in mean reproductive success of females in monogamous and bigamous matings in equivalent environments will affect the likelihood that polygyny will evolve. More specifically, the greater the difference, that is, the farther apart the two curves are, the farther to the left will lie the polygyny threshold, and there must be greater differences in habitat quality for polygyny to pay for the females. Conversely, with less difference, that is, closer curves, a smaller difference in habitat will push part of the population to the left of the polygyny threshold.

*a) Factors promoting occupancy of habitats differing widely in quality*: The existence of individuals to the left of the polygyny threshold will be facilitated by (1) low mortality rates during the nonbreeding season, so that not all individuals can be accommodated in the better areas, and (2) the existence of large differences in the quality of habitats actually occupied by the species.

*b) Factors influencing the difference in average reproductive success of monogamously and bigamously mated females*: A major factor affecting the differences in reproductive success of females will be the role of the male in the care of the offspring. If his role is limited to the provision of genetic information, that is, if the male provides no food, territory, or protection, the success of the females will not be affected by the number of other females that have mated with the males unless his fertility declines with successive matings. Thus, in the limiting case of no male parental care, the two curves become congruent, and the mating choice by the females is based strictly upon the genotype of the male. A high degree of promiscuity is to be expected in such species.

Even if the male also cares for the offspring, in which case the number of other females he already has is a major factor in female choice, there are conditions which tend to minimize the reduction in reproductive success attendant upon polygynous matings. For example, if the successive females of a male are staggered in their breeding so that the periods of dependence of their offspring overlap little or not at all, more support from the male could be available (Verner 1964). This should give selective advantage to two different forms of female behavior. The first is the attempt to exclude additional females from the territory of the male until such time as the overlap in dependency periods of the young would be minimized, and second, the avoidance by newly arriving females of territories in which a prior female was just beginning to breed.

Another factor influencing the difference in reproductive success of monogamously and bigamously mated females is the nature of the food resources present in the area. If the food for the young is not being replenished during the breeding period, the individuals breeding earliest should experience the best foraging conditions, while later-breeding individuals are exploiting an already depressed supply. However, if the recruitment to the food supply is considerable during the breeding period, conditions for later breeders may be no worse (or may even be better) than conditions for earlier breeders. Therefore, other things being equal, species exploiting food supplies that are continually renewed are more likely to cross the polygyny threshold.

The above arguments all assume that the number of offspring being raised by a female (or pair) has evolved under the influence of natural selection to correspond to that which, on the average, is the largest number for which sufficient energy can be mobilized. The theoretical basis for this assumption and the supporting empirical data have been extensively summarized by Lack (1954 pp. 21–44). However, in some species the number of offspring produced is strongly influenced by other factors. In such cases, the existence of male parental care may be of little consequence, and mate choice should be made primarily or strictly on the basis of phenotype and territory quality. Cases in which this situation may be operative will be discussed later.

The model implies the existence of several process which can be directly observed in nature. For example, it should be true that females mate with already mated males when unmated males are readily available and perceived by the females. Evidence that this is true has been gathered for the long-billed marsh wren (Verner 1964), the red-winged blackbird, and the yellow-headed blackbird (Orians, unpublished observations). The great variations in number of females mated with different males in other polygynous species are suggestive of the widespread occurrence of this phenomenon. Verner has also shown that the number of females attracted to a male long-billed marsh wren is correlated with the features of his territory relatable to the available food.

The model also predicts that there should not be a negative correlation between average reproductive success per female and number of females mated with a given male, since females are assumed to enter polygynous matings only when it is advantageous for them to do so. This prediction has been verified for the red-winged blackbird (Haigh 1968).

The model does not require a skewed sex ratio in the breeding population for the initiation and evolution of polygynous mating patterns. This is important because there are theoretical reasons for expecting a sex ratio near equality when the young become independent in most species (Fisher 1958; Kolman 1960) and because sex ratios at the time of fledging are near equality in all polygynous species so far investigated (Haigh 1968; Selander 1960, 1961; Williams 1940).

Using the postulates upon which the polygyny model was erected, a series of seven predictions about mating patterns can be made. These predictions are subject to direct verification or falsification. Current knowledge of mating patterns among mammals, though by no means complete, is extensive enough that I can reasonably assess the goodness of the predictions. Moreover, the predictions from the model serve to draw attention to those cases which would be most rewarding of further study. Though predictions of wide application can be made, I restrict my present consideration to birds and mammals, the groups with which I am most familiar.

1. *Polyandry should be rare among all animal groups.* This prediction follows directly from the basic attributes of maleness and femaleness. A female could presumably increase her reproductive output if several males could be induced to care for her offspring, but such a situation would in most cases be sufficiently disadvantageous to the males to cause the evolution of male behavior patterns that would prevent the system from evolving. However, once a basic sexual role reversal had evolved in a species, males might profit by associating themselves with females on better territories, thus leading to polyandry.

The actual incidence of polyandry among birds and mammals is difficult to assess. Among mammals, Eisenberg (1966) reviewed several polygynous or promiscuous mating patterns but found no good case of a simultaneous association of a female with more than one male. Among birds the situation is confused by the fact that there has been a tendency to assume polyandry in all those species with a reversal of sexual roles. Most such species are tropical and subtropical in distribution and are not well known ecologically, making comparisons even more difficult. For example, it was formerly believed that the phalaropes (Phalaropodidae) were polyandrous, but recent data indicate that, though incubation and care of the young are exclusively the role of the males, the species are nonetheless monogamous (Höhn 1967). However, there is good evidence of polyandry for at least some species in five groups of birds, all with precocial young: the button quails (Turnicidae), painted snipe (Rostratulidae), jacanas (Jacanidae), tinamous (Tinamidae) and rails (Rallidae). Details in all these cases are summarized by Lack (1968).

Among rheas (Rheidae), emus (Dromiceidae), cassowaries (Casuariidae), kiwis (Apterygidae), tinamous and button quails, the males normally incubate the eggs and care for the young, but most species are apparently monogamous. In the best-studied species, the brushland tinamou (*Nothoprocta cinerascens*), the males defend territories while the females travel in small groups. Several females lay eggs in a single nest, but each female may mate with several males in rapid succession (Lancaster 1964). The classification of the type of pattern is somewhat ambiguous since the male may have several females at once, but each female may nevertheless mate with several different males during the laying of a single 'clutch' of eggs. A clearer case of polyandry with a reversal of parental role is provided by the pheasant-tailed jacana (*Hydrophasianus chirurgus*) (Hoffmann 1949).

The evolution of sexual role reversal in birds may have had its origins in a monogamous system with equal sharing of parental care by the two sexes. In many such species, the males incubate first while the females recover the energy lost during egg production. Under such circumstances, especially among species with precocial young, if the female were able to obtain enough energy to produce more eggs, it would be advantageous to mate with another male were one available. It would also be advantageous for the incubating males to induce other females to mate with them and deposit their eggs in the nests, so that the tinamou type of system could readily evolve. A further advantage of such a system is that the length of time the nest is available to be destroyed by predators is reduced, since a full clutch of eggs is placed in the nest in a shorter period of time than if a single female were doing all the laying.

2. *Monogamy should be relatively rare among mammals but should be the predominant mating pattern among birds.* The physiology of mammalian reproduction dictates a minor role of the male in the care of the offspring, whereas among birds the only activity for which males are not equally adept as females is egg laying. This prediction is readily verified. In his extensive review of mammalian social organization, Eisenberg (1966) pointed out that there are very few cases of known monogamy among mammals, the apparent exceptions being the marmosets (Callithricidae), gibbons (Hylobatinae), beavers (Castoridae), the hooded seal (*Cystophora cristata*), the only pinniped known to form a stable family unit, and a number of terrestrial carnivores, such as foxes, badgers, and viverrids. The situation is more difficult to determine among large ungulates, but temporary pair bonds that are apparently monogamous may form in hyraxes (*Dendrohyrax*), rhinoceroses, and some deer (*Capreolus capreolus, Odocoileus*).

The mammals among which monogamy is probably the most prevalent are the terrestrial carnivores, and they provide the most prominent exception to the generalization that the role of the male is limited or nonexistent in the care of offspring. For a carnivore, capturing food is a difficult task, and males can and do make kills and deliver the prey to either the female, who converts it to milk, or to the young once they are old enough to be able to ingest meat. It is difficult to imagine a comparable role for a male herbivore.

In contrast, monogamy is the prevalent mating pattern in the majority of bird species in virtually all families and orders. Assuming that all hummingbirds are promiscuous, Lack (1968) surmises that about 91% of all bird species are monogamous. Given the properties of the model, this is to be expected; but polygyny and promiscuity do exist among birds, and the model, if it is to be generally useful, must provide predictions capable of explaining those cases. Because polygyny should seldom evolve among birds, this group provides a particularly useful test case for the validity of the model. Fortunately, the mating patterns of birds are well enough known to allow tests of the predictions in most cases.

3. *Polygyny should evolve more readily among precocial birds than among altricial species.* This prediction follows from the ability of many precocial young to find their own food and be relatively independent of the provisioning activities of the adults. This decreases the potential role of the male. There are species with precocial young, such as gulls, in which the young, though able to run around actively at birth, are not able to forage for themselves. These species are not included in this prediction.

This prediction is only partially fulfilled in nature. Polygynous and promiscuous species are numerous among upland game birds (Phasianidae, Tetraonidae), and there are a few species among the shorebirds (ruff, *Philomachus pugnax*; buff-breasted sandpiper, *Tryngites subruficollis*; pectoral sandpiper, *Erolia melanotos*; and great snipe, *Capella media*), but most members of the Charadriidae (plovers) and Scolopacidae (sandpipers) are monogamous (Lack 1968, p. 116). Polygyny is also rare in the Anatidae (swans, geese, and ducks). In many of the monogamous species, both sexes take an active role in the care of the young, leading them to suitable foraging areas and keeping on the alert for predators. In geese the male defends the nest and young from predators, but in most species of ducks the female alone cares for the young, and yet monogamous pair bonding seems to be the rule. It is significant that the known exceptions among ducks are all tropical species (Lack 1968, p. 123), suggesting that perhaps the prevalence of monogamy among high latitude species may be the result of the advantage of pair formation on the wintering ground and rapid initiation of breeding which give a stronger advantage to monogamy for the males than would otherwise be the case.

4. *Polygyny is likely to evolve in species with altricial young that nest in marshes.* The marsh environment possesses several

features that make it more likely that the polygyny threshold will be crossed than in any other environment. First, the range in productivity of marshes greatly exceeds that found in upland habitats (Verner and Willson 1966). Differences of over tenfold are not unusual in aquatic environments, whereas the difference between the most productive and least productive woods is much less. Moreover, great differences in productivity in a terrestrial environment are likely to result in a sufficiently altered vegetational profile to cause a change in species rather than the occupation of a broad gradient by one species (L. L. Wolf, personal communication). In marshes, however, productivity differences are not necessarily associated with vegetation structure, and striking changes in vegetational features regularly occur within the span of a few years (Weller and Spatcher 1965), so that opportunities for evolving species that occupy only a small segment of the marsh vegetation pattern are more limited than in terrestrial environments.

The food supply for insectivorous birds in marshes is often rapidly renewed. Many of the breeding passerine birds of marshes exploit primarily the emerging individuals of insects with aquatic larval and terrestrial adult stages. These insects are vulnerable for only a few hours during their lives, and those not taken on the day they emerge are mostly unavailable on subsequent days. Therefore, the supply of food on a given day is not significantly affected by the number of insects removed from the system on previous days but rather upon those factors that regulate the size of the emergence on that particular day. In contrast, the supply of insects on the foliage of trees and shrubs may be seriously depleted by the foraging activities of birds, which lowers the expected reproductive success of later arriving individuals.

In a review of the mating systems of North American passerines, Verner and Willson (1966) demonstrated that, though marsh-nesting species constitute only a small fraction of the total species (about 5%), eight of the 15 polygynous species breed in marshes. Polygyny is also prevalent among the marsh-nesting weaverbirds (Ploceidae) in Africa (Crook 1963, 1964). Also, the only known nonpasserine species with altricial young that is regularly polygynous is the bittern (*Botaurus stellaris*), a marsh-nesting species (Gaukler and Kraus 1965).

5. *Polygyny should be more prevalent among species inhabiting early successional habitats.* Like marshes, early successional terres-

trial vegetation changes rapidly, thus discouraging the evolution of species adapted specifically to minor variants of it. In addition, there are reasons for suspecting that variations in food supply in early successional sites might be considerable. Early successional plants are characterized by rapid growth and the apportionment of large amounts of energy to reproduction. They probably also devote less energy, on the average, to antiherbivore devices than plants or later successional stages. Accordingly, they should be vulnerable to insect attack when found, and may owe their success in part to the fact that many patches escape detection. If this is true, patches of early successional vegetation should consist of some not yet found by insects, therefore containing relatively little food, and others supporting large populations of grazing insects. Five of the 15 regularly polygynous species of North American passerines breed iṇ prairie or savannah habitants (Verner and Willson 1966; Zimmerman 1966), and some of them, notably the dickcissel (*Spiza americana*) and the bobolink (*Dolichonyx oryzivorus*) are restricted to the very early successional stages of grassland vegetation (Zimmerman 1966). Differential food supply and its posssible correlation with the number of females per male in different patches of early successional vegetation has never been measured, but it should not be difficult to do so.

6. *Polygyny should be more prevalent among species in which feeding areas are widespread, but nesting sites are restricted.* If nest sites are restricted and a single male holds several of them, it should be advantageous for females to mate with such males even if they are already mated, particularly if the alternative is accepting an inferior site or no site at all. Two of the polygynous passerines of North America, the house wren (*Troglodytes aedon*) and the winter wren (*T. troglodytes*), nest in cavities but are unable to excavate their own, and the same is true of the polygynous pied flycatcher (*Muscicapa hypoleuca*) of Europe (Curio 1959; von Haartman 1954). This may also be the explanation of the prevalence of polygyny among savannah species of weaverbirds in Africa and Asia (Crook 1962, 1963, 1964), since these are species that feed in grassland but require trees for their nests.

7. *Polygyny and promiscuity should be more prevalent among species in which clutch size is strongly influenced by factors other than the number of offspring that can be supported by the parents.* Clutches smaller than the number of young the parents can feed successfully might occur in species in which the adults feed

primarily on low-energy food sources such as pulpy fruits and nectar which, though sufficient for energy maintenance, are not good for egg production. This supposition is supported by the fact that hummingbirds, unlike most avian species, lay no more than two eggs in all geographical areas and do not show the latitudinal gradient in clutch size characteristic of most birds. In addition, high predation rates may select against high feeding rates in tropical environments, reducing clutches below what the parents could feed (Skutch 1949). Finally, in stable environments competitive ability may demand considerable time expenditure (Cody 1966), so that foraging time is reduced and males may spend all or most of their time at these activities. If any of these factors are operating, the contribution of the male to reproductive success by means of food delivery to the nestlings would be decreased, and conditions favorable for polygyny would be created.

This prediction cannot be tested directly at present, but it is noteworthy, as pointed out by Snow (1963), that polygyny or promiscuity associated with lek displays in the tropics occur only among fruit- and nectar-eating birds  such as hummingbirds, manakins (Snow 1962a, 1962b), cotingas (Snow 1961; Gilliard 1962), birds of paradise (Iredale 1950) and bowerbirds (Marshall 1954; Gilliard 1959a, 1959b), and not among insectivorous species, including insectivorous species in the same families, all of which are apparently monogamous. Only one of the lek species has been studied in detail (Snow 1962a, 1962b), and it was shown that fruits supporting the adult could not have been in short supply during the breeding season, but the effects of this kind of food source on egg production and nature of the food delivered to the young are completely unknown. There is evidence from the bowerbirds that frugivorous species feed their young on insects (Marshall 1954). Conversely, obligatory fruit eaters, such as parrots, have extremely long nestling periods, suggesting that rapid nestling growth and a fruit diet are mutually incompatible.

### Consequences of the evolution of polygyny

If polygynous mating systems evolve from monogamous ones as a result of the existence of choice situations in which it is advantageous for females to select mates already having at least one mate, the very existence of this choice system creates other selective forces that further influence the mating pattern and the morphology of the sexes.

First, in polygynous birds, there should be very keen competition among males for the better-quality territories, because possession of a high-quality territory is likely to result in the attraction of more than one female. The increased intermale competition for good areas should lead to stronger selection for secondary sexual characteristics useful in these contests. The existence of these characteristics, as indicated earlier, is well known among mammals, and Selander (1958) has demonstrated a strong correlation between the amount of sexual dimorphism in size and the degree of polygyny and promiscuity in mating pattern among American blackbirds (Icteridae). Great dimorphism in size is also character-istic of the polygynous marsh-nesting weaverbirds in Africa, the males even showing remarkable convergence toward the plumage patterns shown by males of polygynous species of icterids.

Nevertheless, unless the species are continually evolving to become more highly dimorphic, there must be counterselection against the more dimorphic individuals that stabilizes the degree of dimorphism at its present value. There are two obvious candidates for this counterselection. The first is predation, since the males are rendered exceedingly conspicuous by both their appearance and their behavior patterns. The second derives from the adverse ecological effects of larger size. In polygynous species, females are presumably not normally under selection for size other than that dictated by their basic ecological relationships with their environ-ment (Amadon 1959). The greater the degree to which males depart from the presumably optimal size of the females, the more poorly adapted they should be for general existence, unless by their increased size they are able to exploit food resources not available to the smaller female. To date the only demonstration of a higher mortality rate during the nonbreeding season for males of a highly dimorphic species is that of Selander (1965) for the great-tailed grackle in south-central Texas. Sexing birds returning to communal roosts by examination of greatly enlarged photo-graphs, Selander was able to show that males died at about twice the rate of the females during the winter. He also observed that the large tails of the males interfered with flying in strong winds, and that extremely strong wings completely prevented males from flying, while females were still able to navigate, though with diffi-culty.

Contrary to the theory of Wynne-Edwards, the model developed here predicts delayed maturation on the part of the males but not of females. Unless females are capable of preventing other females from settling in the area, all females should be able to obtain mates and reproduce. In fact, it should be extremely difficult for females to exclude other females from the territory of their mate. During the nest-building and egg-laying periods, defense of the area is easy; but once incubation has begun, eviction of a persistent intruder can only be accomplished at the expense of chilling and possible loss of the clutch of eggs. It is highly unlikely that the adverse effects of the second female could be so great as to select for such behavior. Moreover, by the time the first female is already incubating, the potential period of overlap in time when young are being fed is already minimized.

On the other hand, the strong competition among males for suitable territories and the failure of males with poor territories to obtain mates at all should produce a floating population of non-breeding males. Such floating populations are known to be characteristic of a number of polygynous species of birds. Assuming that older and more experienced birds will be at an advantage in competition for territories, the chances of success for younger birds should be very low. If attempts to obtain territories result in higher mortality rates of the young birds and probability of success is sufficiently low, individuals making vigorous attempts might be selected against and delayed maturation would result. Selander (1965) gives a more detailed development of this argument. In the red-winged blackbird, first-year males do not acquire the full adult plumage, their testes develop later in the spring and do not reach the size characteristic of older males (Wright and Wright 1944), and they usually do not breed, though they may do so if the supply of adult males is reduced in some manner (Orians, unpublished field data).

## Summary

Predictions from a theory assuming mate selection on the part of females, which maximizes reproductive success of individuals, are found to accord closely though not completely, with known mating patterns. These predictions are that (1) polyandry should be rare, (2) polygyny should be more common among mammals than among birds, (3) polygyny should be more prevalent among precocial than among altricial birds, (4) conditions for polygyny should be met in marshes more regularly than among terrestrial environments, (5) polygyny should be more prevalent among species of early successional habitats, (6) polygyny should be more prevalent among species in which feeding areas are widespread but nesting sites are restricted, and (7) polygyny should evolve more readily among species in which clutch size is strongly influenced by factors other than the ability of the adults to provide food for the young. Most cases of polygyny in birds, a group in which monogamy is the most common mating pattern, can be explained on the basis of the model, and those cases not apparently fitting into the predictions are clearly indicated. Thus, there is no need at present to invoke more complicated and restrictive mechanisms to explain the mating patterns known to exist.

# References

Amadon, D. (1959). The significance of sexual differences in size among birds. *Amer. Phil. Soc., Proc.* 103, 531–6.

Cody, M. L. (1966). A general theory of clutch size. *Evolution* 20, 174–84.

Crook, J. H. (1962). The adaptive significance of pair formation types in weaver birds. *Symposia (Zool. Soc. Lond.)* 8, 57–70.

—— (1963). Monogamy, polygamy, and food supply. *Discovery* 24, 35–41.

—— (1964). The evolution of social organization and visual communication in the weaver birds (Ploceinae). *Behaviour (Suppl. 10)*, 1–178.

Curio, E. (1959). Verhaltenstudien am Trauerschnäpper. *Z. Tierpsychol.* 3, 1–118.

Darwin, C. (1871). *The descent of man and selection in relation to sex.* Appleton, New York.

Eisenberg, J. F. (1966). The social organizations of mammals. *Handbuch der Zoologie* 8, (39), 1–92. Walter de Gruyter, Berlin.

Fisher, R. A. (1958). *The genetical theory of natural selection.* Dover, New York.

Gaukler, A., and Kraus, M. (1965). Zur Brutbiologie der grossen Rohrdommel (*Botaurus stellaris*). *Vogelwelt* 86, 129–46.

Gillard, E. T. (1959a). Notes on the courtship behavior of the blue-backed manakin (*Chiroxiphia pareola*). *Amer. mus. Novitates. No.* 1942, 1–19.

—— (1959b). A comparative analysis of courtship movements in closely allied bowerbirds of the genus *Chlamydera. Amer. Mus. Novitates. No.* 1936, 1–8.

—— (1962). On the breeding behaviour of the cock-of-the-rock (Aves, *Rupicola rupicola*). *Amer. Mus. Natur. Hist. Bull.* 124, 31–65.

Haartman, L. von. (1954). Der Trauerfliegenschnäpper. III. Die Nahrungsbiologie. *Acta Zool. Fennica,* 83, 1–96.

Haigh, C. R. (1968). *Sexual dimorphism, sex ratios, and polygyny in the red-winged blackbird.* Ph.D. thesis. Univ. Washington. Seattle.

Hoffmann, A. (1949). Über die Brutpflege des polyandrischen Wasserfasens, *Hydrophasianus chirurgus* (Scop.). *Zool. Jahrb. (Syst.)* 78, 367–403.

Höhn, E. O. (1967). Observations on the breeding of Wilson's phalarope (*Steganopus tricolor*) in central Alberta. *Auk* 84, 220–44.

Iredale, T. (1950). *Birds of paradise and bower birds.* Georgian H., Melbourne.

Kolman, W. A. (1960). The mechanism of natural selection for the sex ratio. *Amer. Natur.* 94, 373–7.

Lack, D. (1954). *The natural regulation of animal numbers.* Clarendon Press, Oxford.

—— (1968). *Ecological adaptations for breeding in birds.* Methuen, London.

Lancaster, D. A. (1964). Biology of the brushland tinamou, *Nothoprocta cinerascens. Amer. Mus. Natur. Hist. Bull.* 127, 270–314.

Marshall, A. J. (1954). *Bower-birds.* Clarendon Press, Oxford.

Maynard Smith, J. (1958). *The theory of evolution.* Penguin, Harmondsworth.

Selander, R. K. (1958). Age determination and molt in the boat tailed grackle. *Condor* 60, 355—76.

—— (1960). Sex ratio of nestling and clutch size in the boat-tailed grackle. *Condor* 62, 34—44.

—— (1961). Supplemental data on the sex ratio of nestling boat-tailed grackles. *Condor* 63, 504.

—— (1965). On mating systems and sexual selection. *Amer. Natur.* 99, 129—41.

Skutch, A. F. (1949). Do tropical birds rear as many young as they can nourish? *Ibis* 91, 430—55.

Snow, B. K. (1961). Notes on the behavior of three contingidae. *Auk* 78, 150—61.

Snow, D. W. (1962a). A field study of the black and white manakin, *Manacus manacus,* in Trinidad. *Zoologica* 47, 60—104.

—— (1962b). A field study of the golden-headed manakin, *Pipra erythrocephala,* in Trinidad, W. I. *Zoologica* 47, 183—98.

—— (1963). *The evolution of manakin displays.* In 13th Int. Ornithological Congr., Proc., Ithaca, N.Y., 1962, p. 553—61.

Verner, J. (1964). Evolution of polygamy in the long-billed marsh wren. *Evolution* 18, 252—61.

—— and Willson, M. F. (1966). The influence of habitats on mating systems of North American passerine birds. *Ecology* 47, 143—7.

Weller, M. W., and Spatcher, C. S. (1965). Role of habitat in the distribution and abundance of marsh birds. *Iowa State Univ. Agr. Home Econ. Exp. Sta. Spec. Rep. No. 43.*

Williams, J. F. (1940). The sex ratio in nestling eastern red-wings. *Wilson Bull.* 52, 267—77.

Willson, M. F., and Pianka, E. R. (1963). Sexual selection, sex ratio, and mating system. *Amer. Natur.* 97, 405—7.

Wright, P. L., and Wright, M. H. (1944). The reproductive cycle of the male red-winged blackbird. *Condor* 46, 46—59.

Wynne-Edwards, V. C. (1962). *Animal dispersion in relation to social behaviour.* Oliver and Boyd, Edinburgh.

Zimmerman, J. L. (1966). Polygyny in the dickcissel. *Auk* 83, 534—46.

# Section 3

# Co-operation and disruption

# Section 3

# Co-operation and disruption

## Introduction

Co-operative or beneficient behaviour poses a problem to evolutionary theorists. We have argued that the evolutionary process is grounded in competition between individuals, yet animals often combine to assist each other: co-operative courtship exists in several bird species (e.g. Watts and Stokes 1971) and co-operative rearing is also common (Skutch 1961; Lack 1968) and occurs in some mammals (Epple 1967; Kühme 1965; Bertram 1976).

There are three ways in which co-operative behaviour is likely to evolve: symbiosis (or mutualism), kin selection, and reciprocal altruism. In a symbiotic relationship both individuals gain immediately from the interaction (Clapham 1973). The more commonly quoted cases are examples of interspecific co-operation (see Barrington 1967), but the same mechanism can favour intra-specific co-operation. One of the best examples is provided by the evolution of group living. Three kinds of advantage of sociality are commonly suggested:

(1) Increased detection, avoidance, or defence against predators (Patterson 1965; Lack 1968; Crook 1970; Vine 1971). While it is obvious that several pairs of eyes are better than one, there is a more complex reason why group living reduces an individual's chances of being eaten. This is described in the paper by Hamilton (1971a) reproduced on p. 142. Hamilton's argument is that an individual that places itself next to another animal reduces its own chances of being eaten if a randomly placed predator selects the nearest individual. This advantage would apply to most group-living amimals, except in cases where the chance of an animal being

taken increases in proportion to group size as a result of increased visibility or reduced opportunity to manoeuvre in groups.

(2)   Increased efficiency of food finding (Ward 1965; Zahavi 1971; Krebs, MacRoberts, and Cullen 1972; Krebs 1974; Thompson, Vertinsky, and Krebs 1974), food handling (Emlen 1973; Schaller 1972; Kruuk 1972), food defence (Alexander 1974; Emlen 1973), or exploitation (Cody 1971; Clutton-Brock 1975). In particular, the reader's attention should be drawn to Krebs' elegant demonstration that group foraging helps individuals to find food supplies in great blue herons, *Ardea herodias* (Krebs 1974). However, it is evident that food finding is not an adequate explanation of group living in all cases. For example, in several species, groups tend to reduce in size when food is less abundant.

(3)   Reproductive advantages including regular access to the opposite sex (Emlen 1973; Brown 1975; Wrangham 1975) and the presence of peers during the period of social learning (Hinde 1974).

The selective forces favouring grouping probably represent a combination of these factors. The important point to note here is that all individuals which form a group benefit from the association. This does not mean that grouping is inevitably advantageous, otherwise we should need to explain why whole populations do not aggregate in single groups. Constraints on increasing group size are probably often set by feeding interference (Alexander 1974).

The second way in which co-operative behaviour can evolve is through kin selection. We have already discussed Hamilton's formulation of the theory (see p. 7) and described its application to the evolution of facultative sterility in the social Hymenoptera. In many cases where animals help each other to mate or to raise offspring, they are closely related (see Skutch 1961; Harrison 1969; Fry 1972; Maynard Smith and Ridpath 1972; Watts and Stokes 1971; Clutton-Brock and Harvey 1976). The paper by Bertram (1976) included in this volume describes the probable role which kin selection has played in the evolution of three types of co-operation between members of lion prides: communal suckling, male tolerance of cubs at kills, and the lack of competition between pride males for access to oestrous females. Perhaps the most likely explanation of alarm calls in birds is that they, too, are the product of kin selection—a view proposed by Maynard Smith in 1965 (see p. 183). However, at least four alternative explanations exist (see also Dawkins 1976):

(1)   Calls decrease the likelihood that a member of the caller's group will be caught, thus discouraging the predator from hunting the same prey in the same area and decreasing the chance that the caller will be caught in future (see Trivers 1971).

(2)   Calls alert other group members, enabling the group to hide or to prepare for co-operative defence or distraction display, and thus reduce the chance that the caller will be caught (Alcock 1975).

(3)   The call announces to the predator that the caller has seen him and is therefore not worth pursuing because it is alert and will be difficult to catch (see Smythe 1970; Alcock 1975).

(4)   Despite appearances, the caller is manipulating other group members to their own disadvantage. Animals who hear the call are informed of the presence of danger but not of its location. In their search for cover, some individuals, at least, are likely to increase their susceptibility to being caught, thereby reducing the caller's own chance of being killed (Charnov and Krebs, 1975).

The third way in which co-operative behaviour can evolve is through 'reciprocal altruism'. This is the term given to cases where a beneficient action by one individual adds more to the recipient's fitness than it detracts from the benefactor's and is likely to be returned in future (thus enhancing the fitness of both benefactor and recipient). It is distinguished from symbiosis by the fact that it involves a time lag between the beneficient act and its reward. The principle of reciprocal altruism and a variety of predictions concerning its distribution are described in Trivers' original (1971) paper (see p. 189). In practice, it is often difficult to distinguish between reciprocal altruism and kin selection since both are most likely to occur under the same conditions (see p. 196) and it is seldom possible to be sure that two individuals are *not* related. As Trivers points out, kin selection can be ruled out where the co-operative relationship involves members of different species. The two examples of reciprocal altruism between conspecifics quoted by Trivers (alarm calls in birds and co-operative relationships in humans) are less convincing, since there are several alternative explanations in both cases. The short paper by Packer (1977) on the formation of supportive coalitions in olive baboons (*Papio anubis*) provides one of the most convincing illustrations of reciprocal altruism yet available. However, even here, some of the co-operating males may have been related.

In any co-operative relationship, selection is likely to favour individuals which extract more than they are prepared to give. In relationships depending on reciprocation, this may lead to selection for various forms of cheating and to concomittant pressures favouring methods for identifying cheats (see p. 205). In co-operative relationships involving kin, similar selection pressures will occur, but the situation is more complex since individuals which short-change their relatives will, to some extent, be acting to the detriment of their own genes. Trivers (1974, p. 233) has investigated the selection pressures operating on parents and offspring in situations where their interests conflict. For example, offspring should favour the continuation of parental care beyond the point at which it is in the parent's interest to discontinue investment and conserve its resources for the next infant. Similarly, it may be advantageous for offspring to attempt to prevent their parents investing in sibs if this is likely to lead to decreased investment in themselves. The extent to which parents and offspring should disagree will depend on the degree of relatedness among offspring, the costs of investment of the parent and the benefits to the offspring. Situations where the interests of parents and offspring conflict, Trivers suggests, may be responsible for periods of agonistic interaction (such as weaning conflicts) in the course of co-operative relationships.

Since competition for resources is common in animal societies, the frequent occurrence of aggressive behaviour is not surprising (Wilson 1971). Descriptive or motivational definitions of aggressive behaviour have proved problematic (Hinde 1974) since elements of aggression are shown in many contexts. It is for this reason that Clutton-Brock and Harvey (1976, see p. 293) consider 'disruptive behaviour' (defined as 'acts which diminish the fitness of the recipient') rather than aggression. 'Maleficent behaviour' would be a more precise description, but the word has gone from current English usage.

Perhaps because disruption poses a less obvious problem to evolutionary theory than beneficence, it has attracted less attention. Nevertheless, the functional significance of many aspects of disruptive behaviour is not obvious and requires careful interpretation. All disruptive actions involve some cost (see Tinbergen 1951; Hutchinson and MacArthur 1959; Geist 1974) which may range from negliglible expenditure of time or energy to risk of serious injury or certain death. Whether an individual is prepared to enter a contest and the risks it will take while fighting should be determined by the benefits of winning, the costs of fighting, and its probability of losing.

Even when fighting for 'valuable' resources, animals often appear to avoid inflicting serious injuries on each other and, in many species, fighting behaviour is extensively ritualized. This has commonly been explained as having evolved for the good of the species (Huxley 1956; Lorenz 1966; Eibl-Eibesfeldt 1970). In fact, it is unnecessary to invoke group selection in this case: individuals should avoid escalating contests whenever the costs of fighting exceed the potential benefits. However, the situation is complex, since the advantages of fighting dangerously (for the individual) will depend on the reactions of other members of the population: in a population of non-escalators, an individual who always fights dangerously would be likely to enhance its fitness. The realization that the advantages of particular social strategies must be considered in relation to the behaviour of other individuals led Maynard Smith to develop the concept of the 'Evolutionarily Stable Strategy' (ESS). This is the strategy which, when adopted by most of the members of a population cannot be bettered by particular 'mutant' strategies. For example, Maynard Smith and Price (1973) have considered the advantages of five different strategies for behaviour in contest situations, ranging from a 'mouse' strategy (where the animal always retreats whenever its opponent escalates) to 'hawk' (where the animal escalates the contest until he is seriously injured or his opponent retreats). The treatment shows that the only single strategy to form an ESS is 'retaliator': a policy of not initiating escalation backed by willingness to fight dangerously if the opponent does so. It has subsequently been shown that a mixed population with a particular frequency of hawks and bullies is also an ESS and cannot be invaded by retaliator mutants (Gale and Eaves 1975). The paper by Maynard Smith reproduced on p. 258 reviews the development of the ESS concept and describes a number of its applications.

This approach to considering the relative advantages of different forms of social behaviour is applicable to almost all cases where conflicts of interest occur between individuals and may represent 'one of the most important advances in evolutionary theory since Darwin' (Dawkins 1976).

Although ritualized fighting is common, recent field studies suggest that the occurence of serious injury during fights is more frequent than was previously thought (see Schaller 1972; Kruuk 1972; Geist 1966, 1971, 1974; Wilkinson and Shank 1976), particularly in polygynous societies with strongly imbalanced sex ratios. Consequently, individuals should assess the probable outcome of fights carefully before they commit themselves, and should avoid fighting when they are unlikely to win. The paper by Parker (1974, see p. 271) considers the factors affecting who wins. Parker argues that the probability that an animal will win a contest depends on its ability to hold the resource and on the amount (of fitness) it is prepared to invest in the contest compared to its opponent. The paper is primarily concerned with the effects of differences in the value of resources to the two contestants. For example, where the initial resource holder must invest time and energy before it can exploit the resource and any successful contestant must expend a similar amount if it replaces the holder, the resource will be worth more to the former than to the latter, who should only attack if its ability to take the resource is greater than the holder's. Similarly, differences in age will affect the potential loss of fitness of the two contestants, and older individuals might be expected to invest more in particular contests than younger ones (see p. 279). It is again important to note that it is not the amount that an individual has already invested in a resource which will determine its value but the cost of exploiting the resource in the future. Parker has also combined his theoretical work with detailed studies of mating competition in dung flies (*Scatophaga stercoraria*) which provide empirical evidence confirming particular predictions of this type of model (Parker 1970).

The two papers on contest strategies are primarily concerned with disputes over resource access. However, competition may take a more subtle form than this. In some situations, it may be to an individual's advantage to 'spitefully' reduce the fitness of its competitors (Hamilton 1970 and 1971). For example, in certain species of parasitic wasps the larvae kill *but do not eat* other larvae inhabiting the same host (Wilson, 1971). Similarly, in some social mammals males which take over a group and expel the previous male systematically kill the younger offspring, thus removing competitors of their own future offspring (though this behaviour may also be directly advantageous because it increases the number of females which can be fertilized in the near future (see Bertram, p. 293; Clutton-Brock and Harvey, p. 000)). Spiteful behaviour may evolve even if the disruptor incurs appreciable costs, a situation termed 'strong spite' (Hamilton 1971*a, b*).

The last paper in this section is an excerpt from a review of our own which interprets some aspects of disruptive behaviour in primate societies in the light of current evolutionary theory. In particular, we believe that it is important to avoid attempts to assign functions to social structures or patterns of interaction and, instead, to consider the advantages to individuals of behaving in different ways (see p. 294). For example, one should not ask 'what is the adaptive significance of a dominance hierarchy?', but 'what are the advantages to the individual of regularly submitting to certain animals?' (see p. 302).

# Geometry for the selfish herd

W. D. HAMILTON†

**Introduction**

This paper presents an antithesis to the view that gregarious behaviour is evolved through benefits to the population or species. Following Galton (1871) and Williams (1964) gregarious behaviour is considered as a form of cover-seeking in which each animal tries to reduce its chance of being caught by a predator.

It is easy to see how pruning of marginal individuals can maintain centripetal instincts in already gregarious species; some evidence that marginal pruning actually occurs is summarized. Besides this, simply defined models are used to show that even in non-gregarious species selection is likely to favour individuals who stay close to others.

Although not universal or unipotent, cover-seeking is a widespread and important element in animal aggregation, as the literature shows. Neglect of the idea has probably followed from a general disbelief that evolution can be dysgenic for a species. Nevertheless, selection theory provides no support for such disbelief in the case of species with outbreeding or unsubdivided populations.

The model for two dimensions involves a complex problem in geometrical probability which has relevance also in metallurgy and communication science. Some empirical data on this, gathered from random number plots, is presented as of possible heuristic value.

## 1. A model of predation in one dimension

Imagine a circular lily pond. Imagine that the pond shelters a colony of frogs and a water-snake. The snake preys on the frogs but only does so at a certain time of day—up to this time it sleeps on the bottom of the pond. Shortly before the snake is due to

† Reprinted with permission from the author and *Journal of Theoretical Biology,* **31** (1971). Copyright © 1971 by Academic Press Inc.

wake up all the frogs climb out onto the rim of the pond. This is because the snake prefers to catch frogs in the water. If it can't find any, however, it rears its head out of the water and surveys the disconsolate line sitting on the rim—it is supposed that fear of terrestial predators prevents the frogs from going back from the rim—the snake surveys this line and snatches *the nearest one.*

Now suppose that the frogs are given opportunity to move about on the rim before the snake appears, and suppose that initially they are dispersed in some rather random way. Knowing that the snake is about to appear, will all the frogs be content with their initial positions? No; each will have a better chance of not being nearest to the snake if he is situated in a narrow gap between two others. One can imagine that a frog that happens to have climbed out into a wide open space will want to improve his position. The part of the pond's perimeter on which the snake could appear and find a certain frog to be nearest to him may be termed that frog's 'domain of danger': its length is half that of the gap between the neighbours on either side. The diagram below shows the best move for one particular frog and how his domain of danger is diminished by it:

But usually neighbours will be moving as well and one can imagine a confused toing-and-froing in which the desirable narrow gaps are as elusive as the croquet hoops in Alice's game in Wonderland. From the positions of the above diagram, assuming the outside frogs to be in gaps larger than any others shown, the following moves may be expected:

What will be the result of this communal exercise? Devious and unfair as usual, natural justice does not, in general, equalize the risks of these selfish frogs by spacing them out. On the contrary, with any reasonable assumptions about the exact jumping behaviour, they quickly collect in heaps. Except in the case of three frogs who start spaced out in an acute-angled triangle I know of no rule of jumping that can prevent them aggregating. Some occupy protected central positions from the start; some are protected only

```
                        10° segments of pool margin (degrees)
        0             90              180             270             360
      1 2 3 3 3   6 1 3 3   4 1 3 5   1 3 4 5 2 9 4   5 6 3 5 4   1 2 2   4 3
      2 5 2 2     8   3 2   6 1 1 7   2 4 7 11 2     7 5 2 5 5   3 2     4 4
      3 6 1 2     8   3 1   8   9     4 9 11         9 4 1 7 5   2 2     4 4
      4 6 1       9   3     9   8     4 11 10        10 3  8 6   1 2     4 5
      5 7         9   2     9   8     5 12  8        12 1  9 6   3       4 5
      6 7         9         9   8     5 14  7        13    9 5   5       3 6
      7 8         7         9   8     5 16  5        15    9 3   6       3 6
      8 9         5         9   9     4 18  3        17    9 1   6       5 5
      9 10        4         8   10    3 20  1        19    8     6       7 4
     10 12        3         7   11    2 22            20   6     6       9 2
     11 13        2         6   12    1 22            22   4     7       9 2
     12 14                  6   13      22            24   3     7       9 2
     13 14                  5   13      22            26   2     8       8 2
     14 15                  3   13      22            28   1     9       7 2
     15 16                  1   13      22            30          10     6 2
     16 17                      12      22            32          9      6 2
     17 17                      11      22            33          8      7 2
     18 18                      9       22            35          6      8 2
     19 19                      7       22            37          5      9 1
```

*(Position number — rows 1–19, read down the left margin)*

Fig. 1. Gregarious behaviour of 100 frogs is shown in terms of the numbers found successively within 10° segments of the margin of the pool. The initial scatter (position 1) is random. Frogs jump simultaneously giving the series of positions shown. They pass neighbours' positions by one-third of the width of the gap. For further explanation, see text.

initially in groups destined to dissolve; some, on the margins of groups, commute wildly from one heap to another and yet continue to bear most of the risk. Fig. 1 shows the result of computer simulation experiment in which 100 frogs are initially spaced randomly round the pool. In each 'round' of jumping a frog stays put only if the 'gap' it occupies is smaller than both neighbouring gaps; otherwise it jumps into the smaller of these gaps, passing the neighbour's position by one-third of the gap-length. Note that at the termination of the experiment only the largest group is growing rapidly.

The idea of this round pond and its circular rim is to study cover-seeking behaviour in an edgeless universe. No apology, therefore, need be made even for the rather ridiculous behaviour that tends to arise in the later stages of the model process, in which frogs supposedly fly right round the circular rim to 'jump into' a gap on the other side of the aggregation. The model gives the hint which I wish to develop: that even when one starts with an edgeless group of animals, randomly or evenly spaced, the *selfish avoidance of a predator can lead to aggregation.*

## 2. Aggregations and predators

It may seem a far cry from such phantasy to the realities of natural selection. Nevertheless I think there can be little doubt that behaviour which is similar in biological intention to that of the hypothetical frogs is an important factor in the gregarious tendencies of a very wide variety of animals. Most of the herds and flocks with which one is familiar show a visible closing-in of the aggregation in the presence of their common predators. Starlings do this in the presence of a sparrowhawk (Baerends and Baerends-van Roon 1950; Hostman 1952; Lorenz 1966; Tinbergen 1951); sheep in the presence of a dog, or, indeed, any frightening stimulus (Scott 1945). Parallel observations are available for the vast flocks of the quelea (Crook 1960), and for deer (Darling 1937). No doubt a thorough search of the literature would reveal many other examples. The phenomenon in fish must be familiar to anyone who has tried to catch minnows or sand eels with a net in British water. Almost any sudden stimulus causes schooling fish to cluster more tightly (Breder 1959), and fish have been described as packing, in the presence of predators, into balls so tight that they cannot swim and such that some on top are thrust above the surface of the water (Springer 1957). A shark has been described as biting mouthfuls from a school of fish 'much in the manner of a person eating an apple' (Bullis 1960).

G. C. Williams, originator of the theory of fish schooling that I am here supporting (Williams 1964, 1966), points out that schooling is particularly evident in the fish that inhabit open water. This fits with the view that schooling is similar to cover-seeking in its motivation. His experiments showed that fish species whose normal environment afforded cover in the form of weeds and rocks had generally less marked schooling tendencies. Among mammals, similarly, the most gregarious species are inhabitants of open grassy plains rather than of forest (Hesse, Allee, and Schmidt 1937). With fish schools observers have noted the apparent uneasiness of the outside fish and their eagerness for an opportunity to bury themselves in the throng (Springer 1957) and a parallel to this is commonly seen in the behaviour of the hindmost sheep that a sheepdog has driven into an enclosure: such sheep try to butt or to jump their way into the close-packed ranks in front. Behaviour of this kind certainly cannot be regarded as showing an unselfish concern for the welfare of the whole group.

With ungulate herds (Galton, 1871; Sdobnikov 1935), with bird flocks (Tinbergen 1951; Wynne-Edwards 1962), and with the dense and sudden-emerging columns of bats that have been described issuing at dusk from great bat caves (Moore 1948; Pryer 1884) observations that predators do often take isolated and marginal individuals have frequently been recorded. Nor are such observations confined to vertebrates or to mobile aggregations. Similar observations have been recorded for locusts (Hudleston 1958), for gregarious caterpillars (Tinbergen 1953a) and, as various entomologists have told me, for aphids.

For the aphids some of the agents concerned are not predators in a strict sense but fatal parasites. Insect parasites of vertebrates are seldom directly fatal but, through transmitted diseases or the weakening caused by the activities of endoparasitic larvae, must often cause death indirectly nevertheless. Thus escape from insect attack is another possible reward for gregarious instincts. From observations in Russia, V. M. Sdobnikov (1935) has stated that when reindeer are standing in dense herds only the outermost animals are much attacked by insects. Among the species which he observed attacking the reindeer, those which produced the most serious affliction were nose flies and warble flies, larvae of which are endoparasites of the nasal passages and the skin respectively. Recently Espmark (1968) has verified and extended most of Sdobnikov's information. His work reinforces the view that such oestrid flies are important and ancient enemies of their various ungulated hosts, as is suggested by the fact that their presence induces a seemly *instinctive* terror. Reindeer seem to be almost as terrified of them as they are of wolves, and cattle react as though they feared the certainly painless egg-laying of warble flies far more than they fear the painful bites of large blood-suckers (Austen 1939).

The occurrence of marginal predation has also been recorded for some of the aggregations formed by otherwise not very gregarious animals for the purpose of breeding. The best data known to me concern nesting black-headed gulls. The work of Kruuk (1964) has been reinforced by further studies, summarized by Lack (1968). The latter seem to have shown that *all* marginal nests failed to rear young, mainly due to predation. Perhaps, nevertheless, the gulls that could not get places in the centre of the colony were right in nesting on the edge rather than in isolation where, for a conspicuous bird like a gull, the chances would have been even worse.

It is perhaps worth digressing here to mention the *temporal* aspects of marginal predation. The 'aggregation' in timing already alluded to for bats issuing from bat caves parallels the marked synchrony in breeding activities which is seen in most aggregations and which has been called the 'Fraser Darling effect'. In explanation of this, Lack (1968) points out that late and early breeders in terns and black-headed gulls do worse in terms of young raised than those best-synchronized with the mass, and he implies that this is mainly due to predation. Individuals coming into breeding condition late or early may also have a problem in sexual selection—that of finding a mate. This point will be touched on later. There are similar influences of temporal selection for flowering plants.

The securing of a nest site in the middle of a colony area is certainly likely to be an achievement of protected position in a sense related to that explained in the story of the frogs, but in the relative immobility of such positions the case diverges somewhat from the initial theme. In all the foregoing examples, except perhaps that of the insect parasites of vertebrates, close analogy to that theme has also been lost through the assumption that the predator is likely to approach from outside the group. This is difficult to avoid.†

When a predator habitually approaches from outside it is comparatively obvious how marginal pruning will at least maintain the centripetal instincts of the prey species. Whether predation could also initiate gregariousness in an originally non-gregarious species is another matter. As mentioned before, this will begin to seem likely if it can be shown that when predators tend to appear within a non-aggregated field of prey geometrical principles of self-protection still orientate towards gregarious behaviour. So far I have only shown this for one highly artificial case: that of jumping organisms in a one-dimensional universe. For the case in two dimensions a more realistic story can be given. The most realistic would, perhaps, take reindeer and warble flies as its subject animals, but in order to follow an interesting historical precedent, which has now to be mentioned, the animals chosen will be cattle and lions.

In 1871 Francis Galton published in *Macmillan's Magazine* an

---

† Cannibalism in gregarious species raises different problems and may help to explain why the nests of gulls do not become very closely aggregated (see 'the gull problem' in Tinbergen (1953b)). Insects have another kind of 'wolf in sheep's clothing': the predator mimicking its prey (Wickler, 1968).

article entitle 'Gregariousness in Cattle and in Men'. In it he out-lined a theory of the evolution of gregarious behaviour based on his own observations of the behaviour of the half-wild herds of cattle owned by the Damaras in South Africa. In spite of the characteristically forceful and persuasive style of his writing, Galton's argument is not entirely clear and consistent. Some specific criticisms will be mentioned shortly. Nevertheless it does contain in embryo the idea of marginal predation as a force of natural selection leading to the evolution of gregarious behaviour. The main predators of the Damaraland cattle, according to Galton, were lions, and he states clearly that these did prefer to take the isolated and marginal beasts. The following passage shows suf-ficiently well his line of thought. After stating that the cattle are unamiable to one another and do not seem to have come together due to any 'ordinary social desires', he writes:

'Yet although the ox has so little affection for, or individual interest in, his fellows, he cannot endure even a momentary severance from his herd. If he be separated from it by strategem or force, he exhibits every sign of mental agony; he strives with all his might to get back again and when he succeeds, he plunges into its middle, to bathe his whole body with the comfort of closest companionship.'

### 3. A model of predation in two dimensions

Although as Galton implies, lions like most other predators, usually attack from outside the herd, it is possible to imagine that in some circumstances a lion may remain hidden until the cattle are feeding on all sides of it. Consider therefore a herd grazing on a plain and suppose that its deep grass may conceal—anywhere—a lion. The cattle are unaware of danger until suddenly the lion is heard to roar. By reason of some peculiar imaginary quality the sound gives no hint of the whereabouts of the lion but it informs the cattle of danger. At any moment, at any point in the terrain the cattle are traversing, the lion may suddenly appear and attack the nearest cow.

As in the case of the 'frog' model, the rule that the predator attacks the nearest prey specifies a 'domain of danger' for each individual. Each domain contains all points nearer to the owner of the domain than to any other individual. In the present case such domains are polygonal (Fig. 2). Each polygon is bounded by lines which bisect at right angles the lines which join the owner to certain neighbours; boundaries meet three at a point and an irreg-

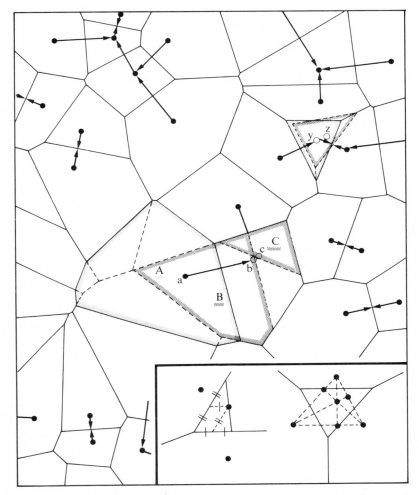

Fig. 2. Domains of danger for a randomly dispersed prey population when a hidden predator will attack the nearest animal. Thin arrowed lines indicate the nearest neighbour of each prey. Thicker arrowed lines show supposed movements of particular prey. Position of prey is given a lower case letter and the domain obtained from each position is given the corresponding upper case letter. The pattern underlining the domain letter corresponds to the pattern used to indicate the boundary of the domain in question. Dashed lines are used for the boundaries of domains that come into existence after the first movement.

Approach to a nearest neighbour usually diminishes the domain of danger (as a → b), but not always (as y → z). On reaching a nieghbour the domain may, in theory, be minimized by moving round to a particular side (as b → c). The inset diagram shows geometrical algorithms for minimum domains attainable by a (on reaching c), and by y. In the case of y the minimum domain is equal to the triangle of neighbours and is obtained from the ortho-centre of that triangle.

ular tesselation of polygons covers the whole plane. On hearing the lion roar each beast will want to move in a way that will cause its polygonal domain to decrease. Not all domains can decrease at once of course: as in the case of the frogs, if some decrease others must grow larger; nevertheless, if one cow moves while others remain stationary the one moving can very definitely improve its position. Hence it can be assumed that inclinations to attempt some adaptive change of position will be established by natural selection. The optimal strategy of movement for any situation is far from obvious, and before discussing even certain better-than-nothing principles that are easily seen it will be a cautionary digression to consider what is already known about a particular and important case of such a tesselation of polygons, that in which the 'centres' of polygons are scattered at random.

Patterns of a more concrete nature which are closely analogous to the tesselation defined certainly exist in nature. In two dimensions they may be seen in the pattern formed by encrusting lichens on rocks, and in the cross-sectional patterns of cracks in columnar basalt. The corresponding pattern in three dimensions, consisting of polyhedra, is closely imitated in the crystal grains of some metals and other materials formed by solidification of liquids. The problem of the statistical description of the pattern in the case where centres are distributed at random was first attacked with reference to the grain structure of metals (Evans 1945). More recently the $N$-dimensional analogue has been studied by E. N. Gilbert (1962) on the incentive of a problem arising in communication science. Yet in spite of great expertise many simple facts about even the two-dimensional case remain unknown. The distribution of the angles of the polygons is known (J. L. Meijering, personal communication); something is known of the lengths of their sides, and Gilbert has found the variance of their areas.† Nothing, apart from the fairly obvious value of six for the mean, is known of the numbers of sides. Since a biologist might be interested to know whether, in the ideal territory system of these polygons, six neighbours is the most likely number, or whether five is more likely than seven, some data gathered from plots of random numbers is given in Table 1.

After glancing at Gilbert's paper a non-mathematical biologist may well despair of finding a theory of *ideal* self-protective movement for the random cattle. Nevertheless experimentation with

---

† Recently R. E. Miles (1970) has greatly extended analysis of this case (*Math. Biosci.* **6**, 85) (footnote added in proof).

Table 1. Distribution of the number of sides of 376 polygons constructed from random number plots

| No. of sides | | 3 | 4 | 5 | 6 | 7 | 8 | 9 | 10 | 11 |
|---|---|---|---|---|---|---|---|---|---|---|
| No. observed | | 3 | 38 | 97 | 123 | 70 | 29 | 11 | 4 | 1 |
| Total no. observed | 376 | | | | | | | | | |

Data for 100 of these polygons were received from E. N. Gilbert. The rarity of triangles indicates that the sample area chosen for Fig. 2 is not quite typical.

ruler and compasses and one small evasion quickly reveal a plausible working principle. As a preliminary it may be pointed out that the problem differs radically from the linear problem in that domains of danger change continuously in size with movements of their owner. (In the linear case no change takes place until an individual actually passes his neighbour; this was the reason for choosing organisms that jump.) Now suppose that one cow alone has sensed the presence of the lion and is hastily moving through an otherwise motionless herd. In most circumstances such a cow will diminish its domain of anger by approaching its nearest neighbour. A direct approach can seldom be the path it would choose if it requires that every step should bring the maximum decrease of danger, and when the nearest neighbour is a rather isolated individual it may even happen that a domain increases during approach to a nearest neighbour (Fig. 2, y → z). Such increases will tend to be associated with polygons with very low side number. When an individual is enclosed in a ring of others and consequently owns a many-sided domain, decrease in area is almost inevitable (Fig. 2, a → b). Since the average number of sides is six and triangles are rare (Table 1), it must be a generally useful rule for a cow to approach its nearest neighbour. This is a rule for which natural selection could easily build the necessary instincts. Behaviour in accord with it has been reported in sheep (Scott 1945). The evasion, then, is to imagine that these imaginary cattle are too slow-witted to do anything better. Most readers will agree, I hope, that there is no need to call them stupid on this account.

If the rest of the herd remains stationary and the alarmed cow reaches its nearest neighbour it will usually further decrease its domain of danger by moving round to a side that gives it a minimal 'corner' of the neighbour's now enlarged domain (Fig. 2, b → c; left inset diagram shows the principle for finding such a minimal 'corner' domain). In this position the cow of the story may be supposed to reach equilibrium; it is some sort of equivalent, albeit not a close one, of the stable non-jump positions that sometimes occur with the frogs. Even more than with the stable positions of the frogs, however, the stable positions of this model must be almost impossible to attain. As in the case of the frogs this is seen as soon as we imagine that all the other cows are moving and have similar aims in view.

Nearest-neighbour relationships connect up points into groups (Fig. 2). Every group has somewhere in it a 'reflexive pair'; that is, a pair for which each is the nearest neighbour of the other. A formula for the frequency of membership of reflexive pairs when the points are randomly dispersed has been derived by Clark and Evans (1955). It is $6\pi/(8\pi + \sqrt{27})$. Halving and inverting this expression gives us the average size of groups at 3·218. No further facts are known mathematically about the relative frequencies of different sizes of groups, but a summary of the forms and sizes found in the random number plots is given in Table 2.

Table 2

| No. in group | 2 | 3 | 4 | 5 | 6 | 7 |
|---|---|---|---|---|---|---|
| No. of cases found | 46 | 37 | 24 | 8 | 5 | 1 |
| Configurations found | 46 | 37 | 9 | 4 | 1 | 1 |
| | | | 11 | 1 | 1 | 1 |
| | | | 4 | 1 | 1 | |
| | | | | 2 | 1 | |
| | | | | 1 | 1 | |
| Mean N in group { observed: 3·107 expected: 3·218 } | | | | | 1 | |

The nearest-neighbour relation allocates points of a random dispersion to groups (see arrowed lines on Fig. 2). Here the configurations found in 121 groups formed by 376 points are recorded together with their frequencies of occurrence. Double connecting lines indicate the reflexive nearest-neighbour relationship.

Clearly, if randomly dispersed cattle condense according to the statistical pattern indicated in Table 2, the groups so formed can hardly be described as herds. Nevertheless there is undoubtedly a gregarious tendency of a kind, and once such primary groups have formed the cattle in each group will see a common advantage in moving through the field of groups using the same principles as before. Thus an indefinite series of such condensations will take place and eventually large herds will appear. Condensation is not prevented by having the points initially more evenly spaced than in the random dispersion. For example, when the points are initially in some lattice formation the inherent instability of the system is actually easier to see than it is for the random dispersion: every point, if moving alone, can definitely diminish its domain of danger by approaching any one of its equidistant neighbours. Thus even for the most unpromising initial conditions it remains evident that predation should lead to the evolution of gregarious behaviour.

No story even as poorly realistic as the foregoing can be given for the case of predation in three dimensions. In the air and in water there is still less cover in which the predator can hide. If there were a hiding place, however, for what it is worth the argument can be repeated almost exactly and we arrive at $3\frac{3}{8}$ for the mean size of primary groups. For celestial cattle in a space of infinitely many dimensions the mean group size attains only to 4.†

Whether or not marginal predation is, as I believe it to be, a common primary cause of the evolution of gregariousness, it is surprising that the idea of such a cause has received so little attention from biologists. The reason for the neglect of Galton's views cannot lie in the non-scientific nature of the journal in which it was originally published. Galton repeated them in his book *Inquiries into human faculty* (1883), and this is one of the best known works. Yet both the hints of the present hypothesis in writings on the schooling of fishes and the admirable study and discussion of this hypothesis by G. C. Williams (1964) were completely independent of Galton's publications. Galton himself is certainly largely to blame for this. With one exception he did not relate his idea to species other than cattle and sheep, and the one exception was

---

† The general formula, where $n$ is the number of dimensions, is

$$\frac{4n}{n+1} + \frac{2}{(n-1)\pi^{\frac{1}{2}}}\left(\frac{3^{\frac{1}{2}}}{2}\right)^{n-1}\frac{\Gamma\{n/2\}}{\Gamma\{(n-1)/2\}}.$$

man. Something of the tenor of his views on the manifestations of gregariousness in man may be gathered from the fact that the relevant passage in the book mentioned above is headed: 'On gregarious and slavish instincts'. Needless to say his analogy between human and bovine behaviour now seems somewhat naive. His views remain, as always, interesting and evocative, but their dogmatic and moralizing tone and their obvious connection with their author's widely distrusted line on eugenics can be imagined to have scared off many potentially interested zoologists. Another probable reason for the neglect of Galton's idea is that he himself presented it mixed up with another really quite separate idea which he treated as if it were simply another aspect of the same thing. This was that every cow, whether marginal or interior, benefited by being part of a herd, and that therefore herding was beneficial to the species. His supporting points are undoubtedly forcible; he mentions mutual warning and the idea that by forming at bay in outward-facing bands the cattle can present a really formidable defence against the lions. However, whether or not the cattle actually do, in the last resort, overcome their centripetal inclination and turn to face the predator (as smaller bands of 'musk ox' certainly do), these points raise a different issue, as do the mass attacks on predators by gregarious nesting birds. My models certainly give no indication that such mutualistic defence is a necessary part of gregarious behaviour. Moreover, mutual defence and warning can hardly be described as 'slavish' behaviour and so seem not to be covered by Galton's heading. There is no doubt, of course, that mutual warning and occasional unselfish defence of others are sometimes shown by gregarious mammals and birds, but where such actions occur they are probably connected on the one hand with the smallness of the risk taken and, on the other, with the closeness of the genetical relationship of the animals benefited (Hamilton 1963). With the musk ox, for example, the bands are small and clearly based on close family relationships. Sheep and cattle also take risks in defending their young (as Galton pointed out), but this is based on recognition of their own offspring. Apart from the forced circumstances, and the unnatural dispositions engendered by domestication, females usually do not associate with, still less defend, young which are not their own (Williams 1966). The ability to recognize particular individuals of the same species is highly developed in mammals and birds and there is no difficulty either practical or theoretical in supposing that the mutualistic behaviour of the adult musk oxen, for instance, is an evolutionary development from the altruism involved in parental care. Positive social relations between members of family groups probably exist submerged in most manifestations of mass gregariousness in animals higher than fish. For example the flock of sheep driven by a shepherd owes its compactness and apparent homogeneity to the presence of the sheep dogs: when it is left undisturbed on the mountain it arranges itself into a loosely clustered and loosely territorial system. The clusters are usually based on kinship and the sheep of a cluster are antagonistic to any strange sheep that attempts to feed on their ground (Hunter 1964). It accords well with the present theory that the breeds of sheep that are most readily driven into large flocks are those which derive from the Merino breed of Spain (Darling 1937), which is an area where, until very recently, predation by wolves was common. Galton likewise gave the relaxation of predation as the reason for European cattle being very much less strongly gregarious than the cattle he observed in Africa. He noted that the centripetal inclinations evolved through predation would be opposed continually by the end to find ungrazed pasture, so that when predation is relaxed gregarious instincts would be selected against. This point has also been made by Williams.

Most writers on the subject of animal aggregation seem to have believed that the evolution of gregarious behaviour must be based on some advantage to the aggregation as a whole. Many well known biologists have subscribed, outspokenly or by implication, to this view. At the same time some, for example Hesse *et al.* (1937), Fisher (1953), and Lorenz (1966), have admitted that the nature of the group advantage remains obscure in many cases. On the hypothesis of gregarious behaviour presented here the apparent absence of a group benefit is not cause for surprise. On the contrary, the hypothesis suggests that the evolution of the gregarious tendency may go on even though the result is a considerable lowering of the overall mean fitness. At the end of our one-dimensional fairy tale, it will be remembered, only the snake lives happily, taking his meal at leisure from the scrambling heaps of frogs which the mere thought of his existence has brought into being. The cases of predators feeding on apparently helpless balls of fish seem to parallel this phantasy: here gregariousness seems much more to the advantage of the species of the predator than to that of the prey. Certain predators of fish appear to have evolved adaptations for exploiting the schooling tendencies of their prey. These cases are not only unfavourable to the hypothesis of a group advantage in schooling but also somewhat unfavourable to the present hypothesis, since it begins to seem that there must be an advantage to fish which do not join the school. The thresher shark is said to use its much elongated tail to round up fish into easily eaten schools (Williams 1964). Even worse, the swordfish is recorded as feeding on schools by first immobilizing large numbers with blows from its sharp-edged sword. This may be its usual method of feeding, and the sawfish may use its weapon similarly. Such predators presumably do best by striking through the middle of a school; lone fish might fail to attract them. It is obvious, however, that such cases should not be assessed on the assumption that a particular predator is the only one (Bullis 1960). At the time of attack by a swordfish there may be other important predators round about that are still concentrating on the isolated and peripheral fish. In the case of locusts the occurrence of such contrary influences from predation is recorded (Hudleston 1957): large bird predators (ravens and hornbills) attacking bands of hoppers of the desert locust did tend to disperse the bands, and to decimate their central members, by settling in their midst, but at the same time smaller birds (chats and warblers) captured only marginal hoppers and stragglers.

Adaptations of the predator to exploit the gregariousness of its prey certainly suggest the possibility of a changeover to a disaggregating phase of selection. In terms of the earliest story, if the snake evolves such a lazy preference for taking its prey from a heap that it comes to overlook a nearer lone frog, then it may come to be the mutant gregariphobe which survives best to propagate its kind. There are in fact many examples of species which are gregarious in only part of their range or their life cycle and possibly this may reflect the differing influences of different predators.

A surprising number of discussions of animal aggregation have mentioned the occurrence of marginal predation and yet shown apparently no appreciation of its possible evolutionary significance. Whether marginal predation or similarly orientated pressure of selection on individuals is sufficiently powerful to account for the gregariousness observed in any particular case cannot be decided by any already existing body of data. Nevertheless it can be claimed that for none of the cases so far discussed does the data exclude the possibility. It can also be claimed that all the rival theories based on the idea of 'group selection' are theoretically insecure. Consider for example the theory of Wynne-Edwards (1962) which, from this basis, attempts to bring many facts of animals aggregation under a common explanation. This theory suggests that aggregations serve to make individuals aware of the current level of population density and that this awareness reacts on reproductive performance in a way that holds off the possibility of a disastrous crash due to over-exploitation of the food supply or other limited resources. Except in cases where groups are exclusive and mutually competitive for territory it is very difficult to see how there can be positive selection of any tendency by an individual to reproduce less effectively than it is able (Hamilton 1971). Wynne-Edwards's cases of massive aggregations, even those having a kin-based substructure as in the gregarious ungulates, are unlikely candidates for the class of exceptions. The alternative theory supported here has no parallel difficulty as to the underlying processes of selection since it can rest firmly on the theory of genetical natural selection.

Certain other versions of the idea of a group benefit, unlike the Wynne-Edwards theory, do not require the concept of self-disciplined restraint by individuals. With such versions, statements that aggregation has evolved because it aids the survival of the species may be treated as merely errors of expression, in which a possibly genuine effect is given the status of a cause. The factor of communal alertness, for instance, may really make life more difficult for a predator and safer for a gregarious prey, especially if the predator is one which relies on stealth rather than speed. In Galton's perhaps over-persuasive words,

'To live gregariously is to become a fibre in a vast sentient web over-spreading many acres; it is to become the possessor of faculties always awake, of eyes that see in all directions, of ears and nostrils that explore a broad belt of air; it is to become the occupier of every bit of vantage ground whence the approach of a lurking enemy might be overlooked.'

But there is of course nothing in the least altruistic in keeping alert for signs of nervousness in companions as well as for signs of the predator itself, and there is, correspondingly, no difficulty in explaining how gregariousness on this basis could be evolved.

Returning to the more interesting and controversial postulate of a population regulatory function for mass aggregation, consider, as a point of detailed criticism, the problem presented by aggregations in which several species are mixed. All the main classes of aggregation that have so far been mentioned—fish schools, birds flocks (both mobile and nesting) and grazing herds—provide numerous examples of species mixture. The species involved may be as widely related as ostriches and antelopes (Hesse *et al.* 1937). If it is difficult to see how the supposed group selection can work within a species it is certainly even more difficult when the groups are mixed. On the other hand the theory that gregariousness is essentially due to the need for cover finds no difficulty over this point provided that the species which mix have at least one important predator in common. None of the vertebrate predators that have been mentioned as possible agents for the moulding of gregarious inclinations in their prey seem so specialized in their hunting as to make it improbable that individuals of one prey species could expect to gain some protection by immersing themselves in an aggregation of another species.

Perhaps most of the examples of mixed aggregation lie outside the class for which Wynne-Edwards and his supporters would claim a population regulatory function. The examples which have been cited in support of the theory are, in general, more directly concerned with reproduction. Apart from the case of nest aggregations, which has been discussed, there remain Wynne-Edwards's numerous citations of nuptial gatherings. With these the case for the effectiveness of marginal predation is admittedly weaker. Marginal predation seems somewhat less likely on general grounds and there is little evidence. However, there are other ways in which selection is likely to favour individuals which are in the nuptial gathering (or at least on it margin) over those that are isolated. Such selection may be sexual; it may work through differences in the chances of obtaining a mate. As one example, consider the swarms of midges that are so common in damp, still, vegetated places in summer. Many species of nematocerous flies have the habit of forming such swarms. Each swarm usually consists of males of a single species and tends to hover in a fixed spot, often near to some conspicuous object. Females come to the swarm and on arrival each is seized by a male. Passing over possible stages in the initiation of such habits, it is at least clear that as soon as proximity to the swarm itself becomes a key to the female's further co-operation in copulation there is likely to be little chance of mating for the male which does not join the swarm. In such a case the optimal position for a male is probably, not, as it is under predation, at the centre of the throng, and considering how males might endeavour to spend the maximum time in relatively favoured positions, downwind or upwind, above or below, possible explanations for the dancelike motion of such swarms become apparent.

# References

Austin, E. E. (1939). In *British bloodsucking flies*. (F. W. Edwards, H. Olroyd and J. Smart, eds.), pp. 149–152. British Museum, London.

Baerends, G. P. and Baerends-van Roon, J. M. (1950). *Behaviour* Suppl. 1, 1.

Breder, G. M. (1959). *Bull. Am. Mus. nat. Hist.* 117, 395.

Bullis, H. R., Jr. (1960). *Ecology* 42, 194.

Clark, P. J. and Evans, F. C. (1955). *Science, N.Y.* 121, 397.

Crook, J. H. (1960). *Behaviour* 16, 1.

Darling, F. F. (1937). *A herd of red deer*. Oxford University Press.

Espmark, Y. (1968). *Zool. Beitr.* 14, 155.

Evans, U. R. (1945). *Trans. Faraday Soc.* 41, 365.

Fisher, J. (1953). In *Evolution as a process*. (J. Huxley *et al.*, eds.), pp. 71–82. Allen and Unwin, London.

Galton, F. (1871). *Macmillan's Mag., Lond.* 23, 353.

—— (1883). *Inquiries into human faculty and its development*. Dent, London; (1928) Dutton, New York.

Gilbert, E. N. (1962). *Ann. math. Statist.* 33, 958.

Hamilton, W. D. (1963). *Amer. Natur.* 97, 354.

—— (1971). *Proc. 3rd int. Symp. Smithson Instn* (in press).

Hesse, R., Allee, W. C., and Schmidt, K. P. (1937). *Ecological animal geography*. Chapman, Wiley and Hall, New York and London.

Horstmann, E. (1952). *Zool. Anz.* Suppl. 17, 153.

Hudleston, J. A. (1958). *Entomologist's mon. Mag.* 94, 210.

Hunter, R. F. (1964). *Advmt. Sci., Lond.* 21 (90) 29.

Kruuk, H. (1964). *Behaviour* Suppl. 11; 1.

Lack, D. (1968). *Ecological adaptations for breeding in birds*. Methuen, London.

Lorenz, K. (1966). *On aggression*. Methuen, London.

Moore, W. G. (1948). *Turtox News* 26, 262.

Pryer, H. (1884). *Proc. zool. Soc. Lond.* 1884, 532.

Scott, J. P. (1945). *Comp. Psychol. Monogr.* 18 (96), 1–29.

Sdobnikov, V. M. (1935). *Trudy arkt. nauchno-issled Inst.* 24, 353.

Springer, S. (1957). *Ecology* 38, 166.

Tinbergen, N. (1951). *The study of instinct*. Clarendon Press, Oxford.

—— (1953a). *Social behaviour in animals*. Methuen, London.

—— (1953b). *The herring gull's world*. Collins, London.

Wickler, W. (1968). *Mimicry in plants and animals*. Weidenfeld and Nicolson, London.

Williams, G. C. (1964). *Mich. St. Univ. Mus. Publ. Biol. Sev.* 2, 351.

—— (1966). *Adaptation and natural selection*. Princeton University Press.

Wynne-Edwards, V. C. (1962). *Animal dispersion in relation to social behaviour*. Oliver and Boyd, Edinburgh.

# Kin selection in lions and in evolution

B. C. R. BERTRAM†

Students of ecology in the field have generally been interested in the way in which a species maintains itself, or in the reproductive success of that species in a broad sense. Students of behaviour in the field, on the other hand, have tended to concentrate more on the behaviour of individuals, or on the relationships between individuals. A result of the relatively recent development of behavioural ecology has been a useful merging of these two approaches, with emphasis being placed on the reproductive success of individuals. It is becoming increasingly clear that there is a great deal of variation in reproductive success among different individuals within a species. It is also being increasingly recognized that, for the larger vertebrates at least, it is these differences in reproductive success between individuals which determine both the direction in which the species is evolving, and the speed at which it is changing. The environment, especially other potentially competing species, sets broad limits within which the species is constrained, but there is a considerable amount of latitude within those limits. Evolutionary change within the limits is caused by selection resulting from differences between conspecifics in the number of offspring they leave; field work is showing that such differences are considerable.

Field work in behavioural ecology has also included studies of the same individuals over long periods of time. For the more social bird and mammal species, these studies have indicated that animals within social groups are often genetically related to one another. This has been demonstrated in primates (Sade 1967; Missakian 1973; Clutton-Brock and Harvey this volume; in carnivores (Mech 1970; Kruuk 1972; Schaller 1972; Bertram 1973, 1975); in ungulates (Klingel 1965; Lincoln and Guiness 1973); in elephants (Douglas-Hamilton 1973); and in birds (Watts and Stokes

1971; Brown 1972; Ridpath 1972; Zahavi 1974). Doubtless there are many other instances.

Hamilton in 1964 introduced the concept of kin selection, pointing out that replicas of an individual's genes were likely to be present not only in that animal's own offspring but also in its close relatives. In effect this was a modification of the way in which natural selection was presumed to work: by the selection of genes rather than by the selection of the individuals carrying those genes. Thus an animal's genes are passed on both via that animal's own young and via its relatives' young. Genes which cause their carrier animal to produce more replicas of themselves increase in frequency in the population, or in other words are selected for. Consequently one would expect selection for genes which cause relatives to be given favoured treatment. Such favours could take the form of co-operating, of refraining from competing or of helping to rear relatives' offspring. But not all relatives can be favoured maximally, partly because the total quantity of these favours is finite (for example food which can be found), and partly because favouring relatives may reduce an animal's own future reproductive potential. Therefore the genes selected for would be expected to be those which cause such favours to be distributed most to those individuals most closely related (generally own offspring), and to a decreasing extent to more distant relatives.

This kin selection pressure is one of the many selective pressures operating on the individual in any species. It will clearly be strongest in those social species where closely related individuals are in close proximity to one another, and where opportunities for co-operation are greatest. In the absence, as yet, of very long-term records of the genealogy of individuals in social groups in the wild, it is rarely known empirically how closely related those individuals are to one another. But given certain assumptions, it is possible to calculate how closely related they are on average, as the following examples indicate.

## Average degrees of relatedness

The *degree of relatedness* ($r$) of two individuals is the probability that a gene in one is a replica of a gene in the other, by virtue of descent from a shared ancestor. Thus between a parent and its offspring (in diploid species) $r = \frac{1}{2}$, because through meiotic division a parent contributes half of its genes to each of its gametes. By the same process, it can be shown (Hamilton 1971*a*) that half-siblings are related by ¼, full siblings by ½, grandparents to grandchild by ¼, full cousins by ⅛, and so on.

*A simple bird case*

Imagine a bird species which is regularly polyandrous, with two equal males both of whom mate with a single female; her offspring are therefore fathered by either male. We want to calculate how closely related to one another the offspring of such trios would be on average. We will take three possible cases:

(*a*)    *The adult males are not related to one another.* Any two off-spring have an equal chance of being either full siblings (so related by ½) or else half siblings (so related by ¼). The average relatedness among any two offspring is therefore (½ + ¼)/2, i.e. ⅜, = 0·38.

(*b*)    *The adult males are full brothers.* Any two offspring have a 50% chance of being full siblings, in which case they are related by ½. Also they have a 50% chance of being half-siblings with the same mother but different full-sibling fathers; in his latter case they are related by ¼ via their mothers plus ¼ × ½ via their fathers. The mean relatedness among any two offspring is thus ½[½ + (¼ + ⅛)], which equals ⁷⁄₁₆, = 0·44.

(*c*)    *The adult males are the offspring of a typical trio, which is how the species has been reproducing for generations.* Let the degree of relatedness between them be $z$. Any two offspring, as above, have a 50% chance of being full sibs and therefore related by ½. They also have a 50% chance of having the same mother, but different fathers related to one another by $z$; in this case they are related by ¼ via their mother, plus ¼ × $z$ via their fathers, i.e. by ¼($z$ + 1). The average relatedness among any two offspring is thus ½[½ + ¼($z$ + 1)]. But if we assume that the reproductive system has been stable for generations, any two male offspring will bear the same average relatedness to one another as their females do (i.e. $z$). Therefore $z$ = ½[½ + ¼($z$ + 1)], which equals ³⁄₇, = 0·43.

Clearly, additional refinements could be included, such as one male getting more than 50% of the successful copulations, or the female occasionally mating with unrelated strangers. And clearly, too, they will not make a great deal of difference to the average degree of relatedness among offspring.

This example has been cited to demonstrate the principles of calculating average degrees of relatedness in a relatively simple case. It is in fact almost exactly the situation found in the Tasmanian native hen (*Tribonyx mortierii*) (Maynard Smith and Ridpath 1972; Ridpath 1972).

*Lion case*

The lion (*Panthera leo*) social system is a great deal more complicated, but sufficient data are now available for the average degrees of relatedness among certain individuals to be calculated. Although the details of the method refer only to lions, the general principles of the method are applicable to a wide variety of social species.

We first make a number of statements about the lion social system, in the form of assumptions (nos. 1 to 11 below) concerning the reproduction of a 'typical' pride from the data of Bertram (1973, 1975) and Schaller (1972). For this typical case, the average relatedness among males and among females is calculated, thus demonstrating the principle of the method simply. There is of course considerable variation in pride size, structure and reproductive performance; I show below that this variation has relatively little effect on the degree of relatedness found.

*Assumptions.* The representative lion pride and its reproduction is illustrated in Fig. 1.

(1)  A pride has two adult males and seven adult females.
(2)  Four of these seven females give birth at about the same time, and rear their cubs together.
(3)  The litter size is three cubs.
(4)  The fathering of litters is shared equally among the adult males.
(5)  Three female cubs when subadult remain in the pride.
(6)  They replace three adult females which died or left during that period, and thus the pride size remains constant.
(7)  The synchronizing of births, survival of cubs, recruitment or departure of female subadults, and death of adults are all independent of relatedness within the pride.
(8)  All male subadults are expelled from the pride.
(9)  The adult males grew up together in another similar pride; they left it before maturity and stayed together; they have taken over, and are reproducing in, a pride which is not the one in which they were born.
(10) The adult males do not retain tenure of a pride for long enough to father more than one batch of young female recruitment to the pride.

Males

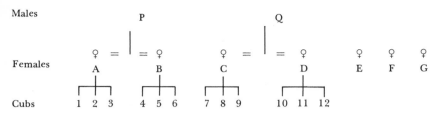

Fig. 1. Diagram of typical case of pride's reproduction, to show possible pairings among cubs.
Pairing of cubs with
    Same father—same mother are
        1—2, 1—3, 2—3, 4—5, 4—6, 5—6, 7—8, . . . (total 12).
    Same father—different mother are
        1—4, 1—5, 1—6, 2—4, 2—5, 2—6, 3—4, 3—5, 3—6, 7—10, . . . (total 18).
    Different father—different mother are
        1—7, 1—8, 1—9, 1—10, 1—11, 1—12, 2—7, 2—8, . . . (total 36).
    Total number of pairings is 66.

(11)    Prides are stable and last for generations, reproducing in this way.

As a result of this system, the adult males are related to one another (by a quantity $x$, to be determined), and the adult females are related to one another (by a quantity $y$); but the males are not related to the females with whom they are reproducing, and therefore no inbreeding occurs.

*Calculations.* Two premises are made. The first is that because males stay together, therefore the relatedness ($x$) between adult males is the same as the average relatedness among any two cubs growing up in the pride together. But this relatedness ($x$, therefore) among the cubs can also be calculated in another way. It is the mean of the following:

    (i)    12 pairings of full-sibling cubs (i.e. same father and same mother; see Fig. 1); these are related by ¼ (via their father) plus ¼ (via their mother).
    (ii)    18 pairings of half-sibling cubs, related by ¼ (via their same father) plus ¼$y$ (via their mothers who are related to one another by $y$).
    (iii)    36 pairings of cubs which are related by ¼$x$ (via their related fathers) plus ¼$y$ (via their related mothers).

Thus for the total of 66 pairings among the 12 cubs, the mean relatedness is

$$x = \frac{(\frac{1}{4} + \frac{1}{4})12 + (\frac{1}{4} + \frac{1}{4}y)18 + (\frac{1}{4}x + \frac{1}{4}y)36}{66}$$

which simplifies to

$$38x - 9y = 7. \tag{1}$$

The second premise is that because the system is stable and repeated over the generations, the mean relatedness among females in a pride remains on average constant with time. Thus $y$ stays the same after replacement of three adult females by subadults. The pride now has four adult and three subadult females in it. The new mean relatedness (still $y$) among all pairs of these seven females is the mean of the following:

(i)   3 pairings of subadult–subadult, who are related by $x$ because they were cubs together.

(ii)   6 pairings of adult–adult, who are related by $y$ by definition.

(iii)   12 pairings of subadult–adult; these are related on average by $(\frac{1}{7})\frac{1}{2} + (\frac{6}{7})\frac{1}{2}y$ (because there is a $\frac{1}{7}$ chance that any adult female was the mother of a particular subadult, and a $\frac{6}{7}$ chance that she was not).

Thus of the total of 21 pairings among the 7 females, the mean relatedness, which still equals

$$y = \frac{(x)3 + (y)6 + \left(\frac{1}{14} + \frac{6y}{14}\right)12}{21} \tag{2}$$

simplifies to

$$7x - 23y = -2.$$

Solving equations 1 and 2 gives the relatedness between adult males $x = 0\cdot22$, and the relatedness between females $y = 0\cdot15$.

*The effects of variation.* To investigate the effects of the variation between prides, it is necessary to put the equations 1 and 2 into general form, and then to apply the same approach. Let $a$ be the number of adult females in the pride; let $b$ be the number of young females recruited into the pride as replacements for the same number of adults; let $c$ be the number of females giving birth at one time; let $d$ be the litter size; and let $e$ be the number of adult males. We now recalculate equations 1 and 2. It can be shown in the same way as before, that the mean relatedness $x$ among cubs also equals

$$\frac{\left(\frac{1}{2}\right)\left[\binom{d}{2}c\right] + (\frac{1}{4} + \frac{1}{4}y)\left[\binom{c/e}{2}d^2e\right] + (\frac{1}{4}x + \frac{1}{4}y)\left[\binom{e}{2}\frac{d^2c^2}{e^2}\right]}{\binom{cd}{2}}$$

which simplifies to

$$x\left(3cd + \frac{cd}{e} - 4\right) + y(1-c)d = d + \frac{cd}{e} - 2. \qquad (1a)$$

Similarly the mean relatedness $y$ among females in the recon-structed pride also equals

$$\frac{x\left[\binom{b}{2}\right] + y\left[\binom{a-b}{2}\right] + \left[\frac{1}{2a} + \left(\frac{a-1}{a}\right)\frac{y}{2}\right](a-b)b}{\binom{a}{2}}$$

which simplifies to

$$x(b-1)b + \left(\frac{b^2}{a} - ab\right) = -\frac{b}{a}(a-b). \qquad (2a)$$

(For integers, $\binom{n}{2}$ is the number of possible combinations of 2 items which can be drawn from a population of $n$ items; it is num-erically equal to $\frac{1}{2}n(n-1)$.)

Equations 1a and 2a can be solved for any values of the para-meters $a$ to $e$. Fig. 2 (i–v) shows that varying any one of these five parameters while keeping the others constant has relatively little effect on the overall degrees of relatedness. $x$ and $y$ rise and fall together. The consequences of departures for most of the assump-tions 1 to 11 can be tested by varying one or more of these five parameters.

For instance, if fatherhood of litters is not shared equally among both adult males (assumption 4), this is equivalent to reducing the number of males to between 2 and 1, which results in an increase in relatedness in both sexes (Fig. 2, v). But males in larger groups (3 or more) reproduce more effectively (Bertram 1975), which counters this effect.

Similarly, it is likely that assumption 8 is not met, and that births, cub survival, recruitment and departure are not independent of relationships in the pride. For example a high cub mortality may tend to remove entire litters rather than individuals at random; but this is equivalent to a reduction in the number of females giving birth (Fig. 2, iii), the effects of which can be seen.

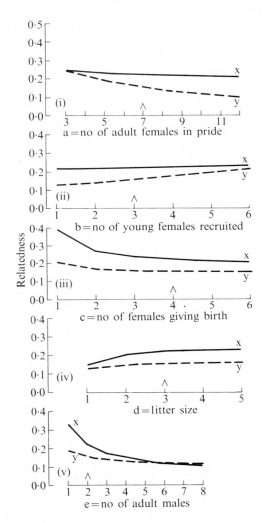

Fig. 2. Relatedness among males (x, solid line) and among females (y, broken line) in lion prides, with parameters a to e varying. ∧ indicates typical case.

The assumptions can be tightened and modified as more data become available.

*Conclusions.* Lack of validity of the different assumptions has different effects on the degrees of relatedness x and y; in some cases they are raised and in other cases lowered, and perhaps to some extent they cancel one another out. But it can be seen that even quite large deviations from the case assumed to be typical produce relatively small effects on these degrees of relatedness, and they do not influence the main conclusions. These are the following:

(i)    Male lions in possession of a pride, and cubs growing up in a pride together, are on average related by about 0·22 (i.e. almost half-siblings).

(ii)    Adult females are related on average by about 0·15 (i.e. a little closer than full cousins).

(iii)    Under all practical circumstances the males are more closely related to one another than are the females.

It must be stressed that the degrees of relatedness calculated are mean figures, made up of a range of differing degrees of relatedness. For example, some animals will be full siblings, related by 0·5. The next most closely related will be related by 0·38; they will be lions who shared a father and whose mothers were full sisters. And so on. Thus the values of relatedness are not normally distributed; instead there is a range of degrees of relatedness, more or less discrete at the higher values, and scattered asymmetrically and discontinuously about a mean of 0·22 or 0·15. This is assuming that the males are totally unrelated to the females, which is unrealistic in practice. However, if they were, for example, their third cousins they would be related to the females by less than 0·01, which is negligible compared with the much closer degrees of relatedness within each sex or peer group of offspring.

*Applicability*

There are a number of cases of animals co-operating with related individuals. Refraining from competition with them is another form of co-operation which may be much commoner although less easily recognizable. At least 60 bird species are known to nest co-operatively (Skutch 1961; Harrison 1969; Brown 1972; Fry 1972). In a variety of mammal species also, individuals other than the parents have been shown to assist in feeding or protecting the young (Kühme 1965; Mech 1970; Schaller 1972; Rood 1974; Hrdy 1976). In many of these cases the individuals are likely to be related to one another. In some cases there are now sufficient data to produce a rough general model of the reproduction of the species, and so to calculate how closely related they are, using methods similar to the one outlined in this paper. To do so would be a valuable first step in quantifying the relative costs and benefits of co-operative behaviour towards other group members, and so understanding the evolution of many aspects of their social behaviour. Examples where such a method would be applicable are the Tasmanian native hen (Maynard Smith and Ridpath 1972; Ridpath 1972), Arabian babblers (*Turdoides squamiceps*) (Zahavi 1974), turkeys (*Meleagris gallopavo*) (Watts and Stokes 1971), wolves (*Canis lupus*) (Mech 1970; E. Zimen, personal communication), wild dogs (*Lycaon pictus*) (van Lawick and van Lawick-Goodall 1971; Schaller 1972), red foxes (*Vulpes vulpes*) (D. W. Macdonald, personal communication), banded mongooses (*Mungos mungo*) (J. P. Rood, personal communication), elephants (*Loxodonta africana*) (Douglas-Hamilton 1973), and a large number of primate species (see e.g. Clutton-Brock and Harvey, this volume).

### Kin selection pressures in lions

The lion social system contains elements of both co-operation and competition among individuals. Lions co-operate in hunting, in driving out intruders, and particularly in rearing their young. They compete to some extent for limited resources: for food, for oestrous females, and for a place in the pride. One of the many factors influencing the compromise between competition and co-operation in any such situation is likely to be the degree of relatedness among those individuals, through the operation of kin selection. We will consider three aspects of this co-operation where a kin selection pressure is likely to have operated in its evolution, and will consider the other selective pressures which are also likely to have been involved.

*Communal suckling*

Adult females in a lion pride generally suckle one another's cubs, if those cubs are of similar age or younger (Schaller 1972, p. 149; Bertram 1975). The suckling is not necessarily indiscriminate, and individuals vary in the extent to which they will allow others' offspring to suckle from them.

The degree of relatedness between a female and the offspring of another female in the pride is $\frac{1}{2} \times 0.15$, or about one-seventh of that between her and her own offspring. If the benefit to the other cub exceeds seven times the cost of her own cub, such communal suckling should evolve through kin selection alone. One can conceive of cases where this disproportion could occur (for example as a result of the deaths of particular females or cubs, or extreme hunger), but it would appear to be unlikely in many cases. Nonetheless, kin selection provides a selective pressure contributing towards communal suckling.

However, it is not the only selective pressure operating. Trivers (1971) showed that reciprocal altruism could be selected for in animals of long-lived species able to detect any failure by companions to reciprocate. It seems reasonable to suppose that a lioness could detect and discriminate against the young of another female who 'cheated' by not allowing communal suckling from her. In a sense, this is a case of a delayed benefit to the individual who feeds another's young: her own offspring will in turn be fed by others later.

There appear to be other advantages to the individual. Cubs born synchronously and reared communally were found to be more likely to survive than cubs born asynchronously (Bertram 1975). It is possible that the presence of other cubs themselves is beneficial to the survival of a female's own cubs (for example by making their mothers more likely to stay together and to hunt co-operatively and therefore more efficiently). When a cub is reared to subadulthood, it benefits in several ways from the presence of companions. And in particular, males have a considerably longer and more effective reproduction life if they have companions (Bertram 1975). Thus it is to the advantage of a lioness for her own cubs' companions to survive, and therefore communal suckling could be selected for even in the absence of kin selection.

*Male tolerance towards cubs at kills*

When food is in short supply, there is a large amount of squabbling and competition for it at kills. Larger animals rob smaller ones of small carcasses or pieces of food, or prevent them from feeding by threatening or attacking them if they come near. The exception which has occasioned frequent surprise is that males more than females will often allow small cubs to feed from a piece of food in their possession (Schaller 1972, p. 152).

A male approached by a cub is on average related to that cub by 0·31 (i.e. there is a 50% chance that it is his own offspring and so related to him by ½, and a 50% chance that it is his companion's offspring and so related to him by ½ × 0·22). It is most unlikely that a male can distinguish which are his own offspring, for a variety of reasons: females come into heat at very irregular intervals, they mate very frequently and in many cases with more than one male in succession, and most oestrous periods do not result in cubs.

A female approached by a cub is related to that cub by either ½ (if it is her own), or ½ × 0·15 (if it is not hers). If we assume that she is one of the four females who gave birth, her average relatedness to the cub is 0·18 (there is a 25% chance that it is her own offspring, and a 75% chance that it is a companion's). It is likely that a female generally can recognize her own offspring, although it may be difficult in practice for her to discriminate among cubs during competition among a number of hungry animals at a kill. In the apparent absence of such discrimination, a male is likely to be considerably more closely related than a female, to a cub which tries to feed from his portion of food. Thus the benefit to a male is greater than to a female from any improved cub survival which results from this tolerance at kills.

One must not forget that there may well be other selective pressures contributing towards this difference in behaviour. For example, males have a larger food intake than females do (Schaller 1972, p. 270). It is possible that the cost to a male of the loss of a small fraction of its food supply is less serious than it would be to a female. (Note that here, as in the previous example, 'cost' and 'benefit' to an animal are defined in terms of that animal's future reproductive success; they are therefore virtually unmeasurable at present.) In addition, a male has a much shorter effective reproductive life than a female does (Bertram 1975). Therefore the cubs fathered by him that are alive at any time represent a much greater proportion of his lifetime's reproductive output than is the case for a female, and so those cubs are more valuable to males than they are to females.

## Competition among males for oestrous females

There is no dominance order among pride males. The first to encounter an oestrous female stays with her and mates with her. He is temporarily dominant while with her, and rarely allows other males to come within 10 metres. Other males make no attempt to do so, but often wait nearby. A female may change males during her oestrous period (which usually lasts two to four days and during which mating takes place at intervals averaging about 15 minutes in length). The lack of a dominance order among males, and the insignificant amount of agonistic behaviour, imply a surprisingly low level of competition among the males in lions as compared with other species.

A male is related on average by 0·22 to his companion, whose offspring thus carry a proportion of his own genes. His own offspring would carry four to five times as many, but the fact that some of his genes are transmitted to the next generation via his related companion must reduce the selective pressure for competition with that companion for matings.

It is interesting to compare the reproductive success of different male lions in terms of offspring produced, with their success as measured by genes in common with the next generation. Table 1 shows for a group of six companion males the relative proportions of offspring presumed to have been fathered by each male; this is on the basis of the number of times each male was seen in possession of an oestrous female. (I appreciate that there are pitfalls in assuming a correlation between the frequencies of copulation and of paternity, but this problem is irrelevant to the present argument.) On the basis of copulations, the most successful male produced 3½ times as many offspring as the least successful. But males are also 'reproducing by proxy' via their companions. One half of a male's gene complement is in one of his own offspring, and one-ninth (i.e. ½ × 0·22) in each cub of one of his companions. Table 1 shows the relative proportions of each male's genes in the next generation. It can be seen that in terms of genes transmitted, the most successful male did only 1½ times as well as the least successful. It is this latter difference (which is still appreciable, but much less than 3½ times) which will determine how strong is the selective pressure for a male to compete with his related companion for an

Table 1. Relative reproductive success of six companion male lions.

| Male lion number | 1 | 3 | . 4 | 5 | 6 | 8 | Total |
|---|---|---|---|---|---|---|---|
| Number of times seen with an oestrous female | 10 | 9 | 4 | 14 | 9 | 6 | 52 |
| Hence % of oestrous females seen with him = presumed proportion of offspring fathered by him. The best does 3½ times as well as the worst | 19·2 | 17·3 | 7·7 | 26·9 | 17·3 | 11·6 | 100 |
| Gene replicas in own offspring (i.e. line above multiplied by ½) | 9·6 | 8·7 | 3·9 | 13·5 | 8·7 | 5·8 | 50·2 |
| % of offspring fathered by his companions: | 80·8 | 82·7 | 92·3 | 73·1 | 82·7 | 88·4 | |
| Gene replicas in offspring of companions (i.e. line above multiplied by ½×0·22) | 9·0 | 9·2 | 10·3 | 8·1 | 9·2 | 9·8 | 55·6 |
| Total of his gene replicas | 18·6 | 17·9 | 14·2 | 21·6 | 17·9 | 15·6 | 105·8 |
| Hence % of his gene replicas in next generation (therefore the best does 1½ times as well as the worst) | 17·6 | 16·9 | 13·4 | 20·4 | 16·9 | 14·8 | 100·0 |

oestrous female. Thus kin selection will be a factor responsible for the low level of competition observed.

However, it is not the only factor. The large number of matings in each oestrous period (about 300), the low probability of cubs being produced as a result (about 20% of oestrous periods result in births), a mean litter size of two to three cubs, and the high cub mortality (75—80%) all mean that on average a very large number of matings are required for each cub reared in the next generation. The probable benefit for a copulation is thus extremely small, which again reduces the selective pressure on a male to compete for it.

The costs of competing, on the other hand, are probably very high. Serious injuries can be inflicted during fights, and such injuries can be fatal. Even if not injured himself, the winner of a fight which leaves him without a male companion is likely to be expelled much sooner by a new group of males (Bertram 1975). When driven out, his reproductive period is effectively terminated; also his offspring in the pride from which he has been expelled suffer an increase in mortality at the takeover of the pride by the new males (Bertram 1975).

## The strength of kin selection pressures

I have dealt at length with the lion example above because it illustrates well the great array of selective pressures operating on animals. Kin selection is only one of these pressures. It has not proved possible so far to demonstrate conclusively that kin selection does operate in the wild, because there are likely always to be a large number of other selective pressures, some operating in the same direction and others pushing in different directions. But it is not reasonable to say that because kin selection cannot be demonstrated to operate it can therefore be ignored. (By analogy, the fact that a lump of iron falls to the ground under the influence of gravity does not mean that there are no other forces acting upon it, such as air resistance, magnetic, and inertial forces, all influencing its descent.) Kin selection pressures provide the most probable explanation for the evolution of non-reproductive castes in social insects (Hamilton 1964; Wilson 1971). It has not been shown on theoretical grounds that genes cannot be selected for regardless of their carrier animals, and therefore Hamilton's (1964) logical modification of the mechanism by which natural selection operates should be accepted. It is surprising that it has not received wider acceptance already.

In the examples above, I have considered only the average degrees of relatedness between individuals interacting with one another. In theory, there should be selection for any gene which enabled its carrier animal to be better at recognizing its degree of relatedness to any other individual; the animal could then discriminate precisely the extent to which it should co-operate (or compete) with any other animal according to how close a relative it was. However, there are a number of complicating factors, at different levels, preventing great precision.

First, there are considerable practical problems involved in the recognition of relatives. Older siblings, for example, can observe directly that younger siblings have come from the same mother; but younger animals cannot distinguish older relatives except by the behaviour of those older siblings or of other individuals.

There can be considerable doubts over paternity, in species where females mate with more than one male. It is difficult to see how the offspring of a female could determine whether they had the same or a different father. Similarly, it is difficult to see how offspring of different females could determine whether they were fathered by the same or by different males.

In the case of two animals which were the offspring of totally different parents, there would seem to be no way in which the offspring could determine whether either parent of one was related to any extent to either parent of the other.

In addition, this is a case where there may be conflict between mother and offspring. It is less to the advantage of the mother for her offspring to be able to recognize their father, because they may then expend resources in co-operating with their relatives on the father's side; there will be kin selection pressures on them to do so. In the absence of inbreeding, the mother is not related to the father's relatives and has no genetic interest in their wellbeing; it is in her interests for her offspring to be altruistic only to her own relatives, not to his. Thus countering any selection on offspring for recognition of relatives, there will be selection on their mother for making difficult the recognition of their father's relatives. Selection pressure on the father will be similar but in the opposite direction, and generally will involve greater practical problems, depending on the social organization of the species.

Further, altruistic behaviour by her offspring towards her

relatives increases the mother's inclusive fitness more than that of the offspring; this is another source of potential conflict between parent and offspring (Trivers 1974). It would be to the mother's advantage to mislead her offspring into treating her relatives as closer relatives than they really are.

Other individuals too, relatives or not, would usually benefit by misleading a potential altruist into treating them as though they were more closely related to him than they really are. Selection for this ability to deceive would make more difficult the task of determining precisely the degree of relatedness of another individual.

Thus for these practical and evolutionary reasons, there is likely to be a considerable range of uncertainty as to the relatedness of any other individual. The degree of relatedness between two animals can be considered as being composed of two portions, one detectable and the other undetectable. The detectable portion of their relatedness comes from such observed facts as a shared mother. The undetectable portion comes from uncertainties over paternity, adoptions, and relatedness between parents and ancestors.

The relative sizes of the detectable and undetectable portions of the relatedness between two animals will depend on a variety of factors, particularly on the rigidity of the social system, the level of perceptive ability of the species, and the degree of cultural transmission of information about relatedness. Between two individuals, the relative sizes of the two portions are not necessarily the same for each animal; for example, older animals are in a position to possess more knowledge of the parentage of younger animals than vice versa.

However, in the course of evolution, the strength of the kin selection pressure depends on the total degree of relatedness, regardless of the relative proportions which the animals are able to detect. For instance, in the lion case, one might expect to find greater co-operation between brothers than the average within the male group, and other types of sub-alliance within the pride. It is not known to science, nor presumably to lions, whether all the cubs in a lion litter produced by a female who mated with more than one male were necessarily fathered by the same or by different males. Litter mates are therefore related by 0·25 (the detectable portion) plus an undetectable portion of between 0 and 0·25 depending on the identities and relatedness of their fathers. Two offspring of different mothers may well have had the same father, and their mothers may well have been related to one another; in consequence, although the degree of relatedness between two such animals is probably composed almost entirely of the undetectable portion, it may sometimes be as great as in the case of the two littermates above. With these large amounts of uncertainty, a lion cannot determine how close its various relatives are. On average, a littermate will be more closely related than other animals, and therefore one would expect that he would receive favoured treatment; but since another peer will on average be appreciably related too, the selective pressure to aid one rather than the other will be small.

A further complication arises in considering the strength of kin selection pressures. The degree of relatedness is not the only factor to be taken into account when behaving altruistically towards another animal. The expected further reproductive success of the relative being helped is also important, and the extent to which that can be improved by co-operation. Clearly it is of little benefit to an individual to give food to a relative who is about to die anyway, or to a relative who has enough food. While the degree of relatedness to another animal may be difficult to determine, it may be relatively easy to determine that animal's age, and to some extent its needs.

## Selective pressures, functions, and adaptiveness

The assumption used to be widespread that every characteristic of an animal species, and every feature of its social system had evolved for 'the good of the species' as a whole. The biologist would try to think of a possible way in which a particular feature benefited the species, and the most plausible one was put forward as the explanation. There are several drawbacks to such an approach.

First, it was not clear what was most to the benefit of the species. To have the greatest possible number of individuals? Or to have animals of the highest possible quality? Or to have animals of as diverse genotypes and therefore greatest adaptability as possible?

Second, different observers might well produce a different explanation according to their subjective expectations. And if the characteristic or feature of the species had been the opposite of that observed, a different explanation might have been satisfactorily found for that too. This subjectiveness is due to the impossibility of testing what were the causes of events which took place in the past; all one can do is to suggest a cause which now would produce the same result. In view of what has been said about the great variety of selective pressures acting on animals, to select a single one as *the* cause seems extremely dubious.

Thirdly, it is nowadays realized that characters may be selected for without necessarily being beneficial to the species as a whole. Hamilton's (1971*a*) model, quoted by Clutton-Brock and Harvey (this volume), showed that a gene responsible for selfish behaviour could be selected for, regardless of its effect on the species, which evolves in the direction of those individuals within it which leave most descendants.

One of the clearest and most striking instances of this is shown by the lion. I have found that male lions taking over a pride are liable to kill cubs they find there (Bertram 1975); by doing so they make the mothers of those cubs give birth sooner to those males' own offspring, which are then also more likely to survive. The new males thus increase their own reproductive output at the expense of the parents of those cubs. Such behaviour does not benefit the species, but the genes responsible for it can be selected for because they make their carriers leave more descendants. In theory, if there were a second lion species living in the same ecological niche and differing from the first only in the lack of cub-killing behaviour, then this second lion species should outbreed the first and eventually eliminate it. However, in practice one does not find such situations, at least in the larger vertebrates. And if one did, it would be only a matter of time before a cub-killing gene appeared as a mutation in the second species and spread rapidly because of the reproductive advantages it would give to the individuals who carried it. Similar cases of infanticide have been reported in several primate studies (summarized by Hrdy 1974, and Clutton-Brock and Harvey, this volume). Infanticide is clearly not adaptive to both participants in the relationship.

It has been suggested (Hamilton 1971*b*) that much gregarious behaviour, such as herding or shoaling, may have been selected for in the same way. The reduced probability of predation on an individual which gets closer to another makes him leave more off-spring on the average, and therefore propagate the genes contributing towards such gregariousness, even if, as could sometimes be the case, this helped the predator to catch more of that prey species.

Instances of behaviour which benefits the individual and is detrimental to the species are not easy to find, but are I think likely to be more common than is generally supposed. Much territorial

behaviour might come into this category, and much of the aggressive behaviour of dominant animals in social groups.

The problem of the evolution of a character or of a behaviour
pattern can be approached from either of two directions.

(1)   The shorthand approach often taken is to say 'This animal
shows a certain behaviour: what is its function?'. A function here
is a beneficial consequences of the behaviour (Hinde 1969, p. 159).
This is an approach most useful at the species level.

(2)   The approach taken throughout this paper has been via
selective pressures. One starts by saying 'This animals shows a
certain behaviour: what selective pressures are acting on it to make
it show this behaviour to a greater or lesser extent?' In my view
this second approach is the more useful when considering selection
at an intra-specific level, and particularly when dealing with complex relationships between individuals. It is easier to see how kin
selection can be incorporated into answers to questions phrased in
terms of selective pressures rather than in terms of functions; also
it helps to draw attention to the diversity of consequences of a
particular behaviour, on all of which natural selection will act.

For example, for a lioness to allow rather than prevent her companions' cubs from suckling from her has a great variety of possible
consequences: some related cubs may get more milk and perhaps
therefore are more likely to survive; her own cubs may get less
milk and perhaps therefore are less likely to survive; her own cubs
if they do survive may be more likely to have companions and all
may therefore do better reproductively; exchange of odour among
animals may make her own cubs able to suckle from other females;
she may have to delay having her next litter; she may therefore
look after her present cubs for better or for longer. Some of these
selection pressures will be stronger than others, and they will act
in different directions, but all play a part in determining through
natural selection the compromise degree of communal suckling in
lions. To label one or two of the beneficial consequences as
functions may lead one to neglect the other consequences.

### Individual selection

It is impossible to test what selective pressures in the past caused
the evolution of a particular character. However, it is possible to
compare the reproductive performance of individuals which possess
that character to a greater or lesser extent. Then if a difference is
found, the direction which evolution will take can be predicted,

and the way in which the selective pressures are operating can be investigated. This is the approach which is increasingly being taken in evolutionary considerations. However, five points need especially to be borne in mind.

(*a*)  We are interested primarily in the frequencies of genes in future generations. Looking at the number of young produced by particular individuals is only a very first stage. We should remember that not all offspring are equal. Young of high-ranking mothers may rank higher than the offspring of lower-ranking mothers (Sade 1967), and this may enhance their reproductive prospects. Larger young may have better survival chances and better competitive ability than smaller young, and the relative advantages of producing many small young or fewer larger young may change in different circumstances (Lack 1968). Young which are driven out have a small but finite chance of founding new colonies; expulsion of offspring to seek new breeding places elsewhere may be a good average strategy for an adult, even if many of those offspring perish as a result. When, in a relatively few years, probably, it becomes possible to determine parentage from examination of blood proteins or by other means, and when long-term studies of vertebrate groups have been continued for much longer, then the reproductive success of individuals will be able to be assessed over a few generations.

(*b*)  If we can detect a difference in reproductive output during a field study, this implies that the selective pressures are strong ones; weak ones will not be detectable.

(*c*)  We must remember that individuals which differ in the characteristic which we are investigating probably differ also in other characteristics which may be much less obvious. For example, in comparing the reproductive performance of peak time versus late breeders, one should bear in mind that the late breeding may be by individuals which are in some way inferior to the majority. In some cases, differences between individuals could be produced by manipulation by the observer, so overcoming this problem.

(*d*)  We must remember that there may be physical or physiological constraints on an animal's reproductive performance. In colonially nesting birds, for example, those at the centre may do better than those at the edge, but there is a physical limit to how many centre nests there can be (Kruuk 1964). Tits nesting earlier in the season do better than those nesting later, but there are physiological and food-supply limits to how early they can nest (Perrins, 1970). Larger animals may compete better, but may need more food, or may wear themselves out sooner.

(*e*)    The operation of kin selection through the reproduction of relatives should not be forgotten.

Eventually it will be possible to understand, and hopefully to measure, the many different selective pressures acting on individuals which have been kept under study: at present the kin selection pressure is the only one which can be quantified. Then, deciding how and why the species evolved to its present day form, behaviour and social system will be slightly less guesswork than it now is.

## Summary

Animals in social groups are often genetically related to one another. Given certain information on their social system and reproduction, it is possible to calculate how closely related they are likely to be.

Such a calculation is carried out for lions (*Panthera leo* L.). The adult males in a typical pride are on average about as closely related as half-siblings, and the females as full cousins. It is shown that variation among prides has little effect on these average degrees of relatedness.

Since these related lions in a pride co-operate with one another, a kin selection pressure is likely to be operating. Its importance is discussed in connection with communal suckling, males' tolerance towards cubs, and the lack of competition for oestrous females. It is shown that individual selective pressures are also operating in these cases.

One would expect selective pressure for enhancing animals' ability to distinguish precisely their degree of relatedness to other individuals; some of the problems hindering this are outlined.

It is emphasized that evolutionary change occurs in response to the net effect of a great array of selective pressures, acting on individuals and exerting their 'pressure' in different directions; kin selection pressure is only one of them.

# References

Bertram, B. C. R. (1973). Lion population regulation. *East African Wildlife Journal* 11, 215–25.
— (1975). Social factors influencing reproduction in wild lions. *Journal of Zoology* 177, 463–82.
Brown, J. L. (1972). Communal feeding of nestlings in the Mexican jay (*Aphelocoma ultramarina*): interflock comparisons. *Animal Behaviour* 20, 395–403.
Douglas-Hamilton, I. (1973). On the ecology and behaviour of the Lake Manyara elephants. *East African Wildlife Journal* 11, 401–3.
Fry, C. H. (1972). The social organisation of bee-eaters (Meropidae) and co-operative breeding in hot-climate birds. *Ibis* 114, 1–14.
Hamilton, W. D. (1964). The genetical evolution of social behaviour. I and II. *Journal of theoretical Biology* 7, 1–16 and 17–52.
— (1971a). Selection of selfish and altruistic behaviour in some extreme models. In *Man and Beast: Comparative Social Behavior,* ed. J. F. Eisenberg and W. S. Dillion, pp. 57–91. Smithsonian Institution Press, Washington, D.C.
— (1971b). Geometry for the selfish herd. *Journal of theoretical Biology* 31, 295–311.
Harrison, C. J. O. (1969). Helpers at the nest in Australian birds. *Emu* 69, 30–40.
Hinde, R. A. (1969). (ed.) *Bird Vocalizations.* Cambridge University Press, London.
Hrdy, S. B. (1974). Male–male competition and infanticide among the langurs (*Presbytis entellus*) of Abu, Rajasthan. *Folia primatologica* 22, 19–58.
— (1976). The care and exploitation of non-human primate infants by conspecifics other than the mother. *Advances in the Study of Behavior* 6, in press.
Klingel, H. (1965). Notes on the biology of the plains zebra, *Equus quagga boehmi* Matschie. *East African Wildlife Journal* 3, 86–8.
Kruuk, H. (1964). Predators and anti-predator behaviour of the black-headed gull (*Larus ridibundus* L.). *Behaviour Supplement* 11, 1–130.
— (1972). *The Spotted Hyena: A Study of Predation and Social Behavior.* University of Chicago Press, Chicago.
Kühme, W. (1965). Communal food distribution and division of labour in African hunting dogs. *Nature, London* 205, 443–4.
Lack, D. (1968). *Ecological Adaptations for Breeding in Birds.* Methuen, London.
Lawick, H. van and Lawick-Goodall, J. van (1971). *Innocent Killers.* Houghton Mifflin Company, Boston.
Lincoln, G. A. and Guiness, F. E. (1973). The sexual significance of the rut in red deer. *Journal of Reproduction and Fertility, Supplement* 19, 475–89.
Maynard Smith, J. and Ridpath, M. G. (1972). Wife-sharing in the Tasmanian native hen, *Tribonyx mortierii*: a case of kin selection? *American Naturalist* 106, 447–52.
Mech, L. D. (1970). *The Wolf: The Ecology and Behaviour of an Endangered Species.* Natural History Press: New York.
Missakian, E. A. (1973). Genealogical mating activity in free-ranging groups of rhesus monkeys (*Macaca mulatta*) on Cayo Santigao. *Behaviour* 45, 225–41.
Perrins, C. M. (1970). The timing of birds' breeding seasons. *Ibis* 112, 242–55.
Ridpath, M. G. (1972). The Tasmanian native hen. *Tribonyx mortierii*. II. The individual, the group, and the population. *CSIRO Wildlife Research* 17, 53–90.
Rood, J. P. (1974). Banded mongoose males guard young. *Nature, London* 248, 176.
Sade, D. S. (1967). Determinants of dominance in a group of free-ranging rhesus monkeys. In *Social Communication Among Primates,* ed. S. A. Altmann, pp. 99–114. University of Chicago Press.
Schaller, G. B. (1972). *The Serengeti Lion: A study of Predator–Prey Relations.* University of Chicago Press.
Skutch, A. F. (1961). Helpers among birds. *Condor* 63, 198–226.

Trivers, R. L. (1971). The evolution of reciprocal altruism. *Quarterly Review of Biology*
    46, 35−57.
−− (1974). Parent-offspring conflict. *American Zoologist* 14, 249−64.
Watts, C. R. and Stokes, A. W. (1971). The social order of turkeys. *Scientific American*
    224, 112−18.
Wilson, E. O. (1971). *Insect Societies*. Harvard University Press, Cambridge, Mass.
Zahavi, A. (1974). Communal nesting by the Arabian babbler: a case of individual selec-
    tion. *Ibis* 116, 84−7.

# The evolution of alarm calls

J. MAYNARD SMITH†

Alarm calls in birds are an often-quoted example of altruistic behavior. The difficulty for an evolutionist is to explain the origin and maintenance by selection of a character which is disadvantageous to the individual. It has usually been supposed that the advantage conferred on relations of the individual giving an alarm call is sufficient to outweigh the disadvantage to the caller, and this is the approach which is pursued further here. But the difficulties of such 'kin selection' (Maynard Smith 1964) are considerable, so it is worth asking first whether the habit could evolve by individual selection, for example because call notes confuse or mislead a predator as to the position of its prey.

It has been pointed out by Marler (1955) that call notes are usually difficult for a predator to locate, being high pitched and with a small range of frequencies. If this is not accidental, it is the consequence of selection; in other words, in the past birds giving alarm notes have been killed by predators, but would not have been killed had they given a less easily locatable note. It follows that there is, or at least that there was in the past, a risk in giving alarm notes. But it is still possible that alarm notes have evolved by individual selection. Thus it may be that, although it is better to give an unlocatable than a locatable alarm, it is better to give a locatable alarm than no alarm. For example, suppose a predator notices the existence of a flock by seeing individual movement, but instead of picking out its quarry immediately approaches closer to the flock before selecting its prey. In this case, to give an alarm call might favor an individual's chances of escape, by causing other members of the flock to remain still, and so causing the predator to overlook the flock altogether. But although such a process is possible, it seems more likely that alarm notes have evolved by kin selection.

The giving of alarm notes probably originated as a signal from

† Reprinted with permission from the author and *American Naturalist* **99** (1965).

parents to their offspring during the breeding season; conditions would be particularly favorable, because the coefficient of relationship between giver and receiver of the signal is high, and, since adults are more mobile and more experienced than young, a comparatively small risk run by an adult might give a larger additional chance of survival to the young. But this does not account for the giving of alarm notes in winter flocks; if it were advantageous, selection could surely produce individuals which gave an alarm note when they saw a hawk in the breeding season, but not in winter.

Kin selection can cause the spread and maintenance of 'alarmism' only if the members of flocks are related. If there is effectively random dispersion between breeding and the establishment of winter flocks, kin selection cannot act, even if individuals subsequently return to breed near where they were themselves raised, and hence tend to mate with close relatives.

The simplest case to analyze, and one particularly favorable to the spread of alarm notes, is that in which winter flocks consist of sets of brothers and sisters. This is probably never the case; the actual degree of relationship between members of winter flocks is unknown but certainly not so close as this, so that the selective advantage to the flock relative to the disadvantage to the individual must be correspondingly greater than that found below for flocks consisting of sibships.

In a flock of $n$ birds, let there be $r$ 'alarmists' and $(n - r)$ 'non-alarmists'. In any given encounter between the flock and a predator, let

$$P_K = \text{probability that a member of the flock is killed}$$
$$P_K \times P_A = \text{probability that an alarmist is killed}$$
$$P_K(1 - P_A) = \text{probability that a non-alarmist is killed.}$$

The larger the number of alarmists in the flock, the earlier is an alarmist likely to see the predator and give the alarm, and so the greater the probability that the flock will escape attention. Hence $P_K$ is a decreasing function of $r$; the simplest assumption is that

$$P_K = P\left(1 - a\frac{r}{n}\right) \qquad (1)$$

where $0 < a < 1$.

It is assumed that an alarm note is given by the first alarmist individual to see the predator. Then if a bird is killed, let the probability that the individual giving the alarm is killed be $b/n$, where

$1 < b < n$, since, if $b \leqslant 1$, giving the alarm does not increase the risk run by the individual, and no problem arises. If $b = n$, the individual giving the alarm is certain to be killed, if any bird is killed.

Then the probability that a specific one of the $(n - 1)$ birds not giving the alarm is killed is

$$\frac{1}{n-1}\left(1 - \frac{b}{n}\right) = \frac{n-b}{n(n-1)}$$

hence

$$P_A = \frac{b}{n} + \frac{(r-1)(n-b)}{n(n-1)}$$

or

$$P_A = 1 - \frac{(n-r)}{n}C$$

$$1 - P_A = \frac{(n-r)}{n}C \tag{2}$$

where $C = (n - b)/(n - 1)$.

In interpreting the results, however, it is convenient to define two parameters, $A$ and $D$, measuring respectively the advantage to the flock and the disadvantage to the individual of giving the alarm note. Similar parameters were defined by Williams and Williams (1957) in their treatment of the evolution of social behavior in insects.

$A$ is defined as the probability that, in a single encounter, a bird will be killed from a wholly non-alarmist flock, divided by the probability that a bird will be killed from a wholly alarmist flock. Then from equation (1),

$$A = \frac{1}{1-a}. \tag{3}$$

$D$, the disadvantage to the individual, is defined as the chance, in a flock consisting of 50 per cent alarmist and 50 per cent non-alarmist individuals, that an alarmist will be killed, divided by the chance that a non-alarmist will be killed. Then from equation (2),

$$D = \frac{2-C}{C}. \tag{4}$$

Consider first the case where alarmism is due to a recessive gene, $gg$ being alarmist and $GG$, $Gg$ non-alarmist.

If $p$ be the frequency of $g$, then the frequencies of the different types of flocks, each of which is a set of siblings, and the probabilities that individuals of various genotypes will be killed, are as follows:

| Parental mating | Frequency | $r/n$ | $P_K$ | $P_A$ | $(1-P_A)$ | Proportion of deaths | | |
|---|---|---|---|---|---|---|---|---|
| | | | | | | $GG$ | $Gg$ | $gg$ |
| $GG \times GG$ | $q^4$ | 0 | $P$ | 0 | 1 | 1 | 0 | 0 |
| $GG \times Gg$ | $4pq^3$ | 0 | $P$ | 0 | 1 | $\frac{1}{2}$ | $\frac{1}{2}$ | 0 |
| $GG \times gg$ | $2p^2q^2$ | 0 | $P$ | 0 | 1 | 0 | 1 | 0 |
| $Gg \times Gg$ | $4p^2q^2$ | $\frac{1}{4}$ | $P\left(1-\frac{a}{4}\right)$ | $1-\frac{3C}{4}$ | $\frac{3C}{4}$ | $\frac{C}{4}$ | $\frac{C}{2}$ | $1-\frac{3C}{4}$ |
| $Gg \times gg$ | $4p^3q$ | $\frac{1}{2}$ | $P\left(1-\frac{a}{2}\right)$ | $1-\frac{C}{2}$ | $\frac{C}{2}$ | 0 | $\cdot\frac{C}{2}$ | $1-\frac{C}{2}$ |
| $gg \times gg$ | $p^4$ | 1 | $P(1-a)$ | 1 | 0 | 0 | 0 | 1 |

Whence the number of deaths are:

$$GG: P\left\{q^4 + 2pq^3 + p^2q^2C\left(1-\frac{a}{4}\right)\right\},$$

$$Gg: P\left\{2pq^3 + 2p^2q^2 + 2p^2q^2\left(1-\frac{a}{4}\right)C + 2p^3q\left(1-\frac{a}{2}\right)C\right\},$$

$$gg: P\left\{4p^2q^2\left(1-\frac{a}{4}\right)\left(1-\frac{3C}{4}\right) + 4p^3q\left(1-\frac{a}{2}\right)\left(1-\frac{C}{2}\right) + p^4(1-a)\right\}.$$

Now at equilibrium,

$$\frac{G \text{ genes lost}}{g \text{ genes lost}} = \frac{q}{q}.$$

Counting each $GG$ death as the loss of two $G$ genes, and so on, this reduces to

$$a = \frac{2(2-p)(1-C)}{2-C} \tag{5}$$

or substituting from (3) and (4)

$$A = \frac{D}{2-D+p(D-1)}, \tag{6}$$

where $p$ is the frequency of the recessive gene for alarmism.

It follows from equation (6) that

(i)   Since $D > 1$, for a given value of $D$, $A$ decreases as $p$ increases. Hence there can be no stable equilibrium between $p = 0$ and $p = 1$.

(ii)   When $p = 1$, $A = D$. Hence, to maintain a recessive gene for alarmism, once it is established as the common allele, it is sufficient that $A$ should be greater than $D$.

(iii)    When $p = 0, A = D/(2 - D)$. Hence if $D > 2$, no selective advantage to the flock will be sufficient to cause an increase of a recessive gene for alarmism if it is initially rare. If the selective advantages are slight, so that $A = 1 + a, D = 1 + \delta$, where $a, \delta$ are small, then $a > 2\delta$ if alarmism is to increase.

The case when alarmism is due to a dominant gene, once more assuming that flocks are sibships, can be analyzed in the same way, and leads to the equilibrium condition:

$$a = \frac{2(3 - q)(1 - C)}{3(2 - C)} \tag{7}$$

or substituting from (3) and (4)

$$A = \frac{3D}{3 + q(D - 1)}, \tag{8}$$

where $q$ is the frequency of the dominant allele for alarmism.

As in the case of the recessive gene, $A$ decreases as the frequency of the allele for alarmism increases; so no stable intermediate frequency is possible.

When alarmism is rare $(q = 0), A = D$.

When alarmism is common $(q = 1), A = 3D/(2 + D)$.

Thus the selective advantage to the flock required to establish or to maintain alarmism is, for a given value of $D$, somewhat smaller than in the case of a recessive allele.

It would be desirable to work out the conditions for the spread of alarmism in the more general case in terms of the mean coefficient of relationship between the members of the flock. I have been unable to do this, but it is clear that the maintenance of alarmism would require $A \gg D$ if the coefficient of relationship is small. It seems likely therefore that alarm notes evolved in the first place as signals from parents to their offspring during the breeding season. When they had acquired properties which made them difficult to locate, $D$ would be small enough for selection to maintain the habit in winter flocks.

The similarity between the alarm notes of different species is probably a case of convergence, although sometimes a similar note may be inherited from a common ancestor. It is also possible that, when mixed flocks are formed, selection analogous to that responsible for Mullerian mimicry may be involved, although it is easier to see how birds would acquire the habit of responding to alarm notes in this way than the habit of giving them.

A final point concerns the possibility of stable genetic polymorphism for altruistic or 'donor' (Williams and Williams, 1957) traits. Polymorphism for such traits in social insects is environmentally induced. In the case of alarm notes, equations (6) and (8) show that there are values of $A$ and $D$ for which intermediate equilibria occur, but these equilibria are always unstable. Kin selection could maintain a stable genetic polymorphism if the ratio of $A$ to $D$ was high when the altruistic or donor trait was rare, and low when it was common. Possible examples are difficult to think of in animals, but relatively easy in human societies.

## Summary

The selective forces which can maintain traits such as alarm calls which are individually disadvantageous but socially advantageous are considered. The particular case in which flocks consist of sets of brothers and sisters is analyzed in detail. It is shown that equilibria with intermediate gene frequencies can exist but are unstable; the conditions for stable polymorphism for altruistic traits are briefly discussed.

## References

Marler, P. (1955). Characteristics of some animal calls. *Nature, Lond.* 176, 6–8.
Maynard Smith, J. (1964). Kin selection and group selection. *Nature, Lond.* 201, 1145–7.
Williams, G. C., and Williams, D. C. (1957). Nature selection of individually harmful social adaptations among sibs with special reference to social insects. *Evolution* 11, 32–9.

# The evolution of reciprocal altruism

ROBERT L. TRIVERS†

## Introduction

Altruistic behavior can be defined as behavior that benefits another organism, not closely related, while being apparently detrimental to the organism performing the behavior, benefit and detriment being defined in terms of contribution to inclusive fitness. One human being leaping into water, at some danger to himself, to save another distantly related human from drowning may be said to display altruistic behavior. If he were to leap in to save his own child, the behavior would not necessarily be an instance of 'altruism'; he may merely be contributing to the survival of his own genes invested in the child.

Models that attempt to explain altruistic behavior in terms of natural selection are models designed to take the altruism out of altruism. For example, Hamilton (1964) has demonstrated that degree of relationship is an important parameter in predicting how selection will operate, and behavior which appears altruistic may, on knowledge of the genetic relationships of the organisms involved, be explicable in terms of natural selection: those genes being selected for that contribute to their own perpetuation, regardless of which individual the genes appear in. The term 'kin selection' will be used in this paper to cover instances of this type—that is, of organisms being selected to help their relatively close kin.

The model presented here is designed to show how certain classes of behavior conveniently denoted as 'altruistic' (or 'reciprocally altruistic') can be selected for even when the recipient is so distantly related to the organism performing the altruistic act that kin selection can be ruled out. The model will apply, for example, to altruistic behavior between members of different species. It will be argued that under certain conditions natural selection favors these altruistic behaviors because in the long run they benefit the organism performing them.

† Reprinted with permission from the author and *Quarterly Review of Biology* **46** (1971). Copyright © by *Quarterly Review of Biology*.

## The model

One human being saving another, who is not closely related and is about to drown, is an instance of altruism. Assume that the chance of the drowning man dying is one-half if no one leaps in to save him, but that the chance that his potential rescuer will drown if he leaps in to save him is much smaller, say, one in twenty. Assume that the drowning man always drowns when his rescuer does and that he is always saved when the rescuer survives the rescue attempt. Also assume that the energy costs involved in rescuing are trivial compared to the survival probabilities. Were this an isolated event, it is clear that the rescuer should not bother to save the drowning man. But if the drowning man reciprocates at some future time and if the survival chances are then exactly reversed, it will have been to the benefit of each participant to have risked his life for the other. Each participant will have traded a one-half chance of dying for about a one-tenth chance. If we assume that the entire population is sooner or later exposed to the same risk of drowning, the two individuals who risk their lives to save each other will be selected over those who face drowning on their own. Note that the benefits of reciprocity depend on the unequal cost/ benefit ratio of the altruistic act, that is, the benefit of the altruistic act to the recipient is greater than the cost of the act to the performer, cost and benefit being defined here as the increase or decrease in chances of the relevant alleles propagating themselves in the population. Note also that, as defined, the benefits and costs depend on the age of the altruist and recipient (see *Age-dependent changes* below). (The odds assigned above may not be unrealistic if the drowning man is drowning because of a cramp or if the rescue can be executed by extending a branch from shore.)

Why should the rescued individual bother to reciprocate? Selection would seem to favor being saved from drowning without endangering oneself by reciprocating. Why not cheat? ('Cheating' is used throughout this paper solely for convenience to denote failure to reciprocate; no conscious intent or moral connotation is implied.) Selection will discriminate against the cheater if cheating has later adverse affects on his life which outweigh the benefit of not reciprocating. This may happen if the altruist responds to the cheating by curtailing all future possible altruistic gestures to this individual. Assuming that the benefits of these lost altruistic acts outweigh the costs involved in reciprocating, the cheater will be selected against relative to individuals who, because neither cheats, exchange many altruistic acts.

This argument can be made precise. Assume there are both altruists and non-altruists in a population of size $N$ and that the altruists are characterized by the fact that each performs altruistic acts when the cost to the altruist is well below the benefit to the recipient, where cost is defined as the degree to which the behavior retards the reproduction of the genes of the altruist and benefit is the degree to which the behavior increases the rate of reproduction of the genes of the recipient. Assume that the altruistic behavior of an altruist is controlled by an allele (dominant or recessive), $a_2$, at a given locus and that (for simplicity) there is only one alternative allele, $a_1$, at that locus and that it does not lead to altruistic behavior. Consider three possiblilities: (1) the altruists dispense their altruism randomly throughout the population; (2) they dispense it nonrandomly by regarding their degree of genetic relationship with possible recipients; or (3) they dispense it nonrandomly by regarding the altruistic tendencies of possible recipients.

### (1) *Random dispensation of altruism*

There are three possible genotypes: $a_1a_1$, $a_2a_1$, and $a_2a_2$. Each allele of the heterozygote will be affected equally by whatever costs and benefits are associated with the altruism of such individuals (if $a_2$ is dominant) and by whatever benefits accrue to such individuals from the altruism of others, so they can be disregarded. If altruistic acts are being dispensed randomly throughout a large population, then the typical $a_1a_1$ individual benefits by $(1/N)\Sigma b_i$, where $b_i$ is the benefit of the $i$th altruistic act performed by the altruist. The typical $a_2a_2$ individual has a net benefit of $(1/N)\Sigma b_i - (1/N)\Sigma c_j$, where $c_j$ is the cost to the $a_2a_2$ altruist of his $j$th altruistic act. Since $-(1/N)\Sigma c_j$ is always less than zero, allele $a_1$ will everywhere replace allele $a_2$.

### (2) *Nonrandom dispensation of reference to kin*

This case has been treated in detail by Hamilton (1964), who concluded that if the tendency to dispense altruism to close kin is great enough, as a function of the disparity between the average cost and benefit of an altruistic act, then $a_2$ will replace $a_1$. Technically, all that is needed for Hamilton's form of selection to operate is that an individual with an 'altruistic allele' be able to distinguish between individuals with and without this allele and discriminate accordingly. No formal analysis has been attempted of the possibilities for selection favoring individuals who increase their chances of receiving altruistic acts by appearing as if they were close kin of altruists, although selection has clearly sometimes favored such parasitism (e.g. Drury and Smith 1968).

(3) *Nonrandom dispensation by reference to the altruistic tendencies of the recipient*

What is required is that the net benefit accruing to a typical $a_2 a_2$ altruist exceed that accruing to an $a_1 a_1$ non-altruist, or that

$$(1/p^2)(\Sigma b_k - \Sigma c_j) > (1/q^2)\Sigma b_m$$

where $b_k$ is the benefit to the $a_2 a_2$ altruist of the $k$th altruistic act performed and toward him, where $c_j$ is the cost of the $j$th altruistic act by the $a_2 a_2$ altruist, where $b_m$ is the benefit of the $m$th altrusitic act to the $a_1 a_1$ nonaltruist, and where $p$ is the frequency in the population of the $a_2$ allele and $q$ that of the $a_1$ allele. This will tend to occur if $\Sigma b_m$ is kept small (which will simultaneously reduce $\Sigma c_j$). And this in turn will tend to occur if an altruist responds to a 'nonaltruistic act' (that is a failure to act altruistically toward the altruist in a situation in which so doing would cost the actor less than it would benefit the recipient) by curtailing future altruistic acts to the nonaltruist.

Note that the above form of altruism does not depend on all altruistic acts being controlled by the same allele at the same locus. Each altruist could be motivated by a different allele at a different locus. All altruistic alleles would tend to be favored as long as, for each allele, the next average benefit to the homozygous altruist exceeded the average benefit to the homozygous nonaltruist; this would tend to be true if altruists restrict their altruism to fellow altruists, regardless of what allele motivates the other individual's altruism. The argument will therefore apply, unlike Hamilton's (1964), to altruistic acts exchanged between members of different species. It is the *exchange* that favors such altruism, not the fact that the allele in question sometimes or often directly benefits its duplicate in another organism.

If an 'altruistic situation' is defined as any in which one individual can dispense a benefit to a second greater than the cost of the act to himself, then the chances of selecting for altruistic behavior, that is, of keeping $\Sigma c_j + \Sigma b_m$ small, are greatest (1) when there are many such altruistic situations in the lifetime of the altruists, (2) when a given altruist repeatedly interacts with the same small set of individuals, and (3) when pairs of altruists are exposed 'symmetrically' to altruistic situations, that is, in such a way that the two are able to render roughly equivalent benefits to each other at roughly equivalent costs. These three conditions can be elaborated into a set of relevant biological parameters affecting the possibility that reciprocally altruistic behavior will be selected for.

(1) *Length of lifetime.* Long lifetime of individuals of a species maximizes the chance that any two individuals will encounter many altruistic situations, and all other things being equal one should search for instances of reciprocal altruism in long-lived species.

(2) *Dispersal rate.* Low dispersal rate during all or a significant portion of the lifetime of individuals of a species increases the chance that an individual will interact repeatedly with the same set of neighbors, and other things being equal one should search for instances of reciprocal altruism in such species. Mayr (1963) has discussed some of the factors that may affect dispersal rates.

(3) *Degree of mutual dependence.* Interdependence of members of a species (to avoid predators, for example) will tend to keep individuals near each other and thus increase the chance they will encounter altruistic situations together. If the benefit of the mutual dependence is greatest when only a small number of individuals are together, this will greatly increase the chance that an individual will repeatedly interact with the same small set of individuals. Individuals in primate troops, for example, are mutually dependent for protection from predation, yet the optimal troop size for foraging is often small (Crook 1969). Because they also meet the other conditions outlined here, primates are almost ideal species in which to search for reciprocal altruism. Cleaning symbioses provide an instance of mutual dependence between members of different species, and this mutual dependence appears to have set the stage for the evolution of several altruistic behaviors discussed below.

(4) *Parental care.* A special instance of mutual dependence is that found between parents and offspring in species that show parental care. The relationship is usually so asymmetrical that few or no situations arise in which an offspring is capable of performing an altruistic act for the parents or even for another offspring, but this is not entirely true for some species (such as primates) in which the period of parental care is unusually long. Parental care, of course, is to be explained by Hamilton's (1964) model, but there is no reason why selection for reciprocal altruism cannot

operate between close kin, and evidence is presented below that such selection has operated in humans.

*Prevent recip altruism* →

(5)  *Dominance hierarchy.* Linear dominance hierarchies consist by definition of asymmetrical relationships: a given individual is dominant over another but not vice versa. Strong dominance hierarchies reduce the extent to which altruistic situations occur in which the less dominant individual is capable of performing a benefit for the more dominant which the more dominant individual could not simply take at will. Baboons (*Papio cynocephalus*) provide an illustration of this. Hall and DeVore (1965) have described the tendency for meat caught by an individual in the troop to end up by preemption in the hands of the most dominant males. This ability to preempt removes any selective advantage that food-sharing might otherwise have as a reciprocal gesture for the most dominant males, and there is no evidence in this species of any food-sharing tendencies. By contrast, Van Lawick-Goodall (1968) has shown that in the less dominance-oriented chimpanzees more dominant individuals often do not preempt food caught by the less dominant. Instead, they besiege the less dominant individual with 'begging gestures', which result in the handing over of small portions of the catch. No strong evidence is available that this is part of a reciprocally altruistic system, but the absence of a strong linear dominance hierarchy has clearly facilitated such a possibility. It is very likely that early hominid groups had a dominance system more similar to that of the modern chimpanzee than to that of the modern baboon (see, for example, Reynolds 1966).

(6)  *Aid in combat.* No matter how dominance-oriented a species is, a dominant individual can usually be aided in aggressive encounters with other individuals by help from a less dominant individual. Hall and DeVore (1965) have described the tendency for baboon alliances to form which fight as a unit in aggressive encounters (and in encounters with predators). Similarly, vervet monkeys in aggressive encounters solicit the aid of other, often less dominant, individuals (Struhsaker 1967). Aid in combat is then a special case in which relatively symmetrical relations are possible between individuals who differ in dominance.

The broad discussion is meant only to suggest the broad conditions that favor the evolution of reciprocal altruism. The most important parameters to specify for individuals of a species are how many altruistic situations occur and how symmetrical they are, and these are the most difficult to specify in advance. Of the three instances of reciprocal altruism discussed in this paper only one, human altruism, would have been predicted from the above broad conditions.

The relationship between two individuals repeatedly exposed to symmetrical reciprocal situations is exactly analogous to what game theorists call the Prisoner's Dilemma (Luce and Raiffa 1957; Rapoport and Chammah 1965), a game that can be characterized by the payoff matrix

*Prisoner's Dilemma*

|       | $A_2$ | $C_2$ |
|-------|-------|-------|
| $A_1$ | R, R  | S, T  |
| $C_1$ | T, S  | P, P  |

where $S < P < R < T$ and where $A_1$ and $A_2$ represent the altruistic choices for the two individuals, and $C_1$ and $C_2$, the cheating choices (the first letter in each box gives the payoff for the first individual, the second letter the payoff for the second individual). The other symbols can be given the following meanings: R stands for the reward each individual gets from an altruistic exchange if neither cheats: T stands for the temptation to cheat: S stands for the sucker's payoff that an altruist gets when cheated; and P is the punishment that both individuals get when neither is altruistic (adapted from Rapoport and Chammah 1965). Iterated games played between the same two individuals mimic real life in that they permit each player to respond to the behavior of the other. Rapoport and Chammah (1965) and others have conducted such experiments using human players, and some of their results are reviewed below in the discussion of human altruism.

W. D. Hamilton (personal communication) has shown that the above treatment of reciprocal altruism can be reformulated concisely in terms of game theory as follows. Assuming two altruists are symmetrically exposed to a series of reciprocal situations with identical costs and identical benefits, then after $2n$ reciprocal situations, each has been 'paid' $nR$. Were one of the two a non-altruist and the second changed to a non-altruistic policy after first being cheated, then the initial altruist would be paid $S + (n - 1)P$ (assuming he had the first opportunity to be altruistic) and the non-altruist would receive $T + (n - 1)P$. The important point here is that unless $T \gg R$, then even with small $n$, $nR$ should exceed $T + (n - 1)P$. If this holds, the nonaltruistic type, when rare, cannot start to spread. But there is also a barrier to the spread of altruism when altruists are rare, for $P > S$ implies $nP > S + (n - 1)P$. As $n$ increases, these two total payoffs tend to equality, so the barrier to the spread of altruism is weak if $n$ is large. The barrier will be overcome if the advantages gained by exchanges between altruists outweigh the initial losses to non-altruistic types.

Reciprocal altruism can also be viewed as a symbiosis, each partner helping the other while he helps himself. The symbiosis has a time lag, however; one partner helps the other and must then wait a period of time before he is helped in turn. The return benefit may come directly, as in human food-sharing, the partner directly returning the benefit after a time lag. Or the return may come indirectly, as in warning calls in birds (discussed below), where the initial help to other birds (the warning call) sets up a causal chain through the ecological system (the predator fails to learn useful information) which rebounds after a time lag to the benefit of the caller. The time lag is the crucial factor, for it means that only under highly specialized circumstances can the altruist be reasonably guaranteed that the causal chain he initiates with his altruistic act will eventually return to him and confer, directly or indirectly, its benefit. Only under these conditions will the cheater be selected against and this type of altruistic behavior evolve.

Although the preconditions for the evolution of reciprocal altruism are specialized, many species probably meet them and display this type of altruism. This paper will limit itself, however, to three instances. The first, behavior involved in cleaning symbioses, is chosen because it permits a clear discrimination between this model and that based on kin selection (Hamilton 1964). The second, warning calls in birds, has already been elaborately analyzed in terms of kin selection; it is discussed here to show how the model presented above leads to a very different interpretation of these familiar behaviors. Finally, human reciprocal altruism is discussed in detail because it represents the best documented case of reciprocal altruism known, because there has apparently been strong selection for a very complex system regulating altruistic behavior, and because the above model permits the functional interpretation of details of the system that otherwise remain obscure.

*Rationale for discussing humans.*

### Altruistic behavior in cleaning symbioses

The preconditions for the evolution of reciprocal altruism are similar to those for the operation of kin selection: long lifetime, low dispersal rate, and mutual dependence, for example, tend to increase the chance that one is interacting with one's close kin. This makes it difficult to discriminate the two alternative hypotheses. The case of cleaning symbiosis is important to analyze in detail because altruistic behavior is displayed that cannot be explained by kin selection, since it is performed by members of one species for the benefit of members of another. It will be shown instead that the behavior can be explained by the model presented above. No elaborate explanation is needed to understand the evolution of the mutually advantageous cleaning symbiosis itself: it is several additional behaviors displayed by the host fish to its cleaner that require a special explanation because they meet the criteria for altruistic behavior outlined above—that is, they benefit the cleaner while apparently being detrimental to the host.

Feder (1966) and Maynard (1968) have recently reviewed the literature on cleaning symbioses in the ocean. Briefly, one organism (e.g. the wrasse, *Labroides dimidiatus*) cleans another organism (e.g. the grouper, *Epinephelus striatus*) of ectoparasites (e.g. caligoid copepods), sometimes entering into the gill chambers and mouth of the 'host' in order to do so. Over forty-five species of fish are known to be cleaners, as well as six species of shrimp. Innumerable species of fish serve as hosts. Stomach analyses of cleaner fish demonstrate that they vary greatly in the extent to which they depend on their cleaning habits for food, some apparently subsisting nearly entirely on a diet of ectoparasites. Likewise, stomach analyses of host fish reveal that cleaners differ in the rate at which they end up in the stomachs of their hosts, some being apparently almost entirely immune to such a fate. It is a striking fact that there seems to be a strong correlation between degree of dependence on the cleaning way of life and immunity to predation by hosts.

Cleaning habits have apparently evolved independently many times (at least three times in shrimps alone), yet some remarkable convergence has taken place. Cleaners, whether shrimp or fish, are distinctively colored and behave in distinctive ways (for example, the wrasse, *L. dimidiatus,* swims up to its host with a curious dipping and rising motion that reminds one of the way a finch flies). These distinctive features seem to serve the function of attracting fish to be cleaned and of inhibiting any tendency in them to feed on their cleaners. There has apparently been strong selection to avoid eating one's cleaner. This can be illustrated by several observations. Hediger (1968) raised a grouper (*Epinephelus*) from

*Cleaner symbiosis.*

infancy alone in a small tank for six years, by which time the fish was almost four feet in length and accustomed to snapping up anything dropped into its tank. Hediger then dropped a small live cleaner (*L. dimidiatus*) into the grouper's tank. The grouper not only failed to snap up the cleaner but opened its mouth and permitted the cleaner free entry and exit.

Soon we watched our second suprise: the grouper made a movement which in the preceeding six years we had never seen him make: he spread the right gill-covering so wide that the individual gill-plates were separated from each other at great distances, wide enough to let the cleaner through (translated from Hediger 1968 p. 93).

When Hediger added two additional *L. dimidiatus* to the tank, all three cleaned the grouper with the result that within several days the grouper appeared restless and nervous, searched out places in the tank he had formerly avoided, and shook himself often (as a signal that he did not wish to be cleaned any longer). Apparently three cleaners working over him constantly was too much for him, yet he still failed to eat any of them. When Hediger removed two of the cleaners, the grouper returned to normal. There is no indication the grouper ever possessed any edible ectoparasites, and almost two years later (in December 1968) the same cleaner continued to 'clean' the grouper (personal observation) although the cleaner was, in fact, fed separately by its zoo-keepers.

Eibl-Eibesfeldt (1959) has described the morphology and behavior of two species (e.g. *Aspidontus tacniatus*) that mimic cleaners (e.g. *L. dimidiatus*) and that rely on the passive behavior of fish which suppose they are about to be cleaned to dart in and bite off a chunk of their fins. I cite the evolution of these mimics, which resemble their models in appearance and initial swimming behavior, as evidence of strong selection for hosts with no intention of harming their cleaners.

Of especial interest is evidence that there has been strong selection not to eat one's cleaner even after the cleaning is over. Eibl-Eibesfeldt (1955) has made some striking observations on the goby, *Elacitinus occanops*:

I never saw a grouper snap up a fish after it had cleaned it. On the contrary, it announced its impending departure by two definite signal movements. First it closed its mouth vigorously, although not completely, and immediately opened it wide again. Upon this intention movement, all the gobies left the mouth cavity. Then the grouper shook its body laterally a few times, and all the cleaners returned to their coral. If one frightened a grouper it never neglected these forewarning movements (translated from Eibl-Eibesfeldt 1955 p. 208).

Randall has made similar observations on a moray eel (*Gymnothorax japonicus*) that signalled with a 'sharp lateral jerk of the eel's head,' after which 'the wrasse fairly flew out of the mouth, and the awesome jaws snapped shut' (Randall 1958, 1962). Likewise, Hediger's Kasper Hauser grouper shook its body when it had enough of being cleaned.

Why does a large fish not signal the end to a cleaning episode by swallowing the cleaner? Natural selection would seem to favor the double benefit of a good cleaning followed by a meal of the cleaner. Selection also operates, of course, on the cleaner and presumably favors mechanisms to avoid being eaten. The distinctive behavior and appearance of cleaners has been cited as evidence of such selection. One can also cite the distinctive behavior of the fish being cleaned. Feder (1966) has pointed out that hosts approaching a cleaner react by 'stopping or slowing down, allowing themselves to assume awkward positions, seemingly in a hypnotic state'. Fishes sometimes alter their color dramatically before and while being cleaned, and Feder (1966) has summarized instances of this. These forms of behavior suggest that natural selection has operated on cleaners to avoid attempting to clean fish without these behaviors, presumably to avoid wasting energy and to minimize the dangers of being eaten. (Alternatively, the behaviors, including color change, may aid the cleaners in finding ectoparasites. This is certainly possible but not, I believe, adequate to explain the phenomenon completely. See, for example, Randall 1962.)

Once the fish to be cleaned takes the proper stance, however, the cleaner goes to work with no apparent concern for its safety: it makes no effort to avoid the dangerous mouth and may even swim inside, which as we have seen, seems particularly foolhardy, since fish being cleaned may suddenly need to depart. The apparent unconcern of the cleaner suggests that natural selection acting on the fish being cleaned does not, in fact, favor eating one's cleaner. No speculation has been advanced as to why this may be so, although some speculation has appeared about the mechanisms involved. Feder advances two possibilities, that of Eibl-Eibesfeldt (1955) that fish come to be cleaned only after their appetite has been satisfied, and one of his own, that the irritation of ectoparasites may be sufficient to inhibit hunger. Both possibilities are contradicted by Hediger's observation, cited above, and seem unlikely on functional grounds as well.

A fish to be cleaned seems to perform several 'altruistic' acts. It desists from eating the cleaner even when it easily could do so and when it must go to special pains (sometimes at danger to itself) to avoid doing so. Furthermore, it may perform two additional behaviors which seem of no direct benefit to itself (and which consume energy and take time), namely, it signals its cleaner that it is about to depart even when the fish is not in its mouth, and it may chase off possible dangers to the cleaner:

> While diving with me in the Virgin Islands, Robert Schroeder watched a Spanish hogfish grooming a bar jack in its bronze color state. When a second jack arrived in the pale color phase, the first jack immediately drove it away. But later when another jack intruded on the scene and changed its pale color to dark bronze it was not chased. The bronze color would seem to mean 'no harm intended; I need service' (Randall, 1962 p. 44).

The behavior of the host fish is interpreted here to have resulted from natural selection and to be, in fact, beneficial to the host because the cleaner is worth more to it alive than dead. This is because the fish that is cleaned 'plans' to return at later dates for more cleanings, and it will be benefited by being able to deal with the same individual. If it eats the cleaner, it may have difficulty finding a second when it needs to be cleaned again. It may lose valuable energy and be exposed to unnecessary predation in the search for a new cleaner. And it may in the end be 'turned down' by a new cleaner or serviced very poorly. In short, the host is abundantly repaid for the cost of its altruism.

To support the hypothesis that the host is repaid its initial altruism, several pieces of evidence must be presented: that hosts suffer from ectoparasites; that finding a new cleaner may be difficult or dangerous; that if one does not eat one's cleaner, the same cleaner can be found and used a second time (e.g. that cleaners are site-specific); that cleaners live long enough to be used repeatedly by the same host; and if possible, that individual hosts do, in fact, reuse the same cleaner.

### (1) *The cost of ectoparasites*

It seems almost axiomatic that the evolution of cleaners entirely dependent on ectoparasites for food implies the selective disadvantage for the cleaned of being ectoparasite-ridden. What is perhaps surprising is the effect that removing all cleaners from a coral reef has on the local 'hosts' (Limbaugh 1961). As Feder (1966) said in his review:

Within a few days the number of fishes was drastically reduced. Within two weeks almost all except territorial fishes had disappeared, and many of these had developed white fuzzy blotches, swellings, ulcerated sores, and frayed fins (p. 366).

Clearly, once a fish's primary way of dealing with ectoparasites is by being cleaned, it is quickly vulnerable to the absence of the cleaners.

### (2) The difficulty and danger of finding a cleaner

There are naturally very few data on the difficulty or danger of finding a new cleaner. This is partially because, as shown below, fish tend repeatedly to return to their familiar cleaners. The only observation of fish being disappointed in their search for cleaners comes from Eibl-Eibesfeldt (1955); 'If the cleaners fail to appear over one coral in about half a minute, the large fishes swim to another coral and wait there a while' (translated from p. 210). It may be that fish have several alternative cleaning stations to go to, since any particular cleaning station may be occupied or unattended at a given moment. So many fish tend to be cleaned at coral reefs (Limbaugh (1961) observed a cleaner service 300 fish in a 6-hour period), that predators probably frequent coral reefs in search of fish being cleaned. Limbaugh (1961) suggested that good human fishing sites are found near cleaning stations. One final reason why coming to be cleaned may be dangerous is that some fish must leave their element to do so (Randall 1962):

Most impressive were the visits of moray eels, which do not ordinarily leave their holes in the reef during daylight hours, and of the big jacks which swam up from deeper water on the reef's edge to be 'serviced' before going on their way (p. 43).

### (3) Site specificity of cleaners

Feder (1966) has reviewed the striking evidence for the site specificity of cleaners and concludes:

Cleaning fishes and cleaning shrimps have regular stations to which fishes wanting to be cleaned can come (p. 367).

Limbaugh, Pederson, and Chase (1961) have reviewed available data on the six species of cleaner shrimps, and say:

The known cleaner shrimps may conveniently be divided into two groups on the basis of behavior, habitat and color. The five species comprising one group are usually solitary or paired. . . . All five species are territorial and remain for weeks and, in some cases, months or possibly years within a meter or less of the same spot. They are omnivorous to a slight extent but seem to be highly dependent upon their hosts for food. This group is tropical, and the individuals are brightly marked. They display themselves to their hosts in a conspicuous manner. They probably rarely serve as prey for fishes. A single species, *Hippolysmata californica*, comprises the second group. . . . This species is a gregarious, wandering, omnivorous animal . . . and is not highly dependent upon its host for survival. So far as is known, it does not display itself to attract fishes (p. 233).

It is *H. californica* that is occasionally found in the stomachs of at least one of its hosts. The striking correlation of territoriality and solitariness with cleaning habits is what theory would predict. The same correlation can be found in cleaner fish. *Labroides,* with four species, is the genus most completely dependent on cleaning habits. No *Labroides* has ever been found in the stomach of a host fish. All species are highly site-specific and tend to be solitary. Randall (1958) reports that an individual *L. dimidiatus* may sometimes swim as much as 60 feet from its cleaning station, servicing fish on the way. But he notes,

This was especially true in an area where the highly territorial damsel fish *Pomacentris nigricans* (Lepede) was common. As one damsel fish was being tended, another nearby would assume a stationary pose with fins erect and the *Labroides* would move on to the latter with little hesitation (p. 333).

Clearly, what matters for the evolution of reciprocal altruism is that the same two individuals interact repeatedly. This will be facilitated by the site specificity of either individual. Of temperate water cleaners, the species most specialized to cleaning is also apparently the most solitary (Hobson 1969).

### (4) Lifespan of cleaners

No good data exist on how long cleaners live, but several observations on both fish and shrimp suggest that they easily live long enough for effective selection against cheaters. Randall (1958) repeatedly checked several ledges and found that different feeding stations were occupied for 'long periods of time', apparently by the same individuals. One such feeding station supported two individuals for over three years. Of one species of cleaner shrimp, *Stenopus hispidus*, Limbaugh, Pederson, and Chase (1961) said that pairs of individuals probably remain months, possibly years, within an area of a square meter.

## (5) *Hosts using the same cleaner repeatedly*

There is surprisingly good evidence that hosts reuse the same cleaner repeatedly. Feder (1966) summarizes the evidence:

> Many fishes spend as much time getting cleaned as they do foraging for food. Some fishes return again and again to the same station, and show a definite time pattern in their daily arrival. Others pass from station to station and return many times during the day; this is particularly true of an injured or infected fish (p. 368).

Limbaugh, Pederson, and Chase (1961) have presented evidence that in at least one species of cleaner shrimp (*Stenopus scutellus*), the shrimp may reservice the same individuals:

> One pair was observed in the same football-sized coral boulder from May through August 1956. During that period, we changed the position and orientation of the boulder several times within a radius of approximately seven meters without disturbing the shrimp. Visiting fishes were momentarily disturbed by the changes, but they soon relocated the shrimps (p. 254).

Randall (1958) has repeatedly observed fish swimming from out of sight directly to cleaning stations, behavior suggesting to him that they had prior acquaintance with the stations. During two months of observations at several feeding stations, Eibl-Eibesfeldt (1955) became personally familiar with several individual groupers (*Epincphelus striatus*) and repeatedly observed them seeking out and being cleaned at the same feeding stations, presumably by the same cleaners.

In summary, it seems fair to say that the hosts of cleaning organisms perform several kinds of altruistic behavior, including not eating their cleaner after a cleaning, which can be explained on the basis of the above model. A review of the relevant evidence suggests that the cleaner organisms and their hosts meet the preconditions for the evolution of reciprocally altruistic behavior. The host's altruism is to be explained as benefiting him because of the advantage of being able quickly and repeatedly to return to the same cleaner.

## Warning calls in birds

Marler (1955, 1957) has presented evidence that warning calls in birds tend to have characteristics that limit the information a predator gets from the call. In particular, the call characteristics do not allow the predator easily to determine the location of the call-giver. Thus, it seems that giving a warning call must result, at least occasionally, in the otherwise unnecessary death of the call-giver, either at the hands of the predator that inspired the call or at the hands of a second predator formerly unaware of the caller's presence or exact location.

Given the presumed selection against call-giving, Williams (1966) has reviewed various models to explain selection for warning cries:

(1)   Warning calls are functional during the breeding season in birds in that they protect one's mate and offspring. They have no function outside the breeding season, but they are not deleted then because 'in practice it is not worth burdening the germ plasm with the information necessary to realize such an adjustment' (Williams 1966, p. 206).

(2)   Warning calls are selected for by the mechanism of group selection (Wynne-Edwards 1962).

(3)   Warning calls are functional outside the breeding season because there is usually a good chance that a reasonably close kin is near enough to be helped sufficiently (Hamilton 1964; Maynard Smith 1964). Maynard Smith (1965) has analyzed in great detail how closely related the benefited kin must be, at what benefit to him the call must be, and at what cost to the caller, in order for selection to favor call-giving.

The first is an explanation of last resort. While it must sometimes apply in evolutionary arguments, it should probably only be invoked when no other explanation seems plausible. The second is not consistent with the known workings of natural selection. The third is feasible and may explain the warning calls in some species and perhaps even in many. But it does depend on the somewhat regular nearby presence of closely related organisms, a matter that may often be the case but that has been demonstated only as a possibility in a few species and that seems very unlikely in some. A fourth explanation is suggested by the above model:

(4)   Warning calls are selected for because they aid the bird giving the call. It is disadvantageous for a bird to have a predator eat a nearby conspecific because the predator may then be more likely to eat him. They may happen because the predator will

(i)    be sustained by the meal,
(ii)   be more likely to form a specific search image of the prey species,
(iii)  be more likely to learn the habits of the prey species and perfect his predatory techniques on it,
(iv)   be more likely to frequent the area in which the birds live, or

(v)    be more likely to learn useful information about the area in which the birds live.

In short, in one way or another, giving a warning call tends to prevent predators from specializing on the caller's species and locality.

There is abundant evidence for the importance of learning in the lives of predatory vertebrates (see, for example, Tinbergen 1960; Leyhausen 1965; Brower and Brower 1965). Rudebeck (1950, 1951) has presented important observations on the tendency of avian predators to specialize individually on prey types and hunting techniques. Owen (1963) and others have presented evidence that species of snails and insects may evolve polymorphisms as a protection against the tendency of their avian predators to learn their appearance. Similarly, Kuyton (1962: cited in Wickler 1968) has described the adaptation of a moth that minimizes the chance of its predators forming a specific search image. Southern (1954), Murie (1944), and numerous others have documented the tendency of predators to specialize on certain localities within their range. Finally, Blest (1963) has presented evidence that kin selection in some cryptic saturnid moths has favored rapid, post-reproductive death to minimize predation on the young. Blest's evidence thus provides an instance of a predator gaining useful information through the act of predation.

It does not matter that in giving a warning call the caller is helping its non-calling neighbors more than it is helping itself. What counts is that it outcompetes conspecifics from areas in which no one is giving warning calls. The non-calling neighbors of the caller (or their offspring) will soon find themselves in an area without any caller and will be selected against relative to birds in an area with callers. The caller, by definition, is always in an area with at least one caller. If we assume that two callers are preferable to one, and so on, then selection will favor the spread of the warning-call genes. Note that this model depends on the concept of *open* groups, whereas 'group selection' (Wynne-Edwards 1962) depends partly on the concept of closed groups.

It might be supposed that one could explain bird calls more directly as altruistic behavior that will be repaid when the other birds reciprocate, but there are numerous objections to this. It is difficult to visualize how one would discover and discriminate against the cheater, and there is certainly no evidence that birds refrain from giving calls because neighbors are not reciprocating.

Furthermore, if the relevant bird groupings are very fluid, with much emigration and immigation, as they often are, then cheating would seem to be favored and no selection against it possible. Instead, according to the model above, it is the mere fact that the neighbor survives that repays the call-giver his altruism.

It is almost impossible to gather the sort of evidence that would discriminate between this explanation and that of Hamilton (1964). It is difficult to imagine how one would estimate the immediate cost of giving a warning call or its benefit to those within earshot, and precise data on the genetic relationships of bird groupings throughout the year are not only lacking but would be most difficult to gather. Several lines of evidence suggest, however, that Hamilton's (1964) explanation should be assumed with caution:

(1)   There exist no data showing a decrease in warning tendencies with decrease in the genetic relationship of those within earshot. Indeed, a striking feature of warning calls is that they are given in and out of the breeding season, both before and after migration or dispersal.

(2)   There do exist data suggesting that close kin in a number of species migrate or disperse great distances from each other (Ashmole 1962; Perdeck 1958; Berndt and Sternberg 1968; Dhont and Hublé 1968).

(3)   One can advance the theoretical argument that kin selection under some circumstances should favor kin dispersal in order to avoid competition (Hamilton 1964, 1969). This would lead one to expect fewer closely related kin near any given bird, outside the breeding season.

The arguments advanced in this section may also apply, of course, to species other than birds.

**Human reciprocal altruism**

Reciprocal altruism in the human species takes place in a number of contexts and in all known cultures (see, for example, Gouldner 1960). Any complete list of human altruism would contain the following types of altruistic behavior:

(1)   helping in times of danger (e.g. accidents, predation, intraspecific aggression);
(2)   sharing food;
(3)   helping the sick, the wounded, or the very young and old;

(4)   sharing implements; and
(5)   sharing knowledge.

All these forms of behavior often meet the criterion of small cost to the giver and great benefit to the taker.

During the Pleistocene, and probably before, a hominid species would have met the preconditions for the evolution of reciprocal altruism; long lifespan; low dispersal rate; life in small, mutually dependent, stable, social groups (Lee and DeVore 1968; Campbell 1966); and a long period of parental care. It is very likely that dominance relations were of the relaxed, less linear form characteristic of the living chimpanzee (Van Lawick-Goodall 1968) and not of the more rigidly linear form characteristic of the baboon (Hall and DeVore 1965). Aid in intraspecific combat, particularly by kin, almost certainly reduced the stability and linearity of the dominance order in early humans. Lee (1969) has shown that in almost all Bushman fights which are initially between two individuals, others have joined in. Mortality, for example, often strikes the secondaries rather than the principals. Tool use has also probably had an equalizing effect on human dominance relations, and the Bushmen have a saying that illustrates this nicely. As a dispute reaches the stage where deadly weapons may be employed, an individual will often declare: 'We are none of us big, and others small; we are all men and we can fight; I'm going to get my arrows' (Lee 1969). It is interesting that Van Lawick-Goodall (1968) has recorded an instance of strong dominance reversal in chimpanzees as a function of tool use. An individual moved from low in dominance to the top of the dominance hierarchy when he discovered the intimidating effects of throwing a metal tin around. It is likely that a diversity of talents is usually present in a band of hunter-gatherers such that the best maker of a certain type of tool is not often the best maker of a different sort or the best user of the tool. This contributes to the symmetry of relationships, since altruistic acts can be traded with reference to the special talents of the individuals involved.

To analyze the details of the human reciprocal altruistic system, several distinctions are important and are discussed here.

## (1) *Kin selection*
The human species also met the preconditions for the operation of kin selection. Early hominid hunter-gatherer bands almost certainly (like today's hunter-gatherers) consisted of many close kin, and

kin selection must often have operated to favour the evolution of some types of altruistic behaviour (Haldane 1955: Hamilton 1964, 1969). In general, in attempting to discriminate between the effects of kin selection and what might be called reciprocal-altuistic selection, one can analyze the form of the altruistic behaviors themselves. For example, the existence of discrimination against non-reciprocal individuals cannot be explained on the basis of kin selection, in which the advantage accruing to close kin is what makes the altruistic behavior selectively advantageous, not its chance of being reciprocated. The strongest argument for the operation of reciprocal-altruistic selection in man is the psychological system controlling some forms of human altruism. Details of this system are reviewed below.

## (2) *Reciprocal altruism among close kin*

If both forms of selection have operated, one would expect some interesting interactions. One might expect, for example, a lowered demand for reciprocity from kin than from nonkin, and there is evidence to support this (e.g. Marshall 1961; Balikci 1964). The demand that kin show some reciprocity (e.g. Marshall 1961; Balikci 1964) suggests, however, that recriprocal-altruistic selection has acted even on relations between close kin. Although interactions between the two forms of selection have probably been important in human evolution, this paper will limit itself to a preliminary description of the human reciprocally altruistic system, a system whose attributes are seen to result only from reciprocal-altruistic selection.

## (3) *Age-dependence changes*

Cost and benefit were defined above without reference to the ages, and hence reproductive values (Fisher 1958), of the individuals involved in an altruistic exchange. Since the reproductive value of a sexually mature organism declines with age, the benefit to him of a typical altruistic act also decreases, as does the cost to him of a typical act he performs. If the interval separating the two acts in an altruistic exchange is short relative to the lifespans of the individuals, then the error is slight. For longer intervals, in order to be repaid precisely, the initial altruist must receive more in return than he himself gave. It would be interesting to see whether humans in fact routinely expect 'interest' to be added to a long overdue altruistic debt, interest commensurate with the intervening decline in reproductive value. In humans reproductive value declines most steeply shortly after sexual maturity is reached (Hamilton 1966), and one would predict the interest rate on altruistic debts to be highest then. Selection might also favor keeping the interval between act and reciprocation short, but this should also be favored to protect against complete non-reciprocation. W. D. Hamilton (personal communication) has suggested that a decline analysis of age-dependent changes in kin altruism and reciprocal altruism should show interesting differences, but the analysis is complicated by the possibility of recriprocity to the kin of a deceased altruist (see *Multiparty interactions* below).

### (4) *Gross and subtle cheating*

Two forms of cheating can be distinguished, here denoted as gross and subtle. In *gross cheating* the cheater fails to reciprocate at all, and the altruist suffers the costs of whatever altruism he has dispensed without any compensating benefits. More broadly, gross cheating may be defined as reciprocating so little, if at all, that the altruist receives less benefit from the gross cheater than the cost of the altruist's acts of altruism to the cheater. That is, $\sum_i c_{ai} > \sum_j b_{aj}$, where $c_{ai}$ is the cost of the $i$th altruistic act performed by the altruist and where $b_{aj}$ is the benefit to the altruist of the $j$th altruistic act performed by the gross cheater; altruistic situations are assumed to have occurred symmetrically. Clearly, selection will strongly favor prompt discrimination against the gross cheater. *Subtle cheating*, by contrast, involves reciprocating, but always attempting to give less than one was given, or more precisely, to give less than the partner would give if the situation were reversed. In this situation, the altruist still benefits from the relationship but not as much as he would if the relationship were completely equitable. The subtle cheater benefits more than he would if the relationship were equitable. In other words,

$$\sum_{i,j}(b_{qi} - c_{qj}) > \sum_{i}(b_{qi} - c_{ai}) > \sum_{i,j}(b_{aj} - c_{ai})$$

where the $i$th altruistic act performed by the altruist has a cost to him of $c_{ai}$ and a benefit to the subtle cheater of $b_{qi}$ and where the $j$th altruistic act performed by the subtle cheater has a cost to him of $c_{qi}$ and a benefit to the altruist of $b_{aj}$. Because human altruism may span huge periods of time, a lifetime even, and because thousands of exchanges may take place, involving many different 'goods' and with many different cost/benefit ratios, the problem of computing the relevant totals, detecting imbalances, and deciding whether they are due to chance or to small-scale cheating is an extremely difficult one. Even then, the altruist is in an awkward position, symbolized by the folk saying, 'half a loaf is better than none', for if attempts to make the relationship equitable lead to the rupture of the relationship, the altruist, assuming other things to be equal, will suffer the loss of the substandard altruism of the subtle cheater. It is the subtlety of the discrimination necessary to detect this form of cheating and the awkward situation that ensues that permit some subtle cheating to be adaptive. This sets up a dynamic tension in the system that has important repercussions, as discussed below.

### (5) *Number of reciprocal relationships*

It has so far been assumed that it is to the advantage of each individual to form the maximum number of reciprocal relationships and that the individual suffers a decrease in fitness upon the rupture of any relationship in which the cost to him of acts dispensed to the partner is less than the benefit of acts dispensed toward him by the partner. But it is possible that relationships are partly exclusive, in the sense that expanding the number of reciprocal exchanges with one of the partners may necessarily decrease the number of exchanges with another. For example, if a group of organisms were to split into subgroups for much of the day (such as breaking up into hunting pairs), then altruistic exchanges will be more likely between member of each subgroup than between members of different subgroups. In that sense, relationships may be partly necessarily decreasing exchanges with others in the group. The importance of this factor is that it adds further complexity to the problem of dealing with the cheater and it increases competition within a group to be members of a favorable subgroup. An individual in a subgroup who feels that another member is subtly cheating on their relationship has the option of attempting to restore the relationship to a completely reciprocal one or of attempting to join another subgroup, thereby decreasing to a minimum the possible exchanges between himself and the subtle cheater and replacing these with exchanges between a new partner or partners. In short, he can switch friends. There is evidence in hunter-gatherers that much movement of individuals from one band to another occurs in response to such social factors as have just been outlined (Lee and DeVore 1968).

## (6) *Indirect benefits or reciprocal altruism?*

Given mutual dependence in a group it is possible to argue that the benefits (non-altruistic) of this mutual dependence are a positive function of group size and that altruistic behaviors may be selected for because they permit additional individuals to survive and thereby confer additional indirect (non-altruistic) benefits. Such an argument can only be advanced seriously for slowly reproducing species with little dispersal. Saving an individual's life in a hunter-gatherer group, for example, may permit non-altruistic actions such as co-operative hunting to continue with more individuals. But if there is an optimum group size, one would expect adaptations to stay near the size, with individuals joining groups when the groups are below this size, and groups splitting up when they are above this size. One would only be selected to keep an individual alive when the group is below optimum and not when the group is above optimum. Although an abundant literature on hunter-gatherers (and also nonhuman primates) suggests that adaptations exist to regulate group size near an optimum, there is no evidence that altruistic gestures are curtailed when groups are above the optimum in size. Instead, the benefits of human altruism are to be seen as coming directly from reciprocality—not indirectly through non-altruistic group benefits. This distinction is important because social scientists and philosophers have tended to deal with human altruism in terms of the benefits of living in a group, without differentiating between non-altruistic benefits and reciprocal benefits (e.g. Rousseau 1954; Baier 1958).

## The psychological system underlying human reciprocal altruism

Anthropologists have recognized the importance of reciprocity in human behavior, but when they have ascribed functions to such behavior they have done so in terms of group benefits, reciprocity cementing group relations and encouraging group survival. The individual sacrifices so that the group may benefit. Recently psychologists have studied altruistic behavior in order to show what factors induce or inhibit such behavior. No attempt has been made to show what function such behavior may serve, nor to describe and interrelate the components of the psychological system affecting altruistic behavior. The purpose of this section is to show that the above model for the natural selection of reciprocally altruistic behavior can readily explain the function of human altruistic behavior and the details of the psychological system underlying such behavior. The psychological data can be organized into functional categories, and it can be shown that the components of the system complement each other in regulating the expression of altruistic and cheating impulses to the selective advantage of individuals. No concept of group advantage is necessary to explain the function of human altruistic behavior.

There is no direct evidence regarding the degree of reciprocal altruism practised during human evolution nor its genetic basis today, but given the universal and nearly daily practice of reciprocal altruism among humans today, it is reasonable to assume that it has been an important factor in recent human evolution and that the underlying emotional dispositions affecting altruistic behavior have important genetic components. To assume as much allows a number of predictions.

### (1) A complex, regulating system

The human altruistic system is a sensitive, unstable one. Often it will pay to cheat: namely, when the partner will not find out, when he will not discontinue his altruism even if he does find out, or when he is unlikely to survive long enough to reciprocate adequately. And the perception of subtle cheating may be very difficult. Given this unstable character of the system, where a degree of cheating is adaptive, natural selection will rapidly favor a complex psychological system in each individual regulating both his own altruistic and cheating tendencies and his responses to these tendencies in others. As selection favors subtler forms of cheating, it will favor more acute abilities to detect cheating. The system that results should simultaneously allow the individual to reap the benefits of altruistic exchanges, to protect himself from gross and subtle forms of cheating, and to practice those forms of cheating that local conditions make adaptive. Individuals will differ not in being altruists or cheaters but in the degree of altruism they show and in the conditions under which they will cheat.

The best evidence supporting these assertions can be found in Kreb's (1970) review of the relevant psychological literature. Although he organizes it differently, much of the material supporting the assertions below is taken from his paper. All references to Krebs below are to this review. Also, Hartshorne and May (1928–1936) have shown that children in experimental situations do not divide bimodally into altruists and 'cheaters' but are distributed normally; almost all the children cheated, but they differed in how much and under what circumstances. ('Cheating' was defined in their work in a slightly different but analogous way).

## (2) *Friendship and the emotions of liking and disliking*

The tendency to like others, not necessarily closely related, to form friendships and to act altruistically toward friends and toward those one likes will be selected for as the immediate emotional rewards motivating altruistic behavior and the formation of altruistic partnerships. (Selection may also favor helping strangers or disliked individuals when they are in particularly dire circumstances). Selection will favor a system whereby these tendencies are sensitive to such parameters as the altruistic tendencies of the liked individual. In other words, selection will favor liking those who are themselves altruistic.

Sawyer (1966) has shown that all groups in all experimental situations tested showed more altruistic behavior toward friends than toward neutral individuals. Likewise, Friedrichs (1960) has shown that attractiveness as a friend was most highly correlated among undergraduates with altruistic behavior. Krebs has reviewed other studies that suggest that the relationship between altruism and liking is a two-way street: one is more altruistic toward those one likes and one tends to like those who are most altruistic (e.g. Berkowitz and Friedman 1967; Lerner and Lichtman 1968).

Others (Darwin 1871; Williams 1966; and Hamilton 1969) have recognized the role friendship might play in engendering altruistic behavior, but all have viewed friendship (and intelligence) as prerequisites for the appearance of such altruism. Williams (1966), who cites Darwin (1871) on the matter, speaks of this behavior as evolving

in animals that live in stable social groups and have the intelligence and other mental qualities necessary to form a system of personal friendships and animosities that transcend the limits of family relationships (p. 93).

This emphasis on friendship and intelligence as prerequisites leads Williams to limit his search for altruism to the Mammalia and to a 'minority of this group'. But according to the model presented above, emotions of friendship (and hatred) are not prerequisites for reciprocal altruism but may evolve *after* a system of mutual altruism has appeared, as important ways of regulating the system.

## (3) *Moralistic aggression*

Once strong positive emotions have evolved to motivate altruistic behavior, the altruist is in a vulnerable position because cheaters will be selected to take advantage of the altruist's positive emotions. This in turn sets up a selection pressure for a protective mechanism. Moralistic aggression and indignation in humans was selected for in order

(a)   to counteract the tendency of the altruist, in the absence of any reciprocity, to continue to perform altruistic acts for his own emotional rewards;

(b)   to educate the unreciprocating individual by frightening him with immediate harm or with the future harm of no more aid; and

(c)   in extreme cases, perhaps, to select directly against the unreciprocating individual by injuring, killing, or exiling him.

Much of human aggression has moral overtones. Injustice, unfairness, and lack of reciprocity often motivate human aggression and indignation. Lee (1969) has shown that verbal disputes in Bushmen usually revolve around problems of gift-giving, stinginess, and laziness. DeVore (personal communication) reports that a great deal of aggression in hunter-gatherers revolves around real or imagined injustices—inequities, for example, in food-sharing (see, for example, Thomas 1958; Balikci 1964; Marshall 1961). A common feature of this aggression is that it often seems out of all proportion to the offenses committed. Friends are even killed over apparently trivial disputes. But since small inequities repeated many times over a lifetime may exact a heavy toll in relative fitness, selection may favor a strong show of aggression when the cheating tendency is discovered. Recent discussions of human and animal aggression have failed to distinguish between moralistic and other forms of aggression (e.g. Scott 1958; Lorenz 1966; Montague 1968; Tinbergen 1968; Gilula and Daniels 1969). The grounds for expecting, on functional grounds, a highly plastic development system affecting moralistic aggression are discussed below.

(4) *Gratitude, sympathy, and the cost/benefit ratio of an altruistic act*

If the cost/benefit ratio is an important parameter in determining the adaptiveness of reciprocal altruism, then humans should be selected to be sensitive to the cost and benefit of an altruistic act, both in deciding whether to perform one and in deciding whether, or how much, to reciprocate. I suggest that the emotion of gratitude has been selected to regulate human response to altruistic acts and that the emotion is sensitive to the cost/benefit ratio of such acts. I suggest further that the emotion of sympathy has been selected to motivate altruistic behavior as a function of the plight of the recipient of such behavior; crudely put, the greater the potential benefit to the recipient, the greater the sympathy and the more likely the altruistic gesture, even to strange or disliked individuals. If the recipient's gratitude is indeed a function of the cost/benefit ratio, then a sympathetic response to the plight of a disliked individual may result in considerable reciprocity.

There is good evidence supporting the psychological importance of the cost/benefit ratio of altruistic ats. Gouldner (1960) has reviewed the sociological literature suggesting that the greater the need state of the recipient of an altruistic act, the greater his tendency to reciprocate; and the scarcer the resources of the donor of the act, the greater the tendency of the recipient to reciprocate. Heider (1958) has analyzed lay attitudes on altruism and finds that gratitude is greatest when the altruistic act does good. Tesser, Gatewood, and Driver (1968) have shown that American undergraduates thought they would feel more gratitude when the altruistic act was valuable and cost the benefactor a great deal. Pruitt (1968) has provided evidence that humans reciprocate more when the original act was expensive for the benefactor. He shows that under experimental conditions more altruism is induced by a gift of 80 per cent of $1·00 than 20 per cent of ·$4·00. Aronfreed (1968) has reviewed the considerable evidence that sympathy motivates altruistic behavior as a function of the plight of the individual arousing the sympathy.

## (5) *Guilt and reparative altruism*

If an organism has cheated on a reciprocal relationship and this fact has been found out, or has a good chance of being found out, by the partner and if the partner responds by cutting off all future acts of aid, then the cheater will have paid dearly for his misdeed. It will be to the cheater's advantage to avoid this, and, providing that the cheater makes up for his misdeed and does not cheat in the future, it will be to his partner's benefit to avoid this, since in cutting off future acts of aid he sacrifices the benefits of future reciprocal help. The cheater should be selected to make up for his misdeed and to show convincing evidence that he does not plan to continue his cheating sometime in  the future. In short, he should be selected to make a reparative gesture. It seems plausible, furthermore, that the emotion of guilt has been selected for in humans partly in order to motivate the cheater to compensate his misdeed and to behave reciprocally in the future, and thus to prevent the rupture of reciprocal relationships.

Krebs has reviewed the evidence that harming another individual publicly leads to altruistic behavior and concludes:

Many studies have supported the notion that public transgression whether intentional or unintentional, whether immoral or only situationally unfortunate, leads to reparative altruism (p. 267).

Wallace and Sadalla (1966), for example, showed experimentally that individuals who broke an expensive machine were more likely to volunteer for a painful experiment than those who did not, but only if their transgression had been discovered. Investigators disagree on the extent to which guilt feelings are the motivation behind reparative altruism. Epstein and Hornstein (1969) supply some evidence that guilt is involved, but on the assumption that one feels guilt even when one behaves badly in private, Wallace and Sadalla's (1966) result contradicts the view that guilt is the only motivating factor. That private transgressions are not as likely as public ones to lead to reparative altruism is precisely what the model would predict, and it is possible that the common psychological assumption that one feels guilt even when one behaves badly in private is based on the fact that many transgressions performed in private are *likely* to become public knowledge. It should often be advantageous to confess sins that are likely to be discovered before they actually are, as evidence of sincerity (see below on detection of mimics).

## (6) *Subtle cheating: the evolution of mimics*

Once friendship, moralistic aggression, guilt, sympathy, and gratitude have evolved to regulate the altruistic system, selection will favor mimicking these traits in order to influence the behavior of others to one's own advantage. Apparent acts of generosity and friendship may induce genuine friendship and altruism in return. Sham moralistic aggression when no real cheating has occurred may nevertheless induce reparative altruism. Sham guilt may convince a wronged friend that one has reformed one's ways even when the cheating is about to be resumed. Likewise, selection will favor the hypocrisy of pretending one is in dire circumstances in order to induce sympathy-motivated altruistic behavior. Finally, mimicking sympathy may give the appearance of helping in order to induce reciprocity, and mimicking gratitude may mislead an individual into expecting he will be reciprocated. It is worth emphasizing that a mimic need not necessarily be conscious of the deception; selection may favor feeling genuine moralistic aggression even when one has not been wronged if so doing leads another to reparative altruism.

Instances of the above forms of subtle cheating are not difficult to find. For typical instances from the literature on hunter-gatherers see Rasmussen (1931), Balikci (1964), and Lee and DeVore (1968). The importance of these forms of cheating can partly be inferred from the adaptations to detect such cheating discussed below and from the importance and prevalence of moralistic aggression once such cheating is detected.

*(7) Detection of the subtle cheater: trustworthiness, trust, and suspicion*

Selection should favor the ability to detect and discriminate against subtle cheaters. Selection will clearly favor detecting and countering sham moralistic aggression. The argument for the others is more complex. Selection may favor distrusting those who perform altruistic acts without the emotional basis of generosity or guilt because the altruistic tendencies of such individuals may be less reliable in the future. One can imagine, for example, compensating for a misdeed without any emotional basis but with a calculating self-serving motive. Such an individual should be distrusted because the calculating spirit that leads this subtle cheater now to compensate may in the future lead him to cheat when circumstances seem more advantageous (because of the unlikelihood of detection, for example, or because the cheated individual is unlikely to survive). Guilty motivation, in so far as it evidences a more enduring commitment to altruism, either because guilt teaches or because the cheater is unlikely not to feel the same guilt in the future, seems more reliable. A similar argument can be made about the trustworthiness of individuals who initiate altruistic acts out of a calculating rather than a generous-hearted disposition or who show either false sympathy or false gratitude. Detection on the basis of the underlying psychological dynamics is only one form of detection. In many cases, unreliability may more easily be detected through experiencing the cheater's inconsistent behavior. And in some cases, third party interactions (as discussed below) may make an individual's behavior predictable despite underlying cheating motivations.

The anthropological literature also abounds with instances of the detection of subtle cheaters (see above references for hunter gatherers). Although I know of no psychological studies on the detection of sham moralistic aggression and sham guilt, there is ample evidence to support the notion that humans respond to altruistic acts according to their perception of the motives of the altruist. They tend to respond more altruistically when they perceive the other as acting 'genuinely' altruistic, that is, voluntarily dispatching an altruistic act as an end in itself, without being directed toward gain (Leeds 1963; Heider 1958). Krebs (1970) has reviewed the literature on this point and notes that help is more likely to be reciprocated when it is perceived as voluntary and intentional (e.g. Goranson and Berkowitz 1966; Lerner and Lichtman 1968) and when the help is appropriate, that is, when the intentions of the altruist are not in doubt (e.g. Brehm and Cole 1966; Schopler and Thompson 1968). Krebs concludes that, 'When the legitimacy of apparent altruism is questioned, reciprocity is less likely to prevail.' Lerner and Lichtman (1968) have shown experimentally that those who act altruistically for ulterior benefit are rated as unattractive and are treated selfishly, whereas those who apparently are genuinely altruistic are rated as attractive and are treated altruistically. Berscheid and Walster (1967) have shown that church women tend to make reparations for harm they have committed by choosing the reparation that approximates the harm (that is, is neither too slight nor too great), presumably to avoid the appearance of inappropriateness.

Rapoport and Dale (1967) have shown that when two strangers play iterated games of Prisoner's Dilemma in which the matrix determines profits from the games played there is a significant tendency for the level of co-operation to drop at the end of the series, reflecting the fact that the partner will not be able to punish for 'cheating' responses when the series is over. If a long series is broken up into subseries with a pause between subseries for totaling up gains and losses, then the tendency to cheat on each other increases at the end of each subseries. These results, as well as some others reported by Rapoport and Chammah (1965), are suggestive of the instability that exists when two strangers are consciously trying to maximize gains by trading altruistic gestures, an instability that is presumably less marked when the underlying motivation involves the emotions of friendship, of liking others, and of feeling guilt over harming a friend. Deutsch (1958), for example, has shown that two individuals playing iterated games of Prisoner's Dilemma will be more co-operative if a third individual, disliked by both, is present. The perceived mutual dislike is presumed to create a bond between the two players.

It is worth mentioning that a classic problem in social science and philosophy has been whether to define altruism in terms of motives (e.g. real vs. 'calculated' altruism) or in terms of behavior, regardless of motive (Krebs 1970). This problem reflects the fact that, wherever studied, humans seem to make distinctions about altruism partly on the basis of motive, and this tendency is consistent with the hypothesis that such discrimination is relevant to protecting oneself from cheaters.

## (8) *Setting up altruistic partnerships*

Selection will favor a mechanism for establishing reciprocal relationships. Since humans respond to acts of altruism with feelings of friendship that lead to reciprocity, one such mechanism might be the performing of altruistic acts toward strangers, or even enemies, in order to induce friendship. In short, do unto others as you would have them do unto you.

The mechanism hypothesized above leads to results inconsistent with the assumption that humans always act more altruistically toward friends than toward others. Particularly toward strangers, humans may initially act more altruistically than toward friends. Wright (1942) has shown, for example, that third grade children are more likely to give a more valuable toy to a stranger than a friend. Later, some of these children verbally acknowledged that they were trying to make friends. Floyd (1964) has shown that, after receiving many trinkets from a friend, humans tend to *decrease* their gifts in return, but after receiving many trinkets from a neutral or disliked individual, they tend to *increase* their gifts in return. Likewise, after receiving few trinkets from a friend, humans tend to increase their gifts in return, whereas receiving few trinkets from a neutral or disliked individual results in a decrease in giving. This was interpreted to mean that generous friends are taken for granted (as are stingy non-friends). Generosity from a non-friend is taken to be an overture to friendship, and stinginess from a friend as evidence of a deteriorating relationship in need of repair. (Epstein and Hornstein 1969, provide new data supporting this interpretation of Floyd 1964.)

## (9) *Multiparty interactions*

In the close-knit social groups that humans usually live in, selection should favor more complex interactions than the two-party interactions so far discussed. Specifically, selection may favor learning from the altruistic and cheating experiences of others, helping others coerce cheaters, forming multiparty exchange systems, and formulating rules for regulating exchanges in such multiparty systems.

(i)    *Learning from others.* Selection should favor learning about the altruistic and cheating tendencies of others indirectly, both through observing interactions of others and, once linguistic abilities have evolved, by hearing about such interactions or hearing characterizations of individuals (e.g. 'dirty, hypocritical, dishonest, untrustworthy, cheating louse'). One important result of this learning is that an individual may be as concerned about the attitude of onlookers in an altruistic situation as about the attitude of the individual being dealt with.

(ii)    *Help in dealing with cheaters.* In dealing with cheaters selection may favor individuals helping others, kin or non-kin, by direct coercion against the cheater or by everyone refusing him reciprocal altruism. One effect of this is that an individual, through his close kin, may be compensated for an altruistic act even after his death. An individual who dies saving a friend, for example, may have altruistic acts performed by the friend to the benefit of his offspring. Selection will discriminate against the cheater in this situation, if kin of the martyr, or others, are willing to punish lack of reciprocity.

(iii)    *Generalized altruism.* Given learning from others and multiparty action against cheaters, selection may favor a multiparty altruistic system in which altruistic acts are dispensed freely among more than two individuals, an individual being perceived to cheat if in an altruistic situation he dispenses less benefit for the same cost than would the others, punishment coming not only from the other individual in that particular exchange but from the others in the system.

(iv)    *Rules of exchange.* Multiparty altruistic systems increase by several-fold the cognitive difficulties in detecting imbalances and deciding whether they are due to cheating or to random factors. One simplifying possibility that language facilitates is the formulation of rules of conduct, cheating being detected as infraction of such a rule. In short, selection may favor the elaboration of norms of reciprocal conduct.

There is abundant evidence for all of the above multiparty interactions (see the above references on hunter-gatherers).

Thomas (1958), for example, has shown that debts of reciprocity do not disappear with the death of the 'creditor' but are extended to his kin. Krebs has reviewed the psychological literature on generalized altruism. Several studies (e.g. Darling and Macker 1966) have shown that humans may direct their altruism to individuals other than those who were hurt and may respond to an altruistic act that benefits themselves by acting altruistically toward a third individual uninvolved in the initial interaction. Berkowitz and Daniels (1964) have shown experimentally, for example, that help from a confederate leads the subject to direct more help to a third individual, a highly dependent supervisor. Freedman, Wallington, and Bless (1967) have demonstrated the suprising result that, in two different experimental situations, humans engaged in reparative altruism only if it could be directed to someone other than the individual harmed, or to the original individual only if they did not expect to meet again. In a system of strong multiparty interactions it is possible that in some situations individuals are selected to demonstrate generalized altruistic tendencies and that their main concern when they have harmed another is to show that they are genuinely altruistic, which they best do by acting altruistic without any apparent ulterior motive, e.g. in the experiments, by acting altruistic toward an uninvolved third party. Alternatively, A. Rapoport (personal communication) has suggested that the reluctance to direct reparative altruism toward the harmed individual may be due to unwillingness to show thereby a recognition of the harm done him. The redirection serves to allay guilt feelings without triggering the greater reparation that recognition of the harm might lead to.

(10) *Developmental plasticity*
The conditions under which detection of cheating is possible, the range of available altruistic trades, the cost/benefit ratios of these trades, the relative stability of social groupings, and other relevant parameters should differ from one ecological and social situation to another and should differ through time in the same small human population. Under these conditions one would expect selection to favor developmental plasticity of those traits regulating altruistic and cheating tendencies and responses to these tendencies in others. For example, developmental plasticity may allow the growing organism's sense of guilt to be educated, perhaps partly by kin, so as to permit those forms of cheating that local conditions make adaptive and to discourage those with more dangerous

consequences. One would not expect any simple system regulating the development of altruistic behavior. To be adaptive, altruistic behavior must be dispensed with regard to many characteristics of the recipient (including his degree of relationship, emotional makeup, past behavior, friendships and kin relations), of other members and of the group, of the situation in which the altruistic behavior takes place, and of many other parameters, and no simple developmental system is likely to meet these requirements.

Kohlberg (1963), Bandura and Walters (1963), and Krebs have reviewed the developmental literature on human altruism. All of them conclude that none of the proposed developmental theories (all of which rely on simple mechanisms) can account for the known diverse developmental data. Whiting and Whiting (in preparation) have studied altruistic behavior directed towards kin by children in six different cultures and find consistent differences among the cultures that correlate with differences in child-rearing and other facets of cultures. They argue that the differences adapt the children to different adult roles available in the cultures. Although the behavior analyzed takes place between kin and hence Hamilton's model (1964) may apply rather than this model, the Whitings' data provide an instance of the adaptive value of developmental plasticity in altruistic behavior. No careful work has been done analyzing the influence of environmental factors on the development of altruistic behavior, but some data exist. Krebs has reviewed the evidence that altruistic tendencies can be increased by the effects of warm, nurturant models, but little is known on how long such effects endure. Rosenhan (1967) and Rettig (1956) have shown a correlation between altruism in parents and altruism in their college age children, but these studies do not separate genetic and environmental influences. Class differences in altruistic behavior (e.g. Berkowitz 1968; Ugurel-Semin 1952; Almond and Verba 1963) may primarily reflect environmental influences. Finally, Lutzker (1960) and Deutsch (1958) have shown that one can predict the degree of altruistic behavior displayed in iterated games of Prisoner's Dilemma from personality typing based on a questionnaire. Such personality differences are probably partly environmental in origin.

It is worth emphasizing that some of the psychological traits analyzed above have applications outside the particular reciprocal altruistic system being discussed. One may be suspicious, for example, not only of individuals likely to cheat on the altruistic system, but of any individual likely to harm oneself; one may be suspicious of the known tendencies toward aldultery of another male or even of these tendencies in one's own mate. Likewise, a guilt-motivated show of reparation may avert the revenge of someone one has harmed, whether that individual was harmed by cheating on the altruistic system or in some other way. And the system of reciprocal altruism may be employed to avert possible revenge. The Bushmen of the Kalahari, for example, have a saying (Marshall 1959) to the effect that, if you wish to sleep with someone else's wife, you get him to sleep with yours, then neither of you goes after the other with poisoned arrows. Likewise, there is a large literature on the use of reciprocity to cement friendships between neighboring groups, now engaged in a common enterprise (e.g. Lee and DeVore 1968).

The above review of the evidence has only begun to outline the complexities of the human altruistic system. The inherent instability of the Prisoner's Dilemma, combined with its importance in human evolution, has led to the evolution of a very complex system. For example, once moralistic aggression has been selected for to protect against cheating, selection favors sham moralistic aggression as a new form of cheating. This should lead to selection for the ability to discriminate the two and to guard against the .latter. The guarding can, in turn, be used to counter real moralistic aggression: one can, in effect, *impute* cheating motives to another person in order to protect one's own cheating. And so on. Given the psychological and cognitive complexity the system rapidly acquires, one may wonder to what extent the importance of altruism in human evolution set up a selective pressure for psychological and cognitive powers which partly contributed to the large increase in hominid brain size during the Pleistocene.

# References

Almond, G. A., and Verba, S. (1963). *The civic culture*. Princeton University Press, Princeton, N.J.

Aronfreed, J. (1968). *Conduct and conscience*. Academic Press, N.Y.

Ashmolf, M. (1962). Migration of European thrushes: a comparative study based on ringing recoveries. *Ibis* 104, 522–59.

Bailr, K. (1958). *The moral point of view*. Cornell University Press, Ithaca, N.Y.

Balikci, A. (1964). Development of basic socioeconomic units in two Eskimo communities. *National Museum of Canada Bulletin No. 202, Ottawa*.

Bandura, A., and Walters, R. H. (1963). *Social learning and personality development*. Holt, Rhinehart and Winston, N.Y.

Berkowitz, L. (1968). Responsibility, reciprocity and social distance in help-giving: an experimental investigation of English social class differences. *J. Exp. Soc. Psychol.* 4, 664–9.

——and Daniels, L. (1964). Affecting the salience of the social responsibility norm: effects of past help on the response to dependency of relationships. *J. Abnorm. Soc. Psychol.* 68, 275–91.

——and Friedman, P. (1967). Some social class differences in helping behavior. *J. Personal. Soc. Psychol.* 5, 217–25.

Berndt, R. and Sternberg, H. (1968). Terms, studies and experiments on the problems of bird dispersion. *Ibis* 110, 256–69.

Berscheid, E., and Walster, E. (1967). When does a harm-doer compensate a victim? *J. Personal. Soc. Psychol.* 6, 435–41.

Blest, A. D. (1963). Longevity, palatability and natural selection in five species of New World Saturuiid moth. *Nature, Lond.* 197, 1183–6.

Brehm, J. W. and Cole, A. H. (1966). Effect of a favor which reduces freedom. *J. Personal. Soc. Psychol.* 3, 420–6.

Brower, J. V. Z., and Brower, L. P. (1965). Experimental studies of mimicy. 8 *Amer. Natur.* 49, 173–88.

Campbell, B. (1966). *Human evolution.* Aldine, Chicago.

Crook, J. H. (1969). The socio-ecology of primates. In J. H. Crook (ed.). *Social behaviour in birds and mammals,* p. 103–66. Academic Press, London.

Darlington, R. B., and Macker, C. E. (1966). Displacement of guilt-produced altruistic behavior. *J. Personal. Soc. Psychol.* 4, 442–3.

Darwin, C. (1871). *The descent of man and selection in relation to sex.* Random House, N.Y.

Deutsch, M. (1958). Trust and suspicion. *J. Conflict Resolution* 2, 267–9.

Dhont, A. A., and Huble, J. (1968). Fledging-date and sex in relation to dispersal in young Great Tits. *Bird Study* 15, 127–34.

Drury, W. H., and Smith, W. J. (1958). Defense of feeding areas of adult herring gulls and intrusion by young. *Evolution* 22, 193–201.

Eibl-Eibesfeldt, I. (1955). Über Symbiosen. Parasitismus und andere besondere zwischenartliche Beziehungen tropische Meeresfische. *Z. f. Tierpsychol.* 12, 203–19.

——(1959). Der Fisch *Aspidontus taeniatus als* Nachahmer des Putzers *Labroides dimidiatus. Z. f. Tierpsychol.* 16, 19–25.

Epstein, Y. M., and Horstein, H. A. (1969). Penalty and interpersonal attraction as factors influencing the decision to help another person. *J. Exp. Soc. Psychol.* 5, 272–82.

Feder, H. M. (1966). Cleaning symbioses in the marine environment. In S. M. Henry (ed.), *Symbiosis,* Vol. 1, p. 327–80. Academic Press, N.Y.

Fisher, R. A. (1958). *The genetical theory of natural selection.* Dover, N.Y.

Floyd, J. (1964). Effects of amount of award and friendship status of the other on the frequency of sharing in children. Unpublished doctoral dissertion. University of Minnesota. (Reviewed in Krebs, 1970.)

Freedman, J. L., Wallington, S. A., and Bless, E. (1967). Compliance without pressure: the effect of guilt. *J. Personal. Soc. Psychol.* 7, 117–24.

Friedrichs, R. W. (1960). Alter versus ego: an exploratory assessment of altruism. *Am. Sociol. Rev.* 25, 496–508.

Gilula, M. F., and Daniels, D. N. (1969). Violence and man's strugle to adapt. *Science* 164; 395–405.

Goranson, R., and Berkowitz (1966). Reciprocity and responsibility reactions to prior help *J. Personal. Soc. Psychol.* 3, 227–32.

Gouldner, A. (1960). The norm of reciprocity: a preliminary statement. *Am. Sociol. Rev.* 47, 73–80.

Haldane, J. B. S. (1955). Population genetics. *New Biology* 18, 34–51.

Hall, K. R. L., and DeVore, I. (1965). Baboon social behavior. In I. DeVore (ed.), *Primate behavior: field studies of monkeys and apes,* p. 53–110. Holt, Rhinehart and Winston, N.Y.

Hamilton, W. D. (1964). The genetical evolution of social behavior. *J. Theoret. Biol.* 7, 1–52.

——(1966). The moulding of senescence by natural selection. *J. Theoret. Biol.* 12, 12–45.

—— (1969). Selection of selfish and altruistic behavior in some extreme models. Paper presented at 'Man and Beast Symposium' (in press, Smithsonian Institution).

Hartshorne, H., and May, M. A. (1928–1930). *Studies in the nature of character. Vol. 1, Studies in deceit; Vol. 2, Studies in self-control; Vol. 3, Studies in the organization of character*. Macmillan, N.Y.

Hediger, H. (1968). Putzer-fische in aquarium. *Natur und Museum* 98, 89–96.

Heider, F. (1958). *The psychology of interpersonal relations*. Wiley, N.Y.

Hobson, E. S. (1969). Comments on certain recent generalizations regarding cleaning symbioses in fishes. *Pacific Science* 23, 35–9.

Kohlberg, L. (1963). Moral development and identification. In H. W. Stevenson (ed.). *Yearbook of the National Society for the Study of Education. Part 1. Child psychology*, p. 277–332. University of Chicago Press.

Krebs, D. (1970). Altruism—an examination of the concept and a review of the literature, *Psychol. Bull.* 73, 258–302.

Kuyton, P. (1962). Verhalten-beobachtungen an der Raupe des Kaiseratlas. *Z. d. Entomol.* 72, 203–7.

Lee R. (1969). !Kung Bushman violence. Paper presented at meeting of American Antropological Association, Nov. 1969.

—— and DeVore, I. (1968). *Man and the hunter*, Aldine, Chicago.

Leeds, R. (1963). Altruism and the norm of living. *Merrill-Pabner Quart.* 9, 229–40.

Lerner, M. J., and Lichfman, R. R. (1968). Effects of perceived norms on attitudes and altruistic behavior toward a dependent other. *J. Personal. Soc. Psychol.* 9, 226–32.

Leyhausen, P. (1965). Über die Funktion der relativen Stimmungshierarchie (dasgestellt am Beispiel der phylogenetischen und ontogenetischen Entwicklung des Beutefangs von Raubtieren. *Z. f. Tierpsychol.* 22, 412–94.

Limbaugh, C. (1961). Cleaning symbioses. *Scient. Amer.* 205, 42–9.

——, Pederson, H., and Chase, F. (1961). Shrimps that clean fishes. *Bull Mar. Sci. Gulf Caribb.* 11, 237–57.

Lorenz, K. (1966). *On aggression*. Harcourt, Brace and Word, N.Y.

Luce, R. D., and Raiffa, H. (1957). *Games and decisions*. Wiley, N.Y.

Lutzker, D. (1960). Internationalism as a predictor of co-operative game behavior. *J. Conflict Resolution* 4, 426–35.

Marler, P. (1955). The characteristics of certain animal calls. *Nature, Lond.* 176, 6–7.

—— (1957). Specific distinctiveness in the communication signals of birds. *Behaviour* 11, 13–30.

Marshall, L. K. (1959). Marriage among !Kung Bushmen. *Africa* 29, 335–65.

—— (1961). Sharing, talking and giving: relief of social tension among !Kung Bushmen. *Africa* 31, 231–49.

Maynard, E. C. L. (1968). Cleaning symbiosis and oral grooming on the coral reef. In P. Person (ed.). *Biology of the Mouth*, p. 79–88. Philip Person, American Association for the Advancement of Science, Wash., D.C.

Maynard Smith, J. (1964). Kin selection and group selection. *Nature, Lond.* 201, 1145–7.

—— (1965). The evolution of alarm calls. *Amer. Natur.* 99, 59–63.

Mayr, E. (1963). *Animal species and evolution*. Belknap Press, Cambridge.

Montagu, F. M. A. (1968). *Man and agression*. Oxford University Press, N.Y.

Murie, A. (1944). *The wolves of Mount McKinley*. Fauna of National Parks, Faunal Series #5; Wash., D.C.

Owen, D. F. (1963). Similar polymorphisms in an insect and a land snail. *Nature, Lond.* 198, 201–3.

Perdfck, A. (1958). Two types of orientation in migrating starlings, *Sturnus vulgaris* L., and chaffinches, *Fringilla coelebs* L., as revealed by displacement experiments. *Ardea* 46, 1–35.

Pruitt, D. G. (1968). Reciprocity and credit building in a laboratory dyad. *J. Personal. Soc. Psychol.* 8, 143–7.

Randall, J. E. (1958). A review of the Labrid fish genus *Labroides* with descriptions of two new species and notes on ecology. *Pacific Science* 12, 327–47.

—— (1962). Fish service stations. *Sea Frontiers*, 8, 40–7.

Rapoport, A., and Chammah, A. (1965). *Prisoner's Dilemma*. University of Michigan Press, Ann Arbor.

—— and Dale, P. (1967). The 'end' and 'start' effects in iterated Prisoner's Dilemma. *J. Conflict Resolution* 10, 363–6.

Rasmussen, K. (1931). The Netslik Eskimos: social life and spiritual culture. Report of the Fifth Thule Expedition 1921–1924, Vol. 8(1,2). Gyldendalske Boghandle, Copenhagen.

Rettig, S. (1956). An explanatory study of altruism. *Dissert. Abstr.* 16, 2220–52.

Reynolds, I. (1966). Open groups in hominid evolution. *Man* 1, 441–52.

Rosenhan, D. (1967). *The origins of altruistic social autonomy*. Educational Testing Service, Princeton N.J.

Rousseau, J. J. (1954). *The social contract*. Henry Regnery Co., Chicago.

Rudebeck, G. (1950). The choice of prey and modes of hunting of predatory birds with special reference to their selective effort. *Oikos* 2, 65–88.

—— (1951). The choice of prey and models of hunting of predatory birds, with special reference to their selective effort (*cont.*) *Oikos* 3: 200–31.

Sawyer, J. (1966). The altruism scale: a measure of cooperative, individualistic, and competitive interpersonal orientation. *Am. J. Sociol.* 71, 407–16.

Schopler, J., and Thompson, V. T. (1968). The role of attribution process in mediating amount of reciprocity for a favor. *J. Personal. Soc. Psychol.* 10, 243–50.

Scott, J. P. (1958). *Aggression*. University of Chicago Press.

Southern, H. N. (1954). Tawny owls and their prey. *Ibis* 96, 384–410.

Struhsaker, T. (1967). Social structure among vervet monkeys (*Cercopithecus aethiops*). *Behaviour* 29, 83–121.

Tesser, A., Gagewood, R., and Driver, M. (1968). Some determinants of gratitude. *J. Personal. Soc. Psychol.* 9, 232–6.

Thomas, E. M. (1958). *The harmless people*. Random House, N.Y.

Tinbergen, L. (1960). The dynamics of insect and bird populations in pine woods. 1. Factors influencing the intensity of predation by song birds. *Arch. Neerl. de Zool.* 13, 265–343.

Tinbergen, N. (1968). On war and peace in animals and man. *Science*, 160, 1411–18.

Ugurel-Semin, R. (1952). Moral behavior and removal judgment of children. *J. Abnrom. Soc. Psychol.* 47, 463–74.

VanLawick-Goodall, J. (1968). A preliminary report on expressive movements and communication in the Gombe Stream chimpanzees. In P. Jay (ed.), *Primates*, p. 313–374 Holt, Rhinehart and Winston, N.Y.

Wallace, J., and Sadalla, E. (1966). Behavioral consequences of transgression: the effects of social recognition. *J. Exp. Res. Personal.* 1, 187–94.

Wickler, W. (1968). *Mimicry in plants and animals*. McGraw-Hill, N.Y.

Williams, C. G. (1966). *Adaptation and natural selection*. Princeton University Press.

Wright, B. (1942). Altruism in children and the perceived conduct of others. *J. Abnorm. Soc. Psychol.* 37, 218–33.

Wynne-Edwards, V. C. (1962). *Animal dispersion in relation to social behavior*. Hafner, N.Y.

# Reciprocal altruism in olive baboons

C. PACKER†

Altruism is behaviour that benefits another individual at some cost to the altruist, costs and benefits being measured in terms of individual fitness. 'Reciprocal altruism'[1] implies the exchange of altruistic acts between unrelated individuals as well as between relatives. If the benefits to the recipient of an altruistic act exceed the costs to the altruist, and if the recipient is likely to reciprocate at a later time, then the cumulative benefits for both individuals will have exceeded the cumulative costs of their altruism. Natural selection would favour individuals that engaged in reciprocal altruism if they distributed their altruism with respect to the altruistic tendencies of the recipient, preferring individuals that were most likely to reciprocate and excluding nonaltruists from the benefits of further altruism. This model has been difficult to test because it is usually impossible to be certain that an example of altruism is not the product of 'kin selection'.[2] The genetic relationship between individuals in animal populations are seldom known and reciprocal altruism can only be cited when it can be found to occur regularly between unrelated individuals. I report here that altruistic behaviour involving the formation of coalitions among male olive baboons (*Papio anubis*) fulfils the criteria for reciprocal altruism.

Eighteen adult males *P. anubis* in three troops at Gombe National Park, Tanzania were studied for more than 1100 h. between May and December 1972 and from June 1974 to May 1975. All data were collected on a focal sample basis.[3] Focal individuals are referred to as 'targets'. Each animal was observed regularly for a fixed period. Observations were made on foot at 5–10 m from the animal. Baboon studies have been in progress since 1967 and data concerning blood relationships and dates of transfer between troops are available. All males leave their natal troop at Gombe, and the males who breed within a particular

†Reprinted with permission from the author and *Nature* 265 (1977). Copyright © 1977 by Macmillan Journals Ltd.

troop are those that have transferred into that troop from else-where.[4] Males known or thought to have transferred into their troop of residence are termed 'adult males'. Sexually mature males still residing in their natal troop are 'natal males'. During the study, there were three, six to eleven, and eight to eleven adult males in the respective study troops.

The effectiveness of temporary coalitions of adult male *Papio* spp. during aggressive interactions against a single opponent has been described in *P. anubis*[5] and in *P. ursinus*.[6] Encounters between coalitions and single opponents in *P. cynocephalus* did not seem to affect subsequent dyadic encounters between a coalition member and the opponent.[7] Ransom noted that coalitions of *P. anubis* at Gombe were generally formed by one male enlisting a partner to help fight against an opponent.[8] The partner had not been involved directly in the encounter with the opponent before his enlistment. Coalitions were sometimes formed in attempts to separate an opponent from an oestrous female. *P. anubis* form exclusive consort pairs lasting for up to several days. If the female becomes available after such an attempt, she could be taken by only one of the two coalition partners.

Attempts at enlisting a coalition partner are referred to as 'soliciting'—a triadic interaction in which one individual, the enlisting animal, repeatedly and rapidly turns his head from a second individual, the solicited animal, towards a third individual (opponent), while continuously threatening the third. The function of headturning by the enlisting animal is to incite the solicited animal into joining him in threatening the opponent. An 'occasion' of soliciting is a bout of the behaviour followed by a gap of more than 10 s. The distribution of the number of 'occasions' of soliciting to the same partner per observation period was tested against a cumulative binomial distribution and did not differ significantly from expected values. Therefore, 'occasions' of soliciting are considered to be statistically independent from each other. For every occasion that soliciting involved the target male during an observation period, the identity and actions (enlisting, solicited, or opponent) of each participant were recorded. 'Occasions' were not recorded consistently in 1972, so data from that period are not included in measures of frequency. Although coalitions occasionally formed spontaneously, only those that resulted from soliciting are considered.

There were 140 examples of one adult male soliciting another during the 1974–75 study period. Soliciting resulted in coalitions on 97 occasions. On 20 occasions the opponent was consorting with an oestrous female and adult males were more likely to join a coalition if the opponent was in consort ($2 \times 2 \chi^2 = 5 \cdot 91, p < 0 \cdot 02, n = 140$). On six occasions during both study periods the formation of a coalition directed against a consorting male resulted in the loss of the female by the opponent. In all six cases the female ended up with the enlisting male of the coalition ($P = 0 \cdot 032$, two-tailed, sign test); the solicited male generally continued to fight the opponent while the enlisting male took over the female. In each of those cases the solicited male risked injury from fighting the opponent while the enlisting male gained access to an oestrous female. In most other examples of soliciting, no resource appeared to be at stake. The greater willingness to join a coalition against a consorting male may be related to the greater benefits that the altruism bestows on the recipient in those cases.

Thirteen different pairs of males reciprocated in joining coalitions at each other's request on separate occasions. These pairs comprised 12 different males. In six of these pairs, both pair members successfully enlisted their partner against the same opponent. Individual males which most frequently gave aid were those which most frequently received aid. There was a strong correlation betweeen the frequency with which adult males joined coalitions and the frequency with which they successfully enlisted coalition partners ($\zeta = 0 \cdot 84, z = 2 \cdot 87, P < 0 \cdot 004$, Kraemer test[9] based on Spearman correlations).

Each male tended to request aid from an individual who in turn requested aid from him. Using one occasion of soliciting per pair per observation period (since spatial patterns are relatively stable), the distribution of soliciting by each of the 18 target males was examined to find which other male each target solicited most often, that is his 'favourite partner'. Only the ten targets who solicited other males on four or more occasions were included. After the favourite partner was found for each target male, the distribution of soliciting by each partner was similarly examined. For nine out of ten target males, the favourite partner in turn solicited the target male more often than the average number of occasions that the partner solicited all adult males in their troop ($P = 0 \cdot 022$, two-tailed, sign test) (Table 1). These results suggest that preferences for particular partners may be partly based on reciprocation.

It is difficult to test whether a nonaltruist is excluded from further altruistic exchanges; a refusal to join a coalition in any given instance may occur because the costs to the potential altruist

Table 1. Soliciting activity of favourite partners of target males

| Target | No. of occasions FP solicited target | Mean no. of occasions FP solicited each male | FP solicited target more; + or less; — than average |
|--------|------|------|------|
| BBB | 3 | 2·7 | + |
| CRS | 3 | 1·2 | + |
| DVD | 0 | 1·0 | — |
| EBN | 5 | 1·0 | + |
| GRN | 1 | 0·7 | + |
| JNH | 2 | 1·1 | + |
| LEO | 4 | 2·7 | + |
| MNT | 1 | 0·4 | + |
| WDY | 5 | 2·7 | + |
| WTH | 4 | 2·8 | + |

FP, favourite partners.

on that occasion are particularly high. There is evidence, however, that adult males are more likely to join coalitions with individuals which in turn would be able to help them. Females and juveniles ('non-males') solicited adult males 12 times, but adult males never solicited 'non-males' ($P < 0.001$, two-tailed, sign test). Adult males were less likely to respond to the solicitings of 'non-males' than they were to other adult males ($2 \times 2\chi^2 = 7.76$, $P < 0.01$, $n = 152$), even though 'non-males' solicited adult males against other 'non-males' far more often than adult males solicited other adult males against 'non-males' ($2 \times 2 \chi^2 = 43.76$, $P < 0.001$, $n = 110$). Adult male baboons are very much larger than either adult females or juveniles, so that even though the aid of an adult male to a 'non-male' would generally incur small costs to the adult male, the benefits to the male from having a 'non-male' partner would be trivial, since a 'non-male' could not provide effective help in an encounter against another adult male and an adult male would not need help in fighting a 'non-male'.

The adult males in the troops studied transferred into their troop of residence singly.[10] The origins of many males entering a study troop were not definitely known, so it was impossible to determine the precise degree of relatedness between most adult males. The prior troop and dates of transfer of each male were known, however. In four of the 13 reciprocating pairs, both partners were first observed as young adults in different troops when they probably had not yet transferred for the first time. They did not reside in the same troop for another 5 yr. One pair not included earlier comprised an adult male and a 'natal' male which were known to have been born in separate troops. Although there is no proof that any of these individuals are completely unrelated, it is unlikely that they are close relatives.

In contests between a coalition and an opponent a previously uninvolved partner may benefit the enlisting male by reducing the latter's risk of injury in fighting the opponent. (Although only one male has been known to die from wounds received in a fight since 1967, non-fatal wounds are common.) By participating in a fight in which he would not have been otherwise involved, his aid will have been at some cost to himself. Whether or not the benefits to the recipient exceed the costs to the altruist (as required by the model) is difficult to determine. But, for the opponent the potential costs in facing two males simultaneously rather than only one might be so much greater that it would often be to his advantage to avoid the coalition without fighting. If so, then the actual costs to the altruist are less than the reduction in costs to the enlisting male since the enlisting male would then be less likely actually to fight the opponent. When the formation of a coalition involves gaining access to an oestrous female further benefits are involved. Coalitions are the most common way of aggressively taking over an oestrous female from a consorting male at Gombe (D. A. Collins, in preparation). During the reproductive life of a male, which extends over 10 yr, there may be a large number of situations where it would be advantageous to enlist a coalition partner in an encounter against a consorting male. The number of offspring that a male sired as a result of participating in reciprocating coalitions would be greater than if he did not, while his lifespan would probably not be appreciably shortened by aiding coalition partners.

There are probably occasions when altruism is not involved. For example, the solicited male may sometimes join a coalition when there is a prospect of immediate benefit to himself. In such cases, explanations of the animal's behaviour are not not necessary; the behaviour is not detrimental to his own fitness. The occurrence of genuine altruism, however, seems to be common enough to demand an alternative explanation.

## References

[1] Trivers, R. L. (1971). *Q. Rev. Biol.* 46, 35–57.
[2] Hamilton, W. D. (1964). *J. Theor. Biol.* 7, 1–52.
[3] Altmann, J. (1974). *Behaviour* 49, 227–67

[4] Packer, C. (1975). *Nature, Lond.* **255**, 219–20.

[5] Hall, K. R., and DeVore, I. (1965). In *Primate behavior* (ed. by DeVore, I.). Holt, New York.

[6] Stoltz, L. P., and Saayman, G. S. (1970). *Ann. Tranvs. Mus.* **26**, 99–143.

[7] Hausfater, G. (1975). *Contrib. Primatol.* 7.

[8] Ransom, T. W. (1971). Thesis. Univ. California, Berkeley.

[9] Kraemer, H. C. (1975). *Psychometrika* **40**, 473–84.

[10] Packer, C. *Proc. Sixth Intl. Congr. Primatol.* (ed. by Harcourt, A.H. and Chivers, D.J.) Academic, London (in press).

# Parent–offspring conflict

ROBERT L. TRIVERS†

In classical evolutionary theory parent-offspring relations are viewed from the standpoint of the parent. If parental investment (PI) in an offspring is defined as anything done by the parent for the offspring that increases the offspring's chance of surviving while decreasing the parent's ability to invest in other offspring (Trivers 1972), then parents are classically assumed to allocate investment in their young in such a way as to maximize the number surviving, while offspring are implicitly assumed to be passive vessels into which parents pour the appropriate care. Once one imagines offspring as *actors* in this interaction, then conflict must be assumed to lie at the heart of sexual reproduction itself—an offspring attempting from the very beginning to maximize its reproductive success (RS) would presumably want more investment than the parent is selected to give. But unlike conflict between unrelated individuals, parent-offspring conflict is expected to be circumscribed by the close genetic relationship between parent and offspring. For example, if the offspring garners more investment than the parent has been selected to give, the offspring thereby decreases the number of its surviving siblings, so that any gene in an offspring that leads to an additional investment decreases (to some extent) the number of surviving copies of itself located in siblings. Clearly, if the gene in the offspring exacts too great a cost from the parent, that gene will be selected against even though it confers some benefit on the offspring. To specify precisely how much cost an offspring should be willing to inflict on its parent in order to gain a given benefit, one must specify how the offspring is expected to weigh the survival of siblings against its own survival.

The problem of specifying how an individual is expected to weigh siblings against itself (or any relative against any other) has been solved in outline by Hamilton (1964), in the context of

† Reprinted with permission from the author and *American Zoologist* 14 (1974).

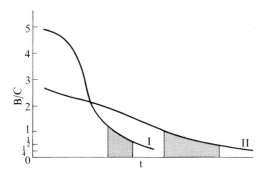

Fig. 1. The benefit/cost ratio (B/C) of a parental act (such as nursing) toward an offspring as a function of time. Benefit is measured in units of reproductive success of the offspring and cost in comparable units of reproductive success of the mother's future offspring. Two species are plotted. In species I the benefit/cost ratio decays quickly; in species II, slowly. Shaded areas indicate times during which parent and offspring are in conflict over whether the parental care should continue. Future sibs are assumed to be full-sibs. If future sibs were half-sibs the shaded areas would have to be extended until B/C = 1/4.

The argument applies to all sexually reproducing species that are not completely inbred, that is, in which siblings are not identical copies of each other. Conflict near the end of the period of PI over the continuation of PI is expected in all such species. The argument applies to PI in general or to any subcomponent of PI (such as feeding the young, guarding the young, carrying the young) that can be assigned a more or less independent cost–benefit function. Weaning conflict in mammals is an example of parent–offspring conflict explained by the argument given here. Such conflict is known to occur in a variety of mammals, in the field and in the laboratory: for example, baboons (DeVore 1963), langurs (Jay 1963), rhesus macaques (Hinde and Spencer-Booth 1971), other macaques (Rosenblum 1971), vervets (Struhsaker 1971), cats (Schneirla *et al.* 1963), dogs (Rheingold 1963) and rats (Rosenblatt and Lehrman 1963). Likewise, I interpret conflict over parental feeding at the time of fledging in bird species as conflict explained by the present argument: for example, Herring Gulls (Drury and Smith 1968), Red Warblers (Elliott 1969), Verreaux's Eagles (Rowe 1947), and White Pelicans (Schaller 1964).

Weaning conflict is usually assumed to occur either because transitions in nature are assumed always to be imperfect or because such conflict is assumed to serve the interests of both parent and offspring by informing each of the needs of the other. In either case, the marked inefficiency of weaning conflict seems the clearest argument in favor of the view that such conflict results from an underlying conflict in the way in which the inclusive fitness of mother and offspring are maximized. Weaning conflict in baboons, for example, may last for weeks or months, involving daily competitive interactions and loud cries from the infant in a species otherwise strongly selected for silence (DeVore 1963). Interactions that are inefficient *within* a multicellular organism would be cause for some surprise, since, unlike mother and offspring, the somatic cells within an organism are identically related.

One parameter affecting the expected length and (intensity) of weaning conflict is the offspring's expected $r_0$ to its future siblings. The lower the offspring's $r_0$ to its future siblings, the longer and more intense the expected weaning conflict. This suggests a simple prediction. Other things being equal, species in which different, unrelated males commonly father a female's successive offspring are expected to show stronger weaning conflict than species in which a female's successive offspring are usually fathered by the same male. As shown below, however, weaning conflict is merely a special case of conflict expected throughout the period of parental investment, so that this prediction applies to the intensity of conflict prior to weaning as well.

## Conflict throughout the period of PI over the amount of PI

In Fig. 1 it was assumed that the amount of investment for each day (or moment in time) had already been established, and that mother and young were only selected to disagree over when such investment should be ended. But it can be shown that, in theory, conflict over the amount of investment that should at each moment be given, is expected throughout the period of PI.

At any moment in the period of PI the female is selected to invest that amount which maximizes the difference between the associated cost and benefit, where these terms are defined as above. The infant is selected to induce that investment which maximizes the difference between the benefit and a cost devalued by the relevant $r_0$. The different optima for a moment in time in a hypothetical species are graphed in Fig. 2. With reasonable assumptions about the shape of the benefit and cost curves, it is clear that the infant will, at each instant in time, tend to favor greater parental

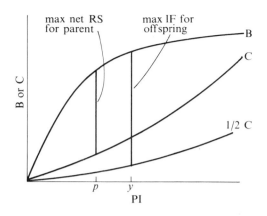

Fig. 2. The benefit, cost, and half the cost of a parental act toward an off-spring at one moment in time as a function of the amount the parent invests in the act (PI). Amount of milk given during one day of nursing in a mammal would be an example of PI. At $p$ the parent's inclusive fitness (B − C) is maxi-mised; at $y$ the offspring's inclusive fitness (B − C/2) is maximized. Parent and offspring disagree over whether $p$ or $y$ should be invested. The offspring's future siblings are assumed to be full-siblings. IF = inclusive fitness.

investment that the parent is selected to give. The period of trans-ition discussed in the previous section is a special case of this continuing competition, namely, the case in which parent and off-spring compete over whether *any* investment should be given, as opposed to their earlier competition over *how much* should be given. Since parental investment begins before eggs are laid or young are born, and since there appears to be no essential distinc-tion between parent–offspring conflict outside the mother (medi-ated primarily by behavioral acts) and parent–offspring conflict inside the mother (mediated primarily by chemical acts), I assume that parent–offspring conflict may in theory begin as early as meiosis.

It must be emphasized that the cost of parental investment referred to above (see Fig. 2) is measured *only* in terms of decreased ability to produce *future* offspring (or, when the brood size is larger than one, decreased ability to produce *other* offspring). To appreciate the significance of this definition, imagine that early in the period of PI the offspring garners more investment than the parent has been selected to give. This added investment may decrease the parent's later investment in the offspring at hand, either through an increased chance of parental mortality during the period of PI, or through a depletion in parental resources, or because parents have been selected to make the appropriate adjustment (that is, to reduce later investment below what otherwise would have been given). In short, the offspring may gain a temporary benefit but suffer a later cost. This self-inflicted cost is subsumed in the benefit function (B) of Fig. 2, because it decreases the benefit the infant receives. It is not subsumed in the cost function (C) because this function refers only to the mother's future offspring.

### The time course of parent–offspring conflict

If one could specify a series of cost–benefit curves (such as Fig. 2) for each day of the period of PI, then the expected time course of parent–offspring conflict could be specified. Where the difference in the offspring's inclusive fitness at the parent's optimum PI ($p$ in Fig. 2) and at the offspring's optimum PI ($y$) is large, conflict is expected to be intense. Where the difference is slight, conflict is expected to be slight or nonexistent. In general, where there is a strong difference in the offspring's inclusive fitness at the two different optima ($p$ and $y$), there will also be a strong difference in the parent's inclusive fitness, so that both parent and offspring will simultaneously be strongly motivated to achieve their respective optimal values of PI. (This technique of comparing cost–benefit graphs can be used to make other predictions about parent-offspring conflict, for example that such conflict should decrease in intensity with increasing age, and hence decreasing reproductive value, of the parent; see Fig. 3.) In the absence of such day-by-day graphs three factors can be identified, all of which will usually predispose parent and offspring to show greater conflict as the period of PI progresses.

(1) *Decreased chance of self-inflicted cost* Due to getting more PI.
As the period of PI progresses, the offspring faces a decreased chance of suffering a later self-inflicted cost for garnering additional investment at the moment. At the end of the period of PI any additional investment forced on the parent will only affect later offspring, so that at that time the interests of parent and offspring are maximally divergent. This time-dependent change in the offspring's chance of suffering a self-inflicted cost will, other things being equal, predispose parent and offspring to increasing conflict during the period of PI.

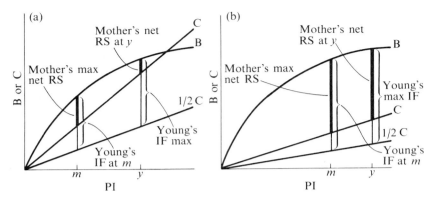

Fig. 3. The benefit and cost of a parental act (as in Fig. 2) towards (a) an off-spring born to a young female and (b) an offspring born to an old female. One assumes that the benefit to the offspring of a given amount of PI does not change with birth order but that the cost declines as a function of the declining reproductive value (Fisher, 1930) of the mother: she will produce fewer future offspring anyway. The difference between the mother's inclusive fitness at $m$ and $y$ is greater for (a) than for (b). The same is true for the offspring. Conflict should be correspondingly more intense between early born young and their mothers than between late born young and their mothers.

## (2) Imperfect replenishment of parental resources

If the parent is unable on a daily basis to replenish resources invested in the offspring, the parent will suffer increasing depletion of its resources, and, as time goes on, the cost of such depletion should rise disproportionately, even if the amount of resources invested per day declines. For example, a female may give less milk per day in the first half of the nursing period than in the second half (as in pigs: Gill and Thomson 1956), but if she is failing throughout to replenish her energy losses, then she is con-stantly increasing her deficit (although at a diminishing rate) and greater deficits may be associated with disproportionate costs. In some species a parent does not feed itself during much of the period of PI and at least during such periods the parent must be depleting its resources (for example, female elephant seals during the nursing period: LeBoeuf et al. 1972). But the extent to which parents who feed during the period of PI fail to replenish their resources is usually not known. For some species it is clear that females typically show increasing levels of depletion during the period of PI (e.g. sheep: Wallace 1948).

## (3) *Increasing size of the offspring*

During that portion of the period of PI in which the offspring receives all its food from its parents, the tendency for the offspring to begin very small and steadily increase in size will, other things being equal, increase the cost to the parent of maintaining and enlarging it. (Whether this is always true will depend, of course, on the way in which the offspring's growth rate changes as a function of increasing size.) In addition, as the offspring increases in size the relative energetic expense to it of competing with its parents should decline.

The argument advanced here is only meant to suggest a general tendency for conflict to increase during the period of PI, since it is easy to imagine circumstances in which conflict might peak several times during the period of PI. It is possible, for example, that weight at birth in a mammal such as humans is strongly associated with the offspring's survival in subsequent weeks, but that the cost to the mother of bearing a large offspring is considerably greater than some of her ensuing investment. In such circumstances, conflict *prior* to birth over the offspring's weight at birth may be more intense than conflict over nursing in the weeks after birth.

Data from studies of dogs, cats, rhesus macaques, and sheep appear to support the arguments of this and the previous section. In these species, parent–offspring conflict begins well before the period of weaning and tends to increase during the period of PI. In dogs (Rheingold 1963) and cats (Schneirla *et al.* 1963) postnatal maternal care can be divided into three periods according to increasing age of the offspring. During the first, the mother approaches the infant to initiate parental investment. No avoidance behaviour or aggression toward the infant is shown by the mother. In the second, the offspring and the mother approach each other about equally, and the mother shows some avoidance behaviour and some aggression in response to the infant's demands. The third period can be characterized as the period of weaning. Most contacts are initiated by the offspring. Open avoidance and aggression characterize the mother.

Detailed quantitative data on the rhesus macaque (Hinde and Spencer-Booth 1967, 1971), and some parallel data on other macaques (Rosenblum 1971), demonstrate that the behavior of both mother and offspring changes during the period of postnatal parental care in a way consistent with theory. During the first weeks after she has given birth, the rhesus mother's initiative in setting up nipple contacts is high but it soon declines rapidly. Concurrently she begins to reject some of the infant's advances, and after her own initiatives toward nipple contact have ceased, she rejects her infant's advances with steadily increasing frequency until at the end of the period of investment all of the offspring's advances are rejected. Shortly after birth, the offspring leaves the mother more often than it approaches her, but as time goes on the initiative in maintaining mother–offspring proximity shifts to the offspring. This leads to the superficially paradoxical result that as the offspring becomes increasingly active and independent, spending more and more time away from its mother, its initiative in maintaining mother–offspring proximity *increases* (that is, it tends to approach the mother more often than it leaves her). According to the theory presented here, this result reflects the underlying tendency for parent-offspring conflict to increase during the period of PI. As the interests of mother and offspring diverge, the offspring must assume a greater role in inducing whatever parental investment is forthcoming.

*[handwritten margin note: Mother keeps kid close → kid stays close.]*

Data on the production and consumption of milk in sheep (Wallace 1948) indicate that during the first weeks of the lamb's life the mother typically produces more milk than the lamb can drink. The lamb's appetite determines how much milk is consumed. But after the fourth week, the mother begins to produce less than the lamb can drink, and from that time on it is the mother who is the limiting factor in determining how much milk is consumed. Parallel behavioral data indicate that the mother initially permits free access by her lamb(s) but after a couple of weeks begins to prevent some suckling attempts (Munro 1956; Ewbank 1967). Mothers who are in poor condition become the limiting factor in nursing earlier than do mothers in good condition, and this is presumably because the cost of a given amount of milk is considerably higher when the mother is in poor condition, while the benefit to the offspring remains more or less unchanged. Females who produce twins permit either twin to suckle on demand during the first three weeks after birth, but in the ensuing weeks they do not permit one twin to suckle unless the other is ready also (Ewbank 1964; Alexander 1960).

## Disagreement over the sex of the offspring

Under certain conditions a potential offspring is expected to disagree with its parents over whether it should become a male or a female. Since one cannot assume that potential offspring are powerless to affect their sex, sex ratios observed in nature should to some extent reflect the offspring's preferred value as well as the parents'.

Fisher (1930) showed that (in the absence of inbreeding) parents are selected to invest as much in the total of their daughters as in the total of their sons. When each son produced costs on average the same as each daughter, parents are selected to produce a sex ratio of 50/50. In such species, the expected reproductive success (RS) of a son is the same as that of a daughter, so that an offspring should be indifferent as to sex. But if (for example) parents are selected to invest twice as much in a typical male as in a typical female, then they will be selected to produce twice as many females as males, and the expected RS of each son will be twice that of each daughter. In such a species a potential offspring would prefer to be a male, for it would then achieve twice the RS it would as a female, without suffering a comparable decrease in inclusive fitness through the cost forced on its parents, because the offspring is selected to devalue that cost by the offspring's expected $r_0$ to the displaced sibling. For the example chosen, the exact gain in the offspring's inclusive fitness can be specified as follows. If the expected RS of a female offspring is defined as one unit of RS, then, in being made male, the offspring gains one unit of RS, but it deprives its mother of an additional daughter (or half a son). This displaced sibling (whether female or half of a male) would have achieved one unit of RS, but this unit is devalued from the offspring's standpoint by the relevant $r_0$. If the displaced sibling would have been a full sibling, then this unit of RS is devalued by ½, and the offspring, in being made a male, achieves a ½ unit net increase in inclusive fitness. If the displaced sibling would have been a half sibling, the offspring, in being made a male, achieves a ¾ unit net increase in inclusive fitness. The parent, on the other hand, experiences initially only a trivial decrease in RS, so that initially any gene in the offspring tending to make it a male against its parents' efforts would spread rapidly.

As a hypothetical gene for offspring control of sex begins to spread, the number of males produced increases, thereby lowering the expected RS of each male. This decreases the gain (in inclusive fitness) to the offspring of being made a male. If the offspring's equilibrial sex ratio is defined as that sex ratio at which an offspring is indifferent as to whether it becomes a male or a female, then this sex ratio can be calculated by determining the sex ratio at which the offspring's gain in RS in being made a male is exactly offset by its loss in inclusive fitness in depriving itself of a sister (or half a brother). The offspring's equilibrial sex ratio will depend on both the offspring's expected $r_0$ to the displaced siblings and on the extent to which parents invest more in males than in females (or vice versa). The general solution is given in the Appendix. Parent and offspring equilibrial sex ratios for different values of $r_0$ and different values of $x$ (PI in a typical son/PI in a typical daughter) are plotted in Fig. 4. For example, where the $r_0$ between siblings is ½ and where parents invest twice as much in a son as in a daughter ($x = 2$), the parents' equilibrial sex ratio is 1:2 (males:females) while that of the offspring is 1:1·414.

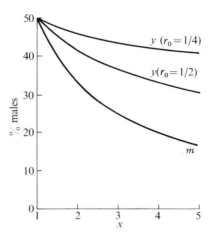

Fig. 4. The optimal sex ratio (per cent males) for the mother (m) and the young (y) where the mother invests more in a son than in a daughter by a factor of $x$ (and assuming no paternal investment in either sex). Two functions are given for the offspring, depending on whether the siblings it displaces are full-siblings ($r_0 = 1/2$) or half-siblings ($r_0 = 1/4$). Note the initial rapid divergence between the mother's and the offspring's preferred sex ratio as the mother moves from equal investment in a typical individual of either sex ($x = 1$) to twice as much investment in a typical male ($x = 2$).

As long as all offspring are fathered by the same male, he will prefer the same sex ratio among the offspring that the mother does. But consider a species such as caribou in which the female produces only one offspring a year and assume that a female's successive offspring are fathered by different, unrelated males. If the female invests more in a son than in a daughter, then she will be selected to produce more daughters than sons. The greater cost of the son is not borne by the father, however, who invests nothing beyond his sperm, and who will not father the female's later off-spring, so the father's equilibrial sex ratio is an equal number of sons and daughters. The offspring will prefer some probability of being a male that is intermediate between its parents' preferred probabilities, because (unlike the father) the offspring is related to the mother's future offspring but (unlike the mother) it is less related to them than to itself.

In a species such as just described (in which the male is hetero-gametic) the following sort of competitive interaction is possible. The prospective father produces more Y-bearing sperm than the female would prefer and she subjects the Y-bearing sperm to differential mortality. If the ratio of the sperm reaching the egg has been reduced to near the mother's optimal value, then the egg preferentially admits the Y-bearing sperm. If the mother ovulated more eggs than she intends to rear, she could then choose which to invest in, according to the sex of the fertilized egg, unless a male egg is able to deceive the mother about its sex until the mother has committed herself to investing in him. Whether such interactions actually occur in nature is at present unknown.

One consequence of the argument advanced here is that there is an automatic selective agent tending to keep maternal investment in a son similar to that in a daughter, for the greater the disparity between the investment in typical individuals of the two sexes, the greater the loss suffered by the mother in competitive interactions with her offspring over their preferred sex and in producing a sex ratio further skewed away from her preferred ratio (see Fig. 4). This automatic selection pressure may partly account for the apparent absence of strongly size-dimorphic young (at the end of PI) in species showing striking adult sexual dimorphism in size.

The argument presented here applies to any tendency of the parent to invest differentially in the young, whether according to sex or some other variable, except that in many species sex is irreversibly determined early in ontogeny and the offspring is expected at the very beginning to be able to discern its own sex and hence the predicted pattern of investment it will receive, so that, unlike other forms of differential investment, conflict is expected very early, namely, at the time of sex determination.

### The offspring as psychological manipulator

How is the offspring to compete effectively with its parent? An offspring cannot fling its mother to the ground at will and nurse. Throughout the period of parental investment the offspring competes at a disadvantage. The offspring is smaller and less experienced than its parent, and its parent controls the resources at issue. Given this competitive disadvantage the offspring is expected to employ psychological rather than physical tactics. (Inside the mother the offspring is expected to employ chemical tactics, but some of the analysis presented below should also apply to such competition.) It should attempt to *induce* more investment than the parent wishes to give.

Since an offspring will often have better knowledge of its real needs than will its parents, selection should favor parental attentiveness to signals from its offspring that apprize the parent of the offspring's condition. In short, the offspring cries when hungry or in danger and the parent responds appropriately. Conversely, the offspring signals its parent (by smiling or wagging its tail) when its needs have been well met. Both parent and offspring benefit from this system of communication. But once such a system has evolved, the offspring can begin to employ it out of context. The offspring can cry not only when it is famished but also when it merely wants more food than the parent is selected to give. Likewise, it can begin to withhold its smile until it has gotten its way. Selection will then of course favor parental ability to discriminate the two uses of the signals, but still subtler mimicry and deception by the offspring are always possible. Parental experience with preceding offspring is expected to improve the parent's ability to make the appropriate discrimination. Unless succeeding offspring can employ more confusing tactics than earlier ones, parent–offspring interactions are expected to be increasingly biased in favor of the parent as a function of parental age.

In those species in which the offspring is more helpless and vulnerable the younger it is, its parents will have been more strongly selected to respond positively to signals of need emitted by the offspring, the younger that offspring is. This suggests that at any stage of ontogeny in which the offspring is in conflict with its parents, one appropriate tactic may be to revert to the gestures and actions of an earlier stage of development in order to induce the investment that would then have been forthcoming. Psychologists have long recognized such a tendency in humans and have given it the name of regression. A detailed functional analysis of regression could be based on the theory presented here.

The normal course of parent–offspring relations must be subject to considerable unpredictable variation in both the condition of the parent and (sometimes independently) the condition of the offspring. Both partners must be sensitive to such variation and must adjust their behavior appropriately. Low investment coming from a parent in poor condition has a different meaning than low investment coming from a parent in good condition. This suggests that from an early age the offspring is expected to be a psychologically sophisticated organism. The offspring should be able to evaluate the cost of a given parental act (which depends in part on the condition of the parent at that moment) and its benefit (which depends in part on the condition of the offspring). When the offspring's interests diverge from those of its parent, the offspring must be able to employ a series of psychological manoeuvres, including the mimicry and regression mentioned above. Although it would be expected to learn appropriate information (such as whether its psychological manoeuvres were having the desired effects), an important feature of the argument presented here is that the offspring cannot rely on its parents for disinterested guidance. One expects the offspring to be pre-programmed to resist some parental teaching while being open to other forms. This is particularly true, as argued below, for parental teaching that affects the altruistic and egoistic tendencies of the offspring.

If one event in a social relationship predicts to some degree future events in that relationship, the organism should be selected to alter its behavior in response to an initial event, in order to change the probability that the predicted events will occur. For example, if a mother's lack of love for her offspring early in its life predicts deficient future investment, then the offspring will be selected to be sensitive to such early lack of love, whether investment at that time is deficient or not, in order to increase her future investment. The best data relevant to these possibilities come from the work of Hinde and his associates on groups of caged rhesus macaques. In a series of experiments, a mother was removed from her 6-month-old infant, leaving the infant in the home cage with other group members. After 6 days, the mother was returned to the home cage. Behavioral data was gathered before, during, and after the separation (see points 1 and 2 below). In a parallel series of experiments, the infant was removed for 6 days from its mother, leaving her in the home cage and the same behavioral data were gathered (see point 3 below). The main findings can be summarized as follows:

(1) *Separation of mother from her offspring affects their relationship upon reunion.* After reunion with its mother, the infant spends more time on the mother than it did before separation—although had the separation not occurred, the infant would have reduced its time on the mother. This increase is caused by the infant, and occurs despite an increase in the frequency of maternal rejection (Hinde and Spencer-Booth 1971). These effects can be detected at least as long as 5 weeks after reunion. These data are consistent with the assumption that the infant has been selected to interpret its mother's disappearance as an event whose recurrence the infant can help prevent by devoting more of its energies to staying close to its mother.

(2) *The mother–offspring relationship prior to separation affects the offspring's behavior on reunion.* Upon reunion with its mother, an infant typically shows distress, as measured by callings and immobility. The more frequently an infant was rejected *prior* to separation, the more distress it shows upon reunion. This correlation holds for at least 4 weeks after reunion. In addition, the more distressed the infant is the greater is its role in maintaining proximity to its mother (Hinde and Spencer-Booth 1971). These data support the assumption that the infant interprets its mother's disappearance in relation to her predeparture behavior in a logical way: the offspring should assume that a rejecting mother who temporarily disappears needs more offspring surveillance and intervention than does a nonrejecting mother who temporarily disappears.

(3) *An offspring removed from its mother shows, upon reunion, different effects than an offspring whose mother has been removed.* Compared to an infant whose mother had been removed, an infant removed from its mother shows, upon reunion, and for up to 6 weeks after reunion, less distress and more time off the mother. In addition, the offspring tends to play a smaller role in maintaining proximity to its mother, and it experiences less frequent maternal rejections (Hinde and Davies 1972*a, b*). These data are consistent with the expectation that the offspring should be sensitive to the *meaning* of events affecting its relationship to its mother. The offspring can differentiate between a separation from its mother caused by its own behavior or some accident (infant removed from group) and a separation which may have been

caused by maternal negligence (mother removed from group). In the former kind of separation, the infant shows less effects when reunited, because, from its point of view, such a separation does not reflect on its mother and no remedial action is indicated. A similar explanation can be given for differences in the mother's behavior.

## Parent–offspring conflict over the behavioral tendencies of the offspring

Parents and offspring are expected to disagree over the behavioral tendencies of the offspring insofar as these tendencies affect related individuals. Consider first interactions among siblings. An individual is only expected to perform an altruistic act toward its full-sibling <u>whenever the benefit to the sibling is greater than twice the cost to the altruist.</u> Likewise, it is only expected to forego selfish acts when $C > 2B$ where a selfish act is defined as one that gives the actor a benefit, B, while inflicting a cost, C, on some other individual, in this case, on a full-sibling. But parents, who are equally related to all of their offspring, are expected to encourage all altruistic acts among their offspring in which $B > C$, and to discourage all selfish acts in which $C > B$. Since there ought to exist altruisitic situations in which $C < B < 2C$, parents and offspring are expected to disagree over the tendency of the offspring to act altruistically toward its siblings. Likewise, whenever for any selfish act harming a full-sibling $B < C < 2B$, parents are expected to discourage such behavior and offspring are expected to be relatively refractory to such discouragement.

*[handwritten margin note: Conflict when B > C but B ≯ 2C]*

This parent–offspring disagreement is expected over behavior directed toward other relatives as well. For example, the offspring is only selected to perform altruistic acts toward a cousin (related through the mother) when $B > 8C$. But the offspring's mother is related to her own nephews and nieces by $r_0 = ¼$ and to her offspring by $r_0 = ½$, so that she would like to see any altruistic acts performed by her offspring toward their maternal cousins whenever $B > 2C$. The same argument applies to selfish acts, and both arguments can be made for more distant relatives as well. (The father is unrelated to his mate's kin and, other things being equal, should not be distressed to see his offspring treat such individuals as if they were unrelated.)

The general argument extends to interactions with unrelated individuals, as long as these interactions have some effect, however remote and indirect, on kin. Assume, for example, that an individual gains some immediate benefit, B, by acting nastily toward some unrelated individual. Asssume that the unrelated individual reciprocates in kind (Trivers 1971), but assume that the reciprocity is directed toward both the original actor and some relative, e.g. his sibling. Assuming no other effects of the initial act, the original actor will be selected to perform the nasty act as long as $B > C_1 + \frac{1}{2}(C_2)$, where $C_1$ is the cost to the original actor of the reciprocal nastiness he receives and $C_2$ is the cost to his sibling of the nastiness the sibling receives. The actor's parents viewing the interaction would be expected to condone the initial act only if $B > C_1 + C_2$. Since there ought to exist situations in which $C_1 + \frac{1}{2}(C_2) < B < C_1 + C_2$, one expects conflict between offspring and parents over the offspring's tendency to perform the initial nasty act in the situation described. A similar argument can be made for altruistic behavior directed towards an unrelated individual if this behavior induces altruism in return, part of which benefits the original altruist's sibling. Parents are expected to encourage such altruism more often than the offspring is expected to undertake on his own. The argument can obviously be extended to behavior which has indirect effects on kin other than one's sibling.

As it applies to human beings, the above argument can be summarized by saying that a fundamental conflict is expected during socialization over the altruistic and egoistic impulses of the offspring. Parents are expected to socialize their offspring to act more altruistically and less egoistically than the offspring would naturally act, and the offspring are expected to resist such socialization. If this argument is valid, then it is clearly a mistake to view socialization in humans (or in any sexually reproducing species) as only or even primarily a process of 'enculturation', a process by which parents teach offspring their culture (e.g. Mussen *et al.* 1969, p. 259). For example, one is not permitted to assume that parents who attempt to impart such virtues as responsibility, decency, honesty, and trustworthiness, generosity, and self-denial are merely providing the offspring with useful information on appropriate behavior in the local culture, for all such virtues are likely to affect the amount of altruistic and egoistic behaviour impinging on the parent's kin, and parent and offspring are expected to view such behavior differently. That some teaching beneficial to the offspring transpires during human socialization can be taken for granted, and one would expect no conflict if socialization involved *only* teaching beneficial to the offspring. According to the theory presented here, <u>socialization is a process by which parents attempt to mold each offspring in order to increase their own inclusive fitness,</u> while each offspring is selected to resist some of the molding and to attempt to mold the behavior of its parents (and siblings) in order to increase its inclusive fitness. Conflict during socialization need not be viewed solely as conflict between the culture of the parent and the biology of the child; it can also be viewed as conflict between the biology of the parent and the biology of the child. Since teaching (as opposed to molding) is expected to be recognized by offspring as being in their own self-interest, parents would be expected to overemphasize their role as teachers in order to minimize resistance in their young. According to this view then, the prevailing concept of socialization is to some extent a view one would expect adults to entertain and disseminate.

Parent–offspring conflict may extend to behavior that is not on the surface either altruistic or selfish but which has consequences that can be so classified. The amount of energy a child consumes during the day, and the way in which the child consumes this energy, are not matters of indifference to the parent when the parent is supplying that energy, and when the way in which the child consumes the energy affects its ability to act altruistically in the future. For example, when parent and child disagree over when the child should go to sleep, one expects in general the parent to favor early bedtime, since the parent anticipates that this will decrease the offspring's demands on parental resources the following day. Likewise, one expects the parent to favor serious and useful expenditures of energy by the child (such as tending the family chickens, or studying) over frivolous and unnecessary expenditures (such as playing cards)—the former are either altruistic in themselves, or they prepare the offspring for future altruism. In short, we expect the offspring to perceive some behavior, that the parent favors, as being dull, unpleasant, moral, or any combination of these. <u>One must at least entertain the assumption that the child would find such behavior more enjoyable if in fact the behavior maximized the offspring's inclusive fitness.</u>

### Conflict over the adult reproductive role of the offspring

As a special case of the preceding argument, it is clear that under certain conditions conflict is expected between parent and offspring over the adult reproductive role of the offspring. To take the extreme case, it follows at once from Hamilton's (1964) work that individuals who choose not to reproduce (such as celibate priests) are not necessarily acting counter to their genetic self-interest. One need merely assume that the nonreproducer thereby increases the reproductive success of relatives by an amount which, when devalued by the relevant degrees of relatedness, is greater than the nonreproducer would have achieved on his own. This kind of explanation has been developed in some detail to explain nonreproductives in the haplodiploid Hymenoptera (Hamilton 1972). What is clear from the present argument, however, is that it is even more likely that the nonreproducer will thereby increase his *parents'* inclusive fitness than that he will increase his own. This follows because his parents are expected to value the increased reproductive success of kin relatively more than he is.

If the benefits of nonreproducing are assumed, for simplicity, to accrue only to full siblings and if the costs of nonreproducing are defined as the surviving offspring the nonreproducer would have produced had he or she chosen to reproduce, then parent–offspring conflict over whether the offspring should reproduce is expected whenever $C < B < 2C$. Assuming it is sometimes possible for parents to predict while an offspring is still young what the cost and benefit of its not reproducing will be, the parents would be selected to mold the offspring toward not reproducing whenever $B > C$. Two kinds of nonreproductives are expected: those who are thereby increasing their own inclusive fitness $(B > 2C)$ and those who are thereby lowering their own inclusive fitness but increasing that of their parents $(C < B < 2C)$. The first kind is expected to be as happy and content as living creatures ever are, but the second is expected to show internal conflict over its adult role and to express ambivalence over the past, particularly over the behavior and influence of its parents. I emphasize that it is not necessary for parents to be conscious of molding an offspring toward nonreproduction in order for such molding to occur and to increase the parent's inclusive fitness. It remains to be explored to what extent the etiology of sexual preferences (such as homosexuality) which tend to interfere with reproduction can be explained in terms of the present argument.

Assuming that parents and offspring agree that the offspring should reproduce, disagreement is still possible over the form of that reproduction. Whether an individual attempts to produce few offspring or many is a decision that affects that individual's opportunities for kin-directed altruism, so that parent and off-spring may disagree over the optimal reproductive effort of the offspring. Since in humans an individual's choice of mate may affect his or her ability to render altruistic behavior toward relatives, mate choice is not expected to be a matter of indifference to the parents. Parents are expected to encourage their offspring to choose a mate that will enlarge the offspring's altruism toward kin. For example, when a man marries his cousin, he increases (other things being equal) his contacts with relatives, since the immediate kin of his wife will also be related to him, and marriage will normally lead to greater contact with her immediate kin. One therefore might expect human parents to show a tendency to encourage their offspring to marry more closely related individuals (e.g. cousins) than the offspring would prefer. Parents may also use an offspring's marriage to cement an alliance with an unrelated family or group, and insofar as such an alliance is beneficial to kin of the parent in addition to the offspring itself, parents are expected to encourage such marriages more often than the off-spring would prefer. Finally, parents will more strongly discourage marriage by their offspring to individuals the local society defines as pariahs, because such unions are likely to besmirch the reputa-tion of close kin as well.

Because parents may be selected to employ parental investment itself as an incentive to induce greater offspring altruism, parent–offspring conflict may include situations in which the offspring attempts to terminate the period of PI *before* the parent wishes to. For example, where the parent is selected to retain one or more offspring as permanent 'helpers at the nest' (Skutch 1961), that is permanent nonreproductives who help their parents raise addi-itional offspring (or help those offspring to reproduce), the parent may be selected to give additional investment in order to tie the offspring to the parent. In this situation, selection on the offspring may favor any urge toward independence which overcomes the offspring's impulse toward additional investment (with its hidden cost of additional dependency). In short, in species in which kin-directed altruism is important, parent–offspring conflict may include situations in which the offspring wants *less* than the parent is selected to give as well as the more common situation in which the offspring attempts to garner *more* PI than the parent is selected to give.

Parent–offspring relations early in ontogeny can affect the later adult reproductive role of the offspring. A parent can influence the altruistic and egoistic tendencies of its offspring whenever it has influence over any variable that affects the costs and benefits associated with altruistic and egoistic behavior. For example, if becoming a permanent nonreproductive, helping one's siblings, is more likely to increase one's inclusive fitness when one is small in size relative to one's siblings (as appears to be true in some polistine wasps: Eberhard 1969), then parents can influence the proportion of their offspring who become helpers by altering the size distribution of their offspring. Parent–offspring conflict over early PI may itself involve parent–offspring conflict over the eventual reproductive role of the offspring. This theoretical possibility may be relevant to human psychology if parental decision to mold an offspring into being a nonreproductive involves differential investment as well as psychological manipulation.

### The role of parental experience in parent–offspring conflict

It cannot be supposed that all parent–offspring conflict results from the conflict in the way in which the parent's and the offspring's inclusive fitnesses are maximized. Some conflict also results, ironically because of an overlap in the interests of parent and young. When circumstances change, altering the benefits and costs associated with some offspring behavior, both the parent and the offspring are selected to alter the offspring's behavior appropriately. That is, the parent is selected to mold the appropriate change in the offspring's behavior and if parental molding is successful, it will strongly reduce the selection pressure on the offspring to change its behavior spontaneously. Since the parent is likely to discover the changing circumstances as a result of its own experience, one expects tendencies toward parental molding to appear, and spread, before the parellel tendencies appear in the offspring. Once parents commonly mold the appropriate offspring behavior, selection still favors genes leading toward voluntary offspring behavior, since such a developmental avenue is presumably more efficient and more certain than that involving parental manipulation. But the selection pressure for the appropriate offspring genes should be weak, and if circumstances change at a faster rate than this selection operates, there is the possibility of continued parent–offspring conflict resulting from the greater experience of the parents.

If the conflict described above actually occurs, then (as mentioned in an earlier section) selection will favor a tendency for parents to overemphasize their experience in all situations, and for the offspring to differentiate between those situations in which greater parental experience is real and those situations in which such experience is merely claimed in order to manipulate the offspring.

### Appendix: the offspring's equilibrial sex ratio

Let the cost of producing a female be one unit of investment, and let the cost of producing a male be $x$ units, where $x$ is larger than one. Let the expected reproductive success of a female be one unit of RS. Let the sex ratio produced be $1:y$ (males: females), where $y$ is larger than one. At this sex ratio the expected RS of a male is $y$ units of RS, so that, in being made a male instead of a female, an offspring gains $y - 1$ units of RS. But the offspring also thereby deprives its mother of $x - 1$ units of investment. The offspring's equilibrial sex ratio is that sex ratio at which the offspring's gain in RS in being made a male $(y - 1)$ is exactly offset by its loss in inclusive fitness which results because it thereby deprives its mother of $x - 1$ units of investment. The mother would have allocated these units in such a way as to achieve a $1:y$ sex ratio, that is, she would have allocated $x/(x + y)$ of the units to males and $y/(x + y)$ of the units to females. In short, she would have produced $(x - 1)/(x + y)$ sons, which would have achieved RS of $y(x - 1)/(x + y)$, and she would have produced $y(x - 1)/(x + y)$ daughters, which would have achieved RS of $y(x - 1)/(x + y)$. The offspring is expected to devalue this loss by the offspring's $r_0$ to its displaced siblings. Hence, the offspring's equilibrial sex ratio results when

$$y - 1 = \frac{r_0 y (x - 1)}{x + y} + \frac{r_0 y (x - 1)}{x + y}$$

$$= (2r_0 y) \frac{x - 1}{x + y}.$$

Rearranging gives

$$y^2 + y(x - 2r_0 x + 2r_0 - 1) - x = 0$$

$$y^2 + (x - 1)(1 - 2r_0)y - x = 0.$$

The general solution for this quadratic equation is

$$y = \frac{-(x - 1)(1 - 2r_0)}{2} + \frac{\sqrt{(x - 1)^2 (1 - 2r_0)^2 + 4x}}{2}.$$

Where $r_0 = \frac{1}{2}$, the equation reduced to $y = \sqrt{x}$. In other words, when the offspring displaces full siblings (as is probably often the case), the offspring's equilibrial sex ratio if $1:\sqrt{x}$, while the parent's equilibrial sex ratio is $1:x$. These values, as well as the offspring's equilibrial sex ratio where $r_0 = \frac{1}{4}$, are plotted in Fig. 4. The same general solution holds if parents invest more in females by a factor of $x$, except that the resulting sex ratios are then reversed (e.g. $\sqrt{x}:1$ instead of $1:\sqrt{x}$).

## References

Alexander, G. (1960). Maternal behaviour in the Merino ewe. *Anim. Prod.* 3, 105–14.

DeVore, I. (1963). Mother-infant relations in free-ranging baboons, p. 305–335. In H. Rheingold (ed.), *Maternal behavior in mammals.* Wiley, N.Y.

Drury, W. H., and Smith, W. J. (1968). Defense of feeding areas by adult Herring Gulls and intrusion by young. *Evolution* 22, 193–201.

Eberhard, M. J. W. (1969). The social biology of polistine wasps, *Misc. Publ. Mus. Zool. Univ. Mich.* 140, 1–101.

Elliott, B. (1969). Life history of the Red Warbler. *Wilson Bull.* 81, 184–95.

Ewbank, R. (1964). Observations on the suckling habits of twin lambs. *Anim. Behav.* 12, 34–7.

—— (1967). Nursing and suckling behaviour amongst Clun Forest ewes and lambs. *Anim. Behav.* 15, 251–8.

Fisher, R. A. (1930). *The genetical theory of natural selection.* Clarendon Press, Oxford.

Gill, J. C., and Thomson, W. (1956). Observations on the behavior of suckling pigs. *Anim. Behav.* 4, 46–51.

Hamilton, W. D. (1964). The genetical evolution of social behavior. *J. Theoret. Biol.* 7, 1–52.

—— (1971). The genetical evolution of social behavior, p. 23–39. Reprinted, with addendum. In G. C. Williams (ed.), *Group selection.* Aldine-Atherton, Chicago.

—— (1972). Altruism and related phenomena, mainly in social insects. *Ann. Rev. Ecol. Syst.* 3, 193–232.

Hinde, R. A., and Spencer-Booth, Y. (1967). The behaviour of socially living rhesus monkeys in their first two and a half years. *Anim. Behav.* 15, 169–96.

—— and —— (1971). Effects of brief separation from mother on rhesus monkeys. *Science* 173, 111–18.

—— and Davies, L. M. (1972a). Changes in mother-infant relationships after separation in rhesus monkeys. *Nature, Lond.* 239, 41–2.

—— and —— (1972b). Removing infant rhesus from mother for 13 days compared with removing mother from infant. *J. Child Psychol. Psychiat.* 13, 227–37.

Jay, P. (1963). Mother–infant relations in langurs, p. 282–304. In H. Rheingold (ed.), *Maternal behavior in mammals.* Wiley, N.Y.

Le Boeuf, B. J., Whiting, R. J. and Gantt, R. F., (1972). Perinatal behavior of northern elephant seal females and their young. *Behaviour* 43, 121–56.

Munro, J. (1956). Observations on the suckling behaviour of young lambs. *Anim. Behav.* 4, 34–6.

Mussen, P. H., Conger, J. J., and Kagan, J. (1969). *Child development and personality.* 3rd ed. Harper and Row, N.Y.

Rheingold, H. (1963). Maternal behavior in the dog, p. 169–202. In H. Rheingold (ed.), *Maternal behavior in mammals.* Wiley, N.Y.

Rosenblatt, J. S., and Lehrman, D. S. (1963). Maternal behavior of the laboratory rat, p. 8–57. In H. Rheingold (ed.), *Maternal behavior in mammals.* Wiley, N.Y.

Rosenblum, L. A. (1971). The ontogeny of mother-infant relations in macaques, p. 315–67. In H. Moltz (ed.), *The ontogeny of vertebrate behavior*. Academic Press, N.Y.

Rowe, E. G. (1947). The breeding biology of *Aquila verreauxi* Lesson. *Ibis* 89, 576–606.

Schaller, G. B. (1964). Breeding behavior of the White Pelican at Yellowstone Lake, Wyoming. *Condor* 66, 3–23.

Schneirla, T. C., Rosenblatt, J. S., and Toback, E. (1963). Maternal behavior in the cat, p. 122–168. In H. Rheingold (ed.), *Maternal behavior in mammals*. Wiley, N.Y.

Skutch, A. F. (1961). Helpers among birds. *Condor* 63, 198–226.

Struhsaker, T. T. (1971). Social behaviour of mother and infant vervet monkeys (*Cercopithecus aethiops*). *Anim. Behav.* 19, 233–50.

Trivers, R. L. (1971). The evolution of reciprocal altruism. *Quart. Rev. Biol.* 46, 35–57.

—— (1972). Parental investment and sexual selection, p. 136–179. In B. Campbell (ed.), *Sexual selection and the descent of man, 1871–1971*. Aldine-Atherton, Chicago.

Wallace, L. R. (1948). The growth of lambs before and after birth in relation to the level of nutrition. *J. Agri. Sci.* 38, 93–153.

# Evolution and the theory of games

J. MAYNARD SMITH†

I want in this article to trace the history of an idea. It is beginning to become clear that a range of problems in evolution theory can most appropriately be attacked by a modification of the theory of games, a branch of mathematics first formulated by Von Neumann and Morgenstern (1944) for the analysis of human conflicts. The problems are diverse and include not only the behavior of animals in contest situations but also some problems in the evolution of genetic mechanisms and in the evolution of ecosystems. It is not, however, sufficient to take over the theory as it has been developed in sociology and apply it to evolution. In sociology, and in economics, it is supposed that each contestant works out by reasoning the best strategy to adopt, assuming that his opponents are equally guided by reason. This leads to the concept of a 'minimax' strategy, in which a contestant behaves in such a way as to minimize his losses on the assumption that his opponent behaves so as to maximize them. Clearly, this would not be a valid approach to animal conflicts. A new concept has to be introduced, the concept of an 'evolutionarily stable strategy'. It is the history of this concept I want to discuss.

## Evolution of the sex ratio

Consider first the evolution of the sex ratio. In most animals and plants with separate sexes, approximately equal numbers of males and females are produced. Why should this be so? Two main kinds of answer have been offered. One is couched in terms of advantage to the population. It is argued that the sex ratio will evolve so as to maximize the number of meetings between individuals of opposite sex. This is essentially a 'group selection' argument. The other, and in my view certainly correct, type of answer was first

†Reprinted with permission from the author and *American Scientist*, Vol. 64, No. 1, January–February 1976, pp. 41–45. Copyright © 1976 by Sigma Xi, The Scientific Research Society of North America, Incorporated.

put forward by Fisher (1930). It starts from the asssumption that genes can influence the relative numbers of male and female off-spring produced by an individual carrying the genes. That sex ratio will be favored which maximizes the number of descendants the individual will have and hence the number of gene copies transmitted. Suppose that the population consisted mostly of females: then an individual which produced only sons would have more grandchildren. In contrast, if the population consisted mostly of males, it would pay to have daughters. If, however, the population consisted of equal numbers of males and females, sons and daughters would be equally valuable. Thus a 1:1 sex ratio is the only stable ratio; it is an 'evolutionarily stable strategy'.

Fisher allowed for the fact that the cost of sons and daughters may be different, so that a parent might have a choice, say, between having one daughter or two sons. He concluded that a parent should allocate equal resources to sons and daughters. Although Fisher wrote before the theory of games had been developed, his theory does incorporate the essential feature of a game—that the best strategy to adopt depends on what others are doing. Since that time, it has been realized that genes can sometimes influence the chromosome or gamete in which they find themselves, so as to make that gamete more likely to participate in fertilization. If such a gene occurs on a sex-determining (X or Y) chromosome, then highly aberrant sex ratios can evolve.

More immediately relevant are the strange sex ratios in certain parasitic hymenoptera (wasps and ichneumonids). In this group of insects, fertilized eggs develop into females and unfertilized eggs into males. A female stores sperm and can determine the sex of each egg she lays by fertilizing it or leaving it unfertilized. By Fisher's argument, it should still pay a female to produce equal numbers of sons and daughters. More precisely, it can be shown that if genes affect the strategy adopted by the female, then a 1:1 sex ratio will evolve. Some parasitic wasps lay their eggs in the larvae of other insects, and the eggs develop within their host. When adult wasps emerge, they mate immediately before dispersal. Such species often have a big excess of females. This situation was analyzed by Hamilton (1967). Clearly, if only one female lays eggs in any given larva, it would pay her to produce one male only, since this one male could fertilize all his sisters on emergence. Things get more complicated if a single host larva is found by two parasitic females, but the details of the analysis do not concern us. The important point is that Hamilton looked for an

'unbeatable strategy'—that is, a sex ratio which would be evolutionarily stable. In effect, he used Fisher's approach but went a step
farther in recognizing that he was looking for a 'strategy' in the
sense in which that word is used by game theorists.

### Animal contests and game theory

A very similar idea was used by the late G. R. Price in an analysis
of animal behavior. Price was puzzled by the evolution of ritualized behavior in animal contests—that is, by the fact that an
animal engaged in a contest for some valuable resource does not
always use its weapons in the most effective way. Examples of
such behavior have been discussed by Lorenz (1966), Huxley
(1966), and others. It seems likely that ethologists have under-
estimated the frequency and importance of escalated, all-out
contests between animals (see particularly Geist 1966). Yet
display and convention are a common enough feature of animal
contests to call for some explanation. Both Lorenz and Huxley
accepted group selectionist explanations: Huxley, for example,
argued that escalated contests would result in many animals being
seriously injured, and 'this would militate against the survival of
the species'. Similar assumptions are widespread in ethology,
although not often so clearly expressed.

Price was reluctant to accept a group selection explanation. It
occurred to him that if animals adopted a strategy of 'retaliation',
in which an animal normally adopts conventional tactics but
responds to an escalated attack by escalating in return, this might
be favored by selection at the individual level. He submitted a
paper to *Nature* arguing this point, which was sent to me to
referee. Unfortunately the paper was some fifty pages in length
and hence quite unsuitable for *Nature*. I wrote a report saying that
the paper contained an interesting idea, and that the author should
be urged to submit a short account of it to *Nature* and/or to
submit the existing manuscript to a more suitable journal. I then
thought no more of the matter until, about a year later, I spent
three months visiting the department of theoretical biology at
Chicago. I decided to spend the visit learning something about the
theory of games, with a view to developing Price's idea in a more
general form and applying it to certain other problems. I was at
that time familiar with Hamilton's work on the sex ratio (indeed,
the work formed part of his Ph.D. thesis, of which I was the

external examiner), but I had not seen its relevance to Price's problem.

While in Chicago, I developed the formal definition of an evolutionarily stable strategy which I will give in a moment and applied it to the 'Dove–Hawk–Retaliator' and 'War of Attrition' games. I also realized the similarity between these ideas and the work of Hamilton (and also MacArthur 1965) on the sex ratio. When I came to write up this work, it was clearly necessary to quote Price. I was somewhat taken aback to discover that he had never published his idea and was now working on something else. When I returned to London I contacted him, and ultimately we published a joint paper (Maynard Smith and Price 1973) in which the concept of an evolutionarily stable stragegy was applied to animal contests.

At this point it will be convenient to describe some ideas from the theory of games. By 'game' or 'contest' is meant an encounter between two individuals (I am not concerned with $n$-person games) in which the various possible outcomes would not be placed in the same order of preference by the two participants: there is a conflict of interest. By 'strategy' is meant a complete specification of what a contestant will do in every situation in which he might find himself. A strategy may be 'pure' or 'mixed'; a pure strategy states 'in situation $A$, always do $X$'; a mixed strategy states 'in situation $A$, do $X$ with probability $P$ and $Y$ with probability $Q$'. Suppose that there are three possible strategies, $A$, $B$, and $C$. A 'payoff matrix' is then a $3 \times 3$ matrix listing the expected gains to a contestant adopting these three strategies, given that his opponent adopts one of the other strategies.

These ideas will be made clearer by an example. Consider the children's game 'Rock–Scissors–Paper'. In each contest, a player must adopt one of these 'strategies' in advance: then Rock blunts Scissors, Scissors cuts Paper, and Paper wraps Rock. Suppose that the winner of each contest receives one dollar from the loser; if both adopt the same strategy, no money changes hands. The payoff matrix is

|   | $R$ | $S$ | $P$ |
|---|---|---|---|
| $R$ | 0 | +1 | −1 |
| $S$ | −1 | 0 | +1 |
| $P$ | +1 | −1 | 0 |

The payoffs are to the player on the left. This particular game is a 'zero-sum' game, in the sense that what one player wins the other loses; in general the games considered below are not of this kind.

Clearly, a player adopting the pure strategy 'Rock' will lose in the long run, because his opponent will catch on and play 'Paper'. A player adopting the mixed strategy '⅓ Rock, ⅓ Scissors, ⅓ Paper' will break even.

How are these ideas to be applied to animal contests? A genotype determines the strategy, pure or mixed, that an animal will adopt. Suppose an animal adopts strategy $I$ and his opponent strategy $J$; then the payoff to $I$ will be written $E_J(I)$, where $E$ stands for 'expected gain'. This payoff is the change in $I$'s fitness as a result of the contest, fitness being the contribution to future generations.

## Evolutionarily stable strategy

We are now in a position to define an evolutionarily stable strategy, or ESS for short. Suppose that a population consists of individuals adopting strategies $I$ or $J$ with frequencies $p$ and $q$, where $p + q = 1$. What is the fitness of an individual adopting strategy $I$?

$$\text{Fitness of } I = p \cdot E_I(I) + q \cdot E_J(I)$$
$$\text{Fitness of } J = p \cdot E_I(J) + q \cdot E_J(J).$$

If a particular strategy, say $I$, is to be an ESS, it must have the following property. A population of individuals playing $I$ must be 'protected' against invasion by any mutant strategy, say $J$. That is *when* $I$ *is common*, it must be fitter than any mutant. That is $I$ is an ESS, if for all $J \neq I$,

$$either \ \ E_I(I) > E_I(J)$$
$$or \ \ E_I(I) = E_I(J)$$
$$and \ \ E_J(I) > E_J(J) \tag{1}$$

If these conditions are satisfied, then a population of individuals playing $I$ is stable; no mutant can establish itself in such a population. This follows from the fact that when $q$ is small, the fitness of $I$ is greater than the fitness of $J$.

It is important to emphasize at this point that the ESS is not necessarily the same as the strategy prescribed by game theorists for human players. There the assumption is that a player will adopt that strategy which minimizes his losses, given that his opponent plays so as to maximize them. Lewontin (1961) applies such 'minimax' strategies to evolution. He was concerned with a contest not between individuals but between a species and 'nature'. The objective of a species is to survive as a species—to avoid extinction. It should therefore adopt that strategy which minimizes its chances of extinction, even if nature does its worst. That is, the species must adopt the minimax strategy. For example, a species should retain sexual reproduction rather than parthenogenesis, because this will enable it to evolve to meet environmental change. This is clearly a group selectionist approach; the advantage is to the species and not to the individual female. In contrast, the concept of an evolutionarily stable strategy is relevant to contests between individuals, not between a species and nature, and is concerned solely with individual advantage.

Let us now apply these ideas to a particular problem. Suppose that two animals are engaged in a contest for some indivisible resource which is worth $+V$ to the victor. An animal can 'display', it can 'escalate'—in which case it may seriously injure its opponent—or it can retreat, leaving its opponent the victor. Serious injury reduces fitness by $-W$ (a 'wound') and forces an animal to retreat. Finally, a long contest costs both animals $-T$. The two simplest strategies are

*Hawk.* Escalate, and continue to do so until injured or until opponent retreats.

*Dove.* Display. Retreat if opponent escalates, before getting injured.

We suppose that two Hawks are equally likely to be injured or to win. We also suppose that two Doves are equally likely to win, but only after a long contest costing both of them $-T$. The payoff matrix is then

$$
\begin{array}{ccc}
 & H & D \\
H & \tfrac{1}{2}(V-W) & V \\
D & 0 & \tfrac{1}{2}V - T
\end{array}
$$

If $W > V$, then there is no pure ESS. Thus $H$ is not an ESS, because $E_H(H) < E_H(D)$, and $D$ is not an ESS, because $E_D(D) < E_D(H)$. The only ESS is

$H$ with probability $(2T + V)/(2T + W)$
$D$ with probability $1 - (2T + V)/(2T + W)$

Thus at an evolutionary equilibrium the population will consist of a mixture of Hawks and Doves. Price's suggestion was that a third stragety, 'Retaliator', $R$, might be an ESS; $R$ plays $D$ against $D$ and $H$ against $H$. The payoff matrix is

|   | $H$ | $D$ | $R$ |
|---|---|---|---|
| $H$ | $\frac{1}{2}(V-W)$ | $V$ | $\frac{1}{2}(V-W)$ |
| $D$ | 0 | $\frac{1}{2}V-T$ | $\frac{1}{2}V-T$ |
| $R$ | $\frac{1}{2}(V-W)$ | $\frac{1}{2}V-T$ | $\frac{1}{2}V-T$ |

It turns out that Price was right. Thus a population of $R$ is stable against invasion by mutant $H$, because $E_R(R) > E_R(H)$. $R$ is not stable against $D$, because in the absence of $H$ they are identical. But a population consisting initially of a mixture of $R$, $D$, and $H$ will evolve to $R$. This game is analyzed further in Maynard Smith and Price (1973) and by Gale and Eaves (1975).

This analysis suggests that we would expect to find retaliation a feature of actual behaviour. One example must suffice: a rhesus monkey which loses a fight will passively accept incisor bites but will retaliate viciously if the winner uses its canines (Bernstein and Gordon 1974).

One assumption made above—that two Doves can settle a contest—needs some justification. Why don't they go on forever? Consider the following game. Two players, $A$ and $B$, can only display. The winner is the one who goes on for longer; the only choice of strategy is how long to go on for. $A$ selects time $T_A$ and $B$ selects $T_B$. The longer the contest actually continues, the more it costs the players; the costs associated with these times are $m_A$ and $m_B$. If $T_A > T_B$, then we have

$$\text{payoff to } A = V - m_B$$
$$\text{payoff to } B = -m_B$$

The cost of $m_A$ which $A$ was prepared to pay is irrelevant, provided that it is greater than $m_B$. Our problem then is: 'How should a player choose a time, and a corresponding value of $m$? More precisely what choice of $m$ is an ESS? For obvious reasons, I have called this the 'War of Attrition'. Clearly, no pure strategy can be an ESS. Any population playing $m$, say, could be invaded by a mutant playing $M$, where $M > m$; if $m > V/2$, it could also be invaded by a mutant playing 0. It can be shown that there is a mixed ESS given by

$$p(x) = \frac{1}{V}e^{-x/V} \tag{2}$$

where $p(x)\delta x$ is the probability of playing $m$ between $x$ and $x + \delta x$.

What does this mean? There are two possible ways in which an ESS of this kind could be realized. First, all members of the population might be genetically identical and have a behavior pattern which varied from contest to contest according to eqn 2. Second, the population might be genetically variable, with each individual having a fixed behavior, the frequencies of different kinds of individuals being given by eqn 2. In either case, the population would be at an ESS.

G. A. Parker (1970) has described a situation which agrees rather well with eqn 2. Female dung flies of the genus *Scatophaga* lay their eggs in cow pats. The males stay close to the cow pat, mating with females as they arrive to lay their eggs. What strategy should a male adopt? Should he stay with a pat once he has found one, or should he move on in search of a fresh pat as soon as the first one begins to grow stale? This is comparable to the choice of a value of $m$ in the War of Attrition. His choice will be influenced by the fact that females arrive less frequently as a pat becomes staler. His best strategy will depend on what other males are doing. Thus if other males leave a pat quickly, it would pay him to stay on, because he would be certain of mating any females which do come. If other males stay on, it would pay him to move.

Parker found that the actual length of time males stayed was given by a distribution resembling eqn 2. By itself this means little, because it is the distribution one would expect if every male had the same constant probability of leaving per unit time. It is the typical negative exponential distribution expected for the 'survivors' of a population suffering a constant 'force of mortality'. What is significant is that Parker was able to show that the expected number of matings was the same for males which left early as for those which stayed on. This means that the males are adopting an ESS; natural selection has adjusted the probability of leaving per unit time to bring this about.

It is not known whether contests between pairs of animals, in which only display is employed, show the appropriate variation in length. It will be interesting to find out.

A major complication in applying these ideas in practice arises because most contests are asymmetrical (Parker 1974; Maynard Smith and Parker 1976), either in the fighting ability of the contestants (i.e. in what Parker has called 'resource-holding power', or RHP), or in the value of the resource to the contestants (i.e. in payoff). Clearly, these asymmetries can only affect the strategies adopted if they are known to the contestants. Thus suppose two animals differ in size, and hence in RHP, but have no way short of escalation of detecting the difference. Then the difference cannot alter their willingness to escalate (i.e. their strategy), although it would affect the outcome of an escalated contest.

In some cases an asymmetry may be clearly perceived by both contestants but have relatively little effect on RHP or on payoff. The obvious example is the asymmetry between the 'owner' of a resource (e.g. a territory, a female, an item of food) and a 'late-comer'. There is no general reason why an owner should have a higher RHP than a latecomer. The value of a resource will often be greater to the owner, but, as I shall show in a moment, no such difference is necessary but an asymmetry can be used to settle a contest conventionally.

To fix ideas, consider the game of Hawks and Doves discussed above, with the arbitrary value $V = 50, W = 100$, and $T = 10$. The payoff matrix for the symmetrical case is then

|   | H | D |
|---|---|---|
| H | −20 | +50 |
| D | 0 | +20 |

Suppose that an animal may be either the owner of the resource or a latecomer, for a particular animal is equally likely to find himself playing either role. Consider now the strategy $I$: 'Play $H$ if you are owner; play $D$ if you are latecomer'. Then, since an animal is owner and latecomer with equal frequency,

$$E_I(I) = \tfrac{1}{2} \times 50 + \tfrac{1}{2} \times 0 = +25.$$

Now let $J$ be a strategy which ignores the asymmetry and plays $H$ with probability $p$ and $D$ with probability $(1 - p)$. Then

$$E_I(J) = \tfrac{1}{2}[50p + 20(1 - p)] + \tfrac{1}{2}[-20p] = 5p + 10.$$

For any value of $p$, $E_I(I) > E_I(J)$. Thus the strategy $I$, which amounts to the conventional acceptance of ownership, is an ESS against any strategy which ignores ownership. (Notice that the game permits the alternative ESS, 'play $H$ if you are latecomer; play $D$ if you are owner'. This raises difficulties which are discussed by Maynard Smith and Parker 1976.)

It follows that conventional acceptance of ownership can be used to settle contests even when there is no asymmetry in payoff or RHP, provided that ownership is unambiguous. Some actual examples will help to illustrate this point.

## The ESS in practice

The hamadryas baboon, *Papio hamadryas*, lives in troops composed of a number of 'one-male groups', each consisting of an adult male, one or more females, and their babies. The male, who is substantially larger than the females, prevents 'his' females from wandering away from his immediate vicinity; a female rapidly comes to recognize this 'ownership'. It is rare for an owning male to be challenged by another. How is this state of affairs maintained?

Kummer (1971) describes the following experiment. Two males, previously unknown to each other, were placed in an enclosure; male *A* was free to move about the enclosure whereas male *B* was shut in a cage from which he could see what was happening but not interfere. A female strange to both males was then loosed into the enclosure. Within 20 minutes male *A* had convinced the female of his ownership, so that she followed him about. Male *B* was then released into the enclosure. He did not challenge male *A*, but kept well away from him, accepting *A*'s ownership.

These observations can be explained in two ways. First, male *B* may have been able to detect that male *A* would win an escalated contest if challenged; second, there may be a conventional acceptance of ownership, for the reasons outlined above. Kummer was able to show that the second explanation is correct. Two weeks later, he repeated the experiment with the same two males but with a different female, but on this occasion male *B* was loose in the enclosure and male *A* confined. Male *B* established ownership of the female and was not challenged by *A*.

One last observation is relevant. If a male is removed from a troop, his females will be taken over by other males. If after some weeks the original male is reintroduced, an escalated fight occurs; both males now behave as 'owners'.

It could be argued that in a hamadryas baboon there is a difference in payoff, because when a male first takes over a new female he has to invest time and energy in persuading her to accept his ownership. This is probably correct, although the theoretical analysis shows that no such difference is required for the establishment of an ESS based on conventional acceptance of ownership. An asymmetry in payoff is less likely in the anubis baboon (Packer 1975). In this species, there is a fairly stable male dominance hierarchy for food but not for females. Females are not the permanent property of particular males; instead, a male 'owns' a female only for a single day—or for several days if he can prevent her from moving away during the night. Once in temporary possession of a female, a male is not challenged, even by those above him in the dominance hierarchy. Why should contests about food and females be settled differently? One possible explanation is that the ownership principle could not be used to settle contests over food, because it must often be the case that two animals see a food item almost simultaneously. Ownership would be ambiguous; two animals would both regard themselves as owners of the same item, and escalated contests would ensue.

This last possibility is beautifully illustrated by the work of L. Gilbert (personal communication) on the swallowtail butterfly. *Papilio zelicaon*. Because this is a relatively rare butterfly, the finding of a sexual partner presents a problem. This problem is solved by 'hilltopping'. Males establish territories at or near the tops of hills, and virgin females fly uphill to mate. There are, however, more males than hilltops, so most males must accept territories lower down the slopes. They attempt to waylay females on their way up and, although they sometimes succeed, the evidence suggests that the male actually at the hill top mates most often. Gilbert marked individual males and observed that a strange male did occasionally arrive at a hilltop and challenge the owner, but the stranger invariably retreated after a brief 'contest'.

As in Kummer's experiments with baboons, we have to choose between two explanations. Either the owner of a hilltop is a particularly strong butterfly, and this fact is perceived by the challenger during a brief contest, or there is again an ESS based on conventional acceptance of ownership. Gilbert showed that the latter explanation is correct by an experiment analogous to that of removing a male hamadryas baboon from a troop and then restoring it. He allowed two male butterflies to occupy a hilltop on alternate days, keeping each in the dark on their off days. After two weeks, when both males had come to regard themselves as owners of the same hilltop, he released them on the same day. A contest lasting many minutes and causing damage to the contestants ensued.

Much has been left out of these simple models. Contests in which only partial information about asymmetries is available to the contestants, or in which information is acquired in the course of a contest, are discussed by Maynard Smith and Parker (1976). The same paper discusses the possibility of 'bluff'—that is, the possession of structures such as manes, ruffs, or crests, which increase apparent RHP without an equivalent increase in actual fighting ability.

I suggested at the beginning of this article that the concept of an ESS is also relevant to the evolution of ecosystems; this idea is developed by Maynard Smith and Lawlor (in press). It is impossible to do more here than indicate the nature of the problem. In nature, animals and plants compete for resources—food, space, light, etc. Genetic changes in an individual can alter its 'choice' of resources: for example, the food items taken by an animal, or the time of year a plant puts out its leaves. Individuals will choose their resources so as to maximize their fitness. The best choice will depend on what other individuals, of the same and other species, are doing. If everyone else is eating spinach it will pay to concentrate on cabbage; since most forest trees put out their leaves late in spring, it pays forest herbs to put out leaves early.

Since the appropriate strategy for an individual depends on what others are doing, we are again concerned with the search for an ESS. Lawlor and I conclude that two competing species will tend to become specialists on different resources even though in isolation each species would be a generalist.

This conclusion is not a new one: it accords with a good deal of observational data and has received several previous theoretical treatments. We would claim, however, that we have clarified a familiar idea and set it in a wider context. That wider context is simply this: whenever the best strategy for an individual depends on what others are doing, the strategy actually adopted will be an ESS.

### References

Berstein, I. S., and Gordon, T. P. (1974). The function of aggression in primate societies. *Amer. Sci.* 62, 304—11.
Fisher, R. A. (1930). *The genetical theory of natural selection.* Oxford University Press.
Gale, J. S., and Eaves, L. J. Logic of animal conflict. *Nature* 254, 463—4.
Geist, V. (1966). The evolution of horn-like organs. *Behaviour* 27, 175—213.
Hamilton, W. D. (1967). Extraordinary sex ratios. *Science,* 156, 477—88.
Huxley, J. S. (1966). Ritualization of behaviour in animals and man. *Phil. Trans. Roy. Soc. B* 251, 249—71.
Kummer, H. (1971). *Primate Societies.* Aldine-Atherton, Chicago.
Lewontin, R. C. (1961). Evolution and the theory of games. *J. Theoret. Biol.* 1, 382—403.
Lorenz, K. (1966). *On aggression.* Methuen, London.
MacArthur, R. H. (1965). In T. Waterman and H. Morowitz, eds., *Theoretical and mathematical biology.* Blaisdell, N.Y.
Maynard Smith, J., and Lawlor, L. R. (in press). The coevolution and stability of competing species. *Amer. Natur.*

—— and Parker, G. A. (1976). The logic of asymmetric contests. *Anim. Behav.* **24**, 159—75.

—— and Price, G. R. (1973). The logic of animal conflicts. *Nature* **246**, 15—18.

Packer, C. (1975). Male transfer in olive baboons, *Nature* **255**, 219—20.

Parker, G. A. (1970). The reproductive behaviour and the nature of sexual selection in *Scatophaga stercoraria* L. II. *J. Anim. Ecol.* **39**, 205—28.

—— (1974). Assessment strategy and the evolution of animal conflicts. *J. Theoret. Biol.* **47**, 223—43.

Von Neumann, J., and Morgenstern, O. (1944). *Theory of games and economic behaviour.* Princeton University Press.

# Assessment strategy and the evolution of fighting behaviour

G. A. PARKER†

## Introduction

There is much in favour of viewing a great deal of animal behaviour as optimum strategies for maximizing the rate of extraction of 'fitness gain' from the available series of 'fitness gain parameters' (resources) present in its environment. One consequence of the occurrence of discontinuously distributed resources is that they may be in short supply. Animal aggression (in the form of resource guarding) will be favoured by selection when there are less resources than competitors and where an individual can achieve an immediate gain in fitness by forcibly ousting one of its fellows. Selection for aggression will be more intense the more discrete the resource (i.e. the easier it is to guard) and the higher its yield as a fitness gain parameter (a function both of its absolute effect and its shortness of supply). It is not surprising therefore that most of animal aggression relates to food fighting and especially to mating. Territoriality is often merely an adjunct to these two situations— e.g. an area is guarded because it has a high probable yield of food or mates, or both. Fighting tendency will be much modified by the probable relatedness of the two competing individuals, an effect studied by Hamilton (1964).

Darwin (1871) was very well aware of the individual advantages of aggression when he founded the theory of sexual selection. Since then it has become fashionable amongst certain ecologists and ethologists to view aggression and territory in terms of advantages it may confer upon groups or species, rather than on individuals (see Wynne-Edwards 1962). The fact that much aggression is highly ritualized (as displays, pushing contests, etc.) and does not involve damage (termed 'conventional fighting', Maynard Smith 1972) has fuelled 'group selectionism' because it can be argued that an immediate advantage would be conferred on any individual which indulged in damaging or escalated fighting. Group selection

† Reprinted with permission from the author and *Journal of Theoretical Biology*, **47** (1974). Copyright. © by Academic Press Inc.

poses major problems in terms of modern population genetics, and it seems likely to be a very weak selective agent compared to individual selection (see Williams 1971). An excellent account of the position of fighting in relation to group selection and individual selection can be found in Maynard Smith (1972).

Recently (Maynard Smith 1972; Maynard Smith and Price 1973) it has become abundantly clear that there is no conflict between observed fighting strategy and individual selection. Of a number of possible strategies, it can be shown that the only one to form an evolutionary stable strategy (ESS—i.e. where, if most of the individuals in the population adopt it there is no other strategy which would give higher reproductive fitness) is one where individuals start conventionally but escalate to damaging fights later, especially when the opponent escalates. These 'limited war' strategies appear stable against 'total war' or 'total peace' strategies when analyzed in relation to game theory.

In the present paper, fighting strategy is again considered in relation to individual selection. The view that the 'retaliate if opponent escalates' will initially form an ESS is accepted. Further adaptations, once this strategy has stabilized, are examined, in particular the theory that relative strengths of combatants are estimated during displays; a suggestion which recurs continually in the literature (see Ewer 1968) but which has attracted very little consideration in evolutionary terms.

## 2. Conventional fighting as assessment of RHP (Resource Holding Power)

Once 'retaliator' has stabilized as an ESS, any mutant individual able to assess from the conventional fighting stage how its own RHP (resource holding power) compares with that of its opponent would have a selective advantage, since it could withdraw without damage when the RHP of its opponent exceeds its own by a sufficiently large amount. It is assumed that RHP is a measure of the absolute fighting ability of a given individual. If this character spreads, we may end up with a 'total peace' strategy, where all disputes are settled conventionally. In this case, provided that the characteristic of retaliation is not lost, a mutant deficient in responding to the signals of RHP during conventional fighting will not spread—it will be disadvantageous since it will not gain any extra resources and will be beaten in encounters with individuals of higher RHP. Thus our 'conventional assessor/retaliator' becomes the ESS.

It has certain problems to face, however. Firstly, there is the obvious difficulty that selection will immediately favour exaggeration of those cues used to assess RHP. The selective advantage of this form of 'evolutionary cheating' is simple; if (for example) size is used as the cue for RHP, then where for other reasons it is disadvantageous to increase absolute size (and RHP), what will be favoured are mechanisms to increase apparent size (and therefore apparent RHP). That this has happened often seems very likely. The canid threat posture involves raising the neck hair and standing erectly (Darwin 1872), so does that of many other groups including rodents (Eibl-Eibesfeldt 1970). Lions have manes, fish often raise fins, birds fluff out feathers, certain species have inflatable pouches (e.g. the lizard *Phenacosaurus richteri*; Kästle 1963). Examples are legion. Another cue very commonly used could be weaponry. Much of threat display involves exaggeration and display of teeth, antlers, claws (e.g. in crabs; Crane 1966) or even hind legs, which are the main defensive weapons in locusts (Parker, Hayhurst, and Bradley 1974). It seems quite likely that these features might initially have given good indices of RHP. Where there is this type of drive for 'evolutionary cheating', a counter-selective compensatory adustment of the assessment mechanism would continually follow in its wake.

More reliable measures of RHP might be provided by direct trials of strength between combatants. Pushing and pulling contests, on head and/or tail beating clashes abound in all groups from fish to ungulates (for many classic examples see Eibl-Elbesfeldt 1970). Very often conventional fighting consists of combinations of 'unreliable' display and 'reliable' contests of strength, implying that many cues may be used to assess relative RHP. An independent analysis has been made of contests carrying no information of RHP by Maynard Smith (1974). He shows that selection will here mainly favour persistence during displays. Persistence durations should stabilize at a negative exponential distribution, but should all be of constant intensity. The present paper mainly concerns contests which provide RHP information. Though the 'conventional assessor–retaliator' theory ascribes a definite selective advantage to the display behaviour, it poses an alternate problem— why do damaging fights ever occur? With a nearly perfect assessment mechanism one would predict escalation only where combatants are very closely matched, and there is no clear-cut predictive outcome. We shall consider this problem later (section 4) but first let us consider how assessment might operate and whether the simple comparison of RHPs provides a satisfactory solution for aggressive behaviour.

Clearly, this model is a naive one, and certain behavioural observations do not conform to its prediction. Very often the odds appear heavily weighted in favour of the resource holder, and the absolute RHP (as judged by human eyes) of the attacker apparently has to exceed that of the holder very considerably before a take-over occurs. There are several possible reasons why this effect operates:

(1)   Suppose that the 'resource structuring' is near perfect; i.e. by the assorting action of disputes, the population is perfectly truncated with the highest RHP individuals occupying all the resources. In this case, restructuring can only occur by inputs and outputs of competitors and resources or by changes in RHP status of individuals in one or both of the two groups (holders or non-holders). Here most of the observed disputes would obviously be won by the holders.

(2)   The tenure of the resource may well itself increase RHP, especially where the resource is a food source. Also the outcome of a fight may involve experience of the local environment, hence tenure of the resource may increase RHP in this way. Position of the holder in guarding the resource may be very important, for instance, in the female-guarding behaviour where the male clings to the female (see Parker 1974) because the attacker must prise the holder off before a take-over can occur.

(3)   Pay-offs may be different for the holder and attacker. Suppose that the holder will lose more than the attacker will gain, it might be expected that the holder could afford to sacrifice more units of fitness in the fights than the attacker could afford to expend. Hence where the combatants are of equal RHP, the attacker should withdraw because it will run out of expendable fighting units before the holder. Hence an attacker must be of higher RHP before it can win.

The last suggestion causes us to modify our suggestion about the type of assessment favoured by selection, and leads to a less naive model. Individuals will be favoured which respond appropriately to the correct threshold of RHP prediction before they withdraw. This threshold, which will be set by evolution, will not be simply 'does his RHP exceed mine?'; rather, it will be 'given that his RHP is $x$ and mine $y$, and that in this situation I have $a$ units available to expend and he has $b$ units, will I run out of expendable fitness units before he does?' It will be the probable relative rates at which the combatants will expend fitness during an interaction which will (via selection) set the appropriate thresholds for withdrawal, since this will determine which individual will expend its fighting budget first (and hence lose). We shall consider this less naive model more fully in section 4, but first we shall consider how pay-offs depend on whether one is a holder or attacker, and circumstances in which imbalances in pay-offs arise.

## 3. Pay-off imbalances between holder and attacker

Let us assume for the present section that we are considering two average individuals of equal RHP, and that one has held a resource for a certain time ($t$) before it is encountered by an attacker. What we wish to estimate is the change in fitness sustained by each individual as a result of a possible change in state (i.e. a take-over or a withdrawal). Where the holder's loss in fitness exceeds the attacker's gain, then the attacker should show a greater tendency to withdraw because it will have less fitness units available to expend in the fight. This model is compatible with an adaptive interpretation of motivation. It is suggested that motivational state will be a function of the 'fitness change effect' achieved by shifting from one motivational state to another.

Take a very simple case where a resource can be extracted at a constant rate of gain $g$ fitness units through time 1, summating to $g$ by the end of the extraction. The interaction occurs at time $t$ during extraction, $(1 - t)$ time units before the resource will finish. If there is an escalated fight, the individuals will be damaged and the winner's gain rate reduced to $w$ for the rest of the time. (This arises mainly because the RHP will have declined, though other reproductive disadvantages may also have been incurred.) Let the search cost before finding the resource $= s_h$ for the holder and $s_a$ for the attacker (these values really summarize the fitness of each individual before it encountered the resource). If withdrawal occurs before an escalated fight, the withdrawing individual achieves a probable gain rate of $p_g$, if it occurs after an escalated fight, the probable gain rate is $p_w$. We can therefore summarize all the possible fitness outcomes to each individual (by the end of resource extraction) in the following way:

|  | Fitness if fights | Fitness if withdraws without damaging fight |
|---|---|---|
| **Holder** | | |
| Wins: | $gt + w(1-t) - s_h$    (1) | $gt + p_g(1-t) - s_h$    (3) |
| Loses: | $gt + p_w(1-t) - s_h$    (2) | |
| **Attacker** | | |
| Wins: | $w(1-t) - s_a$    (4) | $p_g(1-t) - s_a$    (6) |
| Loses: | $p_w(1-t) - s_a$    (5) | |

Suppose that the combatants escalate. The change in fitness of the holder if it loses $= \Delta h = (1)-(2) = w(1-t) - p_w(1-t)$; and that for the attacker if it wins $= \Delta a = (4)-(5) = w(1-t) - p_w(1-t)$. Hence $\Delta h = \Delta a$, there is no imbalance in pay-offs and therefore no clearcut predictive outcome. If motivation to fight is proportional to $\Delta h$ or $\Delta a$, both should be equally motivated.

We can ask a second question: are both individuals equally motivated to withdraw without fighting? Do both experience the same change in fitness for the choice between withdrawal without escalation rather than escalating? That is, what is the fitness change effect due to withdrawing during conventional fighting. For the holder, this is

$$\Delta hr = (1)/2 + (2)/2 - (3) = w(1-t)/2 + p_w(1-t)/2 - p_g(1-t),$$

and for the attacker $\Delta ar = (4)/2 + (5)/2 - (6)$, which is exactly the same value. Hence again the pay-offs are equal and there is no clearcut predictive outcome; motivations to withdraw will be equal. Both these conclusions are intuitively obvious from this simple model where the pay-off remaining after the interaction is of equal value to both combatants. For estimates of change in fitness, it is obvious that the fitness of the individual *before* the interaction is irrelevant (values $s_h$, $s_a$, and $gt$ cancel out). The best estimate of a 'fitness budget for fighting' is undoubtedly $\Delta hr - \Delta ar$ since this considers the disparity between the alternative strategies of *escalate* versus *withdraw without escalation*. We can now consider some more realistic imbalance situations.

*(A) Case 1: investment occurs before gain; fixed investment period*
Many examples of guarding involve a period of investment (expenditure of fitness) before gain can be extracted from the guarded resource. Males of many groups show precopulatory female-guarding phases until the female becomes receptive and mates. These are especially common in crustaceans (e.g. *Asellus, Gammarus, Talitrus, Orchestria, Artemia*, and copepods) and

in insects (reviewed by Parker 1970a). In the vertebrates, a territory or mate may often be guarded for some time before any obvious gain can be recognized. A holder can be said to have entered the gain phase when it has a probability of offsetting some of its investment (e.g. when it has begun to transfer sperm—so it will have a probable fertilization gain). Take-over (when it occurs) usually happens during the investment phase, possibly because it is generally longer and because take-over is often easier then.

(a) Suppose that the investment rate is constant. If take-over occurs, the attacker simply supplies the remaining investment necessary before gain ( = that which the holder would have put in after the interaction, had he won). It can be shown simply analytically after the method above that there is no imbalance here; $\Delta h = \Delta a$ and $\Delta hr = \Delta ar$, irrespective of whether take-over occurs during investment or gain. This may seem odd because the holder loses his existing investment as well as his possible gain. However, the attacker's net gain is correspondingly greater because he has to invest correspondingly less. Again the holder's possible loss = the attacker's possible gain.

(b) This will not be true where the investment rate is not constant. Suppose that if it won the holder would invest a total of $j_i$ fitness units after the interaction and the attacker would invest a total of $k_i$ units, before gain extraction. In this case, $\Delta h - \Delta a = k_i - j_i$ and $\Delta hr - \Delta ar =$ half this value. Clearly, when investment rate is highest at the start, $k_i > j_i$ and the holder has more expendable fitness units. This is not unfeasible for certain situations, e.g. where a territory must be elaborately marked and its characteristics learned, or where nests must be rebuilt, or females re-courted etc. In the reverse case (investment rate increasing with time) then $j_i > k_i$ and the odds will be weighted in favour of the attacker. This is an interesting possibility in cases where, for example, female-guarding or territory-holding means that males are impaired in their feeding activity. If this is so they may lose condition (and increase investment) at an accelerating rate with time. In the kob, males holding TGs (small territories, continually contested, to which the females come for mating) sustain a higher takeover rate than males on STs (much larger territories, little contested, where females are reluctant to mate)—probably because food is in very short supply in the TGs (Leuthold 1966). This is probably best regarded as an example of case 4 with $j$(net gain, $g - i$) < $k$(net gain, $g - i$) because the gain rate (probability of insemination) will be constant but the investment rate accelerating. Note that when $g = i$, the male should leave anyhow, with or without any contest. In locusts there is a precopulatory female-guarding phase and the take-over rate is much higher during oviposition than earlier in guarding (Parker, Hayhurst, and Bradley 1974). Oviposition is usually the last stage of guarding investment before copulation. It is very difficult, however, to determine whether or not this effect results from a decline of absolute RHP, or merely from the change in posture during oviposition.

For arthropod precopulatory guarding phases, it seems very likely that the main feature heavily weighting the odds in favour of the holder is that it initially has a major postural advantage.

It seems unlikely that the $j_i > k_i$ case could ever exert a major effect. If the probability of take-over before gain gets too high then the existing investment strategy becomes disadvantageous. This is one of the features which may be expected to stabilize drive for investment earlier and earlier before the pay-off (Parker 1974); in other words the $j_i:k_i$ relationship can never become heavily $j_i > k_i$ biassed because this is not evolutionarily stable. Note that this will not apply to the $j_i > k_i$ case.

*(B) Case 2: investment occurs before gain; attacker must reinvest*

Here there is an obvious imbalance in favour of the holder. If we assume a roughly constant investment rate summating to $i$ by the end of investment, and that the attacker must complete a full $i$ before collecting the gain, we find that $\Delta h - \Delta a = it + p_w t$ and $\Delta hr - \Delta ar = (it + p_w t)/2$. That is the holder has an extra number of expendable fitness units, equivalent to the difference between how much more he would have to invest until pay-off ($i(1 - t)$ units) and the full investment ($i$ units), *plus* the value of searching during the time ($t$) that the attacker is reinvesting ($p_w t$ units). If the interaction occurs during the gain phase, the imbalance is even greater, simply $i + p_w$ and $(i + p_w)/2$.

This case appears particularly relevant to postcopulatory guarding in insects (see Parker 1970a). Here the male copulates and then guards the female during oviposition. In the dung fly the sperm transfer phase can really be considered the investment phase because the last male to mate fertilizes most of the eggs (80%) and the female would certainly be quickly remated if left unguarded during oviposition. Hence the gain phase is the pay-off of ensuring precedence of one's ejaculate while eggs are being laid. Here a full reinvestment (full copulation) occurs if there is a take-over. In interactions, struggles (escalated fights?) are rare (about 7%) and usually the holder wins. However, it is again difficult to estimate the positional advantage of the holder. Take-overs are more frequent during oviposition than copulation, although this model would predict the opposite, because the imbalance is most favourable to the holder during the gain phase. This effect certainly relates to a relatively greater positional advantage during copulation (Parker 1970$b$). Positional advantage must be minimal in the non-contact post-copulatory guarding phases shown by many dragonflies, and though take-overs are relatively common, the odds are weighted in favour of the holder (Jacobs 1955), as we would predict. A full recopulation occurs after take-over, and guarding continues until the end of oviposition, as in dung flies.

## (C) Case 3: damage from fight permanent; combatants different ages

For many prolonged resource guarding situations (e.g. males in certain primate troops, lions, and certain ungulates, etc.) it seems very likely that the average age of holders will exceed that of attackers. In this case if the combatants fight and the damage persists to some extent throughout life, then it seems likely that the overall fitness of the younger combatant will be reduced more than that of the older one; it will have a longer part of its reproductive life in the reduced RHP condition. Suppose that $y$ and $z$ represent the proportions of reproductive life spent before the interaction by the holder and attacker respectively. They have therefore $(1 - y)$ and $(1 - z)$ left. $p_g$ and $p_w$ again represent the gain rates due to searching time in the undamaged and damaged states respectively. We know by the present analysis that what happens beforehand is irrelevant (i.e. $p_g y$, $p_g z$) because it cannot alter the *change* in fitness arising from the interaction. If we add $p_w(1 - y)$ to equations (1) and (2) and $p_w(1 - z)$ to (4) and (5) though $\Delta h = \Delta a$, there is a clear imbalance with $\Delta hr - \Delta ar = p_w(1 - y) - p_w(1 - z) + p_g(1 - z) - p_g(1 - y)$. This example indicates why $\Delta hr - \Delta ar$ gives the best indication of disparity in fitness budgets; it takes into account the value of the alternative strategy, *withdrawal without escalation*. If we signify the difference between the attacker's and holder's remaining reproductive life as $a$ [i.e. $(1 - z) - (1 - y)$], then $\Delta hr - \Delta ar$ is simply $p_g a - p_w a$ which is clearly positive if the attacker is younger than the holder. Thus the odds are again weighted in favour of the holder, because of the more disastrous effect of damage on the younger attacker. This certainly fits observed data if we assume that there will generally be older holders and younger attackers. Note that there will be no imbalance if the damage does not persist until the death of the older individual.

*(D) Case 4: non-constant extraction rate from time of start of extraction*

This is the converse of case 2(b). Suppose that if it won the holder would extract a total of $j_g$ fitness units after the interaction and the attacker would extract $k_g$ units. Here $\Delta h - \Delta a = j_g - k_g$, $\Delta hr - \Delta ar = (j_g - k_g)/2$, and so where $j_g > k_g$ the holder has the higher budget (i.e. where gain increases with time a given individual has been extracting). If the rate of gain is greatest at the start $(j_g < k_g)$, the attacker has the edge.

Now, for prolonged guarding as for example in certain feeding territories, it seems likely that learning the characteristics of the resource will increase the rate of uptake from the resource. Hence the remainder of the resource may be worth more to the holder than to the attacker, and so $j_g > k_g$ and the holder wins. However, this will not be true for temporary feeding resources, e.g. food-fighting situations. It seems highly likely here that the first few units of food intake will affect fitness disproportionately more than subsequent units. Thus the value of the extractable remainder of the resource will be higher to the attacker than to the holder $(j_g < k_g)$. This time the odds should be weighted towards the attacker. In many species there is a clear cut ordering of feeding with dominant male feeding first, then females, then young. This ordering corresponds exactly to the expected absolute RHPs of the individuals—highest first. There is some evidence that where the disparity between RHPs is not so obvious, the 'holder has precedence' effect does not apply. An example is the intra-specific food-fighting found in some birds. Often, if it is a single food item that is being contested, there is no obvious precedence and the interaction is a mixture of fighting and snatching. We would predict this if neither individual has yet extracted from the resource and both are equally hungry $(j_g = k_g)$. Even more interesting is the case of sparrows around pieces of bread (D. Barnes and G. A. Parker, unpublished observations). A 'holder' (the hungriest?) guards all or part of the resource some time while feeding and giving threat displays to contestants. However, even after a few pecks the odds become heavily weighted in favour of an attacker—which usually takes over the resource for a time until it itself is ousted. The same sort of pattern also appears to apply to starlings.

Group selectionists would probably interpret this as a mechanism which overall gives an even share out of the resource to all individuals and which therefore has an adaptive advantage at the species level. The individual selection interpretation is favoured since it is likely to form the greater selective agent.

*(E) Summary of imbalances*

Most of the cases considered are very inadequate for a given natural situation, which often consists of elaborate combinations of the above effects. Apart from the last example ($j_g < k_g$), holders should generally maintain tenure of their resources unless the attacker is of sufficiently higher RHP to offset the imbalance. Predictions from holder: attacker imbalances appear to fit observed data reasonably well, suggesting that a combination of assessment of absolute RHP (during conventional fighting) with an appropriate imbalance threshold, could be operative in aggressive decisions. The model as proposed so far depends on an assessment from these two parameters of the relative rates at which each combatant will lose fitness and therefore which will run out first. Because there is a single 'I will win—he will win' outcome to this problem, then the model based on relative rates of loss of fitness predicts 'never escalate' for a perfect assessment mechanism. Let us now consider a more realistic model.

## 4. Why escalate?

Clearly, if the combatants can predict exactly their relative rates of loss of fitness during a subsequent damaging fight, they should never escalate. This is unrealistic as a model because there will be a strong element of chance involved. Assessment will give only a probabilistic prediction of winning, not an absolute one. Instead of a precisely ordered rate of fitness loss, a much more valid description of observed escalations would be a series of bouts in which either combatant can score an injury inflicted upon his opponent. Injuries will occur as discrete events. Let us now revise the model as follows:

(1)    As before, the function of conventional fighting is to assess relative RHPs. This will give an absolute probability ($c_{abs}$) for each individual to win the first bout of an escalated fight (score the first injury against his opponent).

(2)    Suppose that the loss in fitness due to an injury in the first bout would be $l$. For this possible loss, there will be a critical

minimum probability of winning the first bout ($c_{crit}$) below which retreat (rather than escalation) is the more favourable strategy. $c_{crit}$ is greater the greater the search cost for an alternative resource.

(3)    Now, only where $c_{abs} > c_{crit}$ for both individuals does escalation occur. Where the sign is reversed for one individual, it retreats rather than be damaged. Where the sign is reversed for both, the winner is the one which the lesser negative score; it can afford to persist without escalation longer.

(4)    Should escalation occur, a reassessment should occur immediately after the end of the first bout (first injury) because the RHP of the loser will have decreased and so will its $c_{abs}$ for the next bout (also $c_{abs}$ for the winner has actually increased). Thus the chance that withdrawal (before the next bout) will be advantageous to the loser is likely to be considerably increased, depending on $l$ (greater $l$, greater the probability that withdrawal will be favourable before the next bout). It is a common feature of damaging fights that as soon as a combatant sustains an obvious injury, it retreats; an observation which fits the model.

(5)    The 'game' being played is that of reversing the opponent's $c_{abs} > c_{crit}$ to become $c_{abs} < c_{crit}$; i.e. playing for the withdrawal of one's opponent. Thus fighting in disputes over resources is regarded as a form of resource assessment strategy in that the probable gain from a given resource is weighed against the probable search cost for an alternative. In the present case the withdrawal point is defined by the changing nature of probability of winning and its cost, measured against the cost of withdrawal for searching for an alternative resource (i.e. one which is unguarded or has a holder with a lower RHP, more favourable to attack).

It is interesting to attempt a quantification of the above model to examine its characteristics. A rigorous examination becomes extremely complex, so only a first order approximation will be attempted here. Evolution's job is in a sense a much simpler one— selection merely favours individuals showing the optimum withdrawal/escalate thresholds, out of a series of 'threshold variants'. Let us assume a normal distribution of RHP in the competing population, so that frequencies of individuals in relation to RHP will therefore be summarized by:

$$f(r) = \frac{1}{\sigma\sqrt{2\pi}} \, e^{\frac{-(r-\mu)^2}{2\sigma^2}}$$

where the variable $r$ = RHP, $\sigma$ = standard deviation and $\mu$ = mean

RHP for the competitors. Integrating this distribution between $r = 0$ and $r_x$ ($r_x$ is a given RHP individual being considered) we get

$$F(r_x) = \int_0^{r_x} f(r) \, \mathrm{d}r$$

which gives the proportion of the competing population with an absolute RHP below $r_x$ if we set $F(r_\infty) = 1 \cdot 0$. $F(r_x)$ can be calculated from tables of integrals of variable $\pm\sigma$ limits of the normal distribution.

Consider an individual without a resource searching in a locality where *all* resources are guarded, but by a *random* sample of the competing population. The time taken to come across a resource held by an individual of lower RHP will be on average $t/F(r_x)$, where $t =$ the mean time between successive encounters of different resources. Obviously, where $t$ is low relative to the resource life, an 'imperfect structuring' of the resources (resources held randomly) cannot persist; there will be a change towards 'perfect structuring' (resources held by truncated top end of RHP distribution) at a rate inversely related to $t$. To make a gross simplification, we could assume that a proportion $s$ of the resources are perfectly structured, and that $(1 - s)$ are held randomly. There are many more competitors than resources. A less naive approach (e.g. where there is an increase in the average level of structuring with encounters through resource life) is complex and is not justified until other aspects of the model are also elaborated. All that is required of $s:(1 - s)$ is that it gives an approximate index of structuring so that we can assess roughly how long it will take a given individual to find a takeable resource. For the present model we will assume that no holder:attacker imbalance operates, other than disparity in RHP.

*(A) Value of search time if withdraw without fighting*

We shall assume that the advantage of high RHP is related to the lower search time to find a resource, and that a 'takeable' resource (from the viewpoint of estimating the value of search time) is simply one occupied by an individual of lower RHP. This is obviously a reasonable approximation only when the range of opponent RHP which will result in escalation with a given searcher (the 'escalation range'—see later) is narrow. On average a searcher will come across a lower RHP holder once in every $t/F(r_x)(1 - s)$ time units from the non-structured resources, i.e. $t/F(r_x) \times 1/(1 - s)$. Of all structured resources, a proportion $[F(r_x) - (1 - s)]/s$ will be occupied by lower RHP individuals. Hence the time to take a 'structured' resource will be:

$$\frac{t}{[F(r_x) - (1-s)]/s} \times \frac{1}{s} = \frac{t}{F(r_x) - (1-s)}.$$

If $F(r_x) - (1-s)$ is negative, the value is taken as 0 (none of the structured resources held by lower RHP individuals).

Now, in one (long) time unit of searching, the total number of encounters of takeable resources is therefore

$$\{1/[t/F(r_x)(1-s)]\} + \{1/[t/F(r_x) - (1-s)]\} = \frac{F(r_x)(1-s) + [F(r_x) - (1-s)]}{t}$$

Thus the mean search time taken to find a resource held by a lower RHP individual will be the reciprocal of this value (i.e. total time divided by total fruitful encounters), and hence the overall gain rate for a gain $G$ with gain extraction time $h$

$$= \frac{G}{\left\{ \frac{t}{F(r_x)(1-s) + [F(r_x) - (1-s)]} \right\} + h}. \qquad (7)$$

*(B) Probability of winning a bout if escalate*

In a bout between any two given combatants the probability of winning ($c_{abs}$) will be assumed to be directly proportional to the relation between their RHPs. For individual $x$ fighting $y$, this is $r_x/(r_x + r_y)$.

*(C) Fitness budgets for fighting*

A withdrawal point in resource assessment strategy is set by the stage in investment where

| (a) | | (b) |
|---|---|---|
| Probable future fitness gain rate due to continued investment in the resource (in gain extraction, fighting, courtship persistence, etc.) | = | Probable future fitness gain rate due to withdrawal for resumption of searching for an alternative resource. |

(see Parker 1974). Obviously the optimum strategy is to continue investment when (a) > (b), but to withdraw when the sign is reversed. The theoretical withdrawal point for each combatant depends on how much fitness it can afford to lose (during fighting) before withdrawal becomes the favourable strategy; i.e. on its 'fitness budget' for fighting. For the present analysis, we shall measure fitness loss entirely in terms of reduction in RHP; the actual fitness loss will be greater than this for a variety of reasons, but (especially for sexually selected fighting) RHP loss may often form the major component. (The model to be developed can be modified quite simply to include, say, an increased probability of mortality as a result of fight damage; however, it is interesting to examine whether RHP loss on its own can account for observed behaviour). We shall consider the effect of a loss $l$ in RHP. Hence an individual of RHP $r_x$ falls to RHP $= r_x - l$ if it loses a bout of escalated fighting.

Now, the winner will gain from the resource at a rate $G/h$. When he leaves the resource (after extraction is complete), the value of search time is equivalent to (7). For the loser, the search time gain rate will be reduced to

$$\frac{G}{\left\{\dfrac{t}{F(r_x - l)(1 - s) + [F(r_x - l) - (1 - s)]}\right\} + h}. \tag{8}$$

Supposing that loss $l$ persists for time $n$, and then the individual recovers to RHP $r_x$. For each combatant, the fitness budget (maximum permissible loss $l$) can now be calculated roughly as:

$$
\underset{\text{(a)}}{c_{abs}\{hG/h + (n - h)(7)\} + (1 - c_{abs})n(8)} = \qquad \underset{\text{(b)}}{n(7)} \tag{9a}
$$

for the condition where $n > h$, the evaluation is roughly:

$$
\underset{\text{(a)}}{c_{abs}hG/h + (1 - c_{abs})\{n(8) + (h - n)(7)\}} = \qquad \underset{\text{(b)}}{h(7)} \tag{9b}
$$

In (9a) and (9b) above, the parts (a) and (b) correspond quantitatively to (a) and (b) in the descriptive equation for withdrawal point.

## (D) Calculation of $c_{crit}$: when should escalation occur?

We can make a major simplification if we ensure that the model operates as a 'one step game', i.e. there is a definite solution after one bout—the loser withdraws. This can be done by adjusting the stake played for by the combatants. We can find the minimum value of $l$ in a combat which will ensure that the loser will withdraw.

We know that the withdrawal point at any stage in the interaction is set by (9a) and (9b). Stake $l$ is determined as follows. Assume a given individual loses at the first bout. On reassessment of $c_{abs2}$ ($c_{abs}$ for the second bout), will it fight for a further bout assuming that the stake played for will be the same as in the first bout (i.e. escalation is maintained at the same level)? A second bout will be favourable if

$$c_{abs2} > c_{crit2} = \frac{n(7) - n(8)}{[G + (n-h)(7)] - n(8)} \qquad (10a)$$

where $n > h$, or if

$$c_{abs2} > c_{crit2} = \frac{h(7) - [n(8) + (h-n)(7)]}{G - [n(8) + (h-n)(7)]} \qquad (10b)$$

for the condition $n < h$. The value of $c_{crit}$ above follows from substituting $c_{crit2}$ for $c_{abs}$ in (9a) or (9b). Thus by supplying a range of values for $l$ in (10a) or (10b) we can plot $c_{crit2}$ against $l$. The intercept of this curve with one for $c_{abs2}$ against $l$ gives the minimum injury ($l_{crit}$) which will ensure withdrawal if the individual loses the first bout (i.e. where $c_{abs2} = c_{crit2}$). We calculate two stakes, one for each combatant, and use the higher value to determine whether or not escalation should occur after a period of 'conventional assessment'. Escalation should occur only when $c_{abs1} - c_{crit1}$ is positive for both opponents; $c_{crit1}$ is calculated from (10a) or (10b) above using the higher value for $l_{crit}$ as the stake this time for both opponents for the *first* bout.

The above procedure for deciding the stake $l_{crit}$ is by no means as arbitrary as it may at first appear. It relies entirely on the relative fitness budgets of the two opponents; an integral part of the proposed model. It assumes that each combatant is playing for the retreat of its opponent and that this demands the infliction of a certain critical level of injury. It also assumes that the level of escalation necessary (higher $l_{crit}$) to ensure the withdrawal of one's opponent renders oneself vulnerable to the same possible danger. The opponent with the lower fitness budget must play for a higher stake than would be necessary to ensure his own withdrawal if he lost; this automatically escalates the fight to the same level for both opponents.

### (E) Some predictions of the model

Obviously escalation tendency will be inversely related to damage cost. Where the effects of damage persist less than the encounter time ($n < t$), damage costs nothing in the present model because the loser will have recovered before the next resource is encountered. Hence RHP loss as the sole fitness cost of fighting damage can operate only where $n > t$. Note that in (9) and (10) $n$ is used to apportion relative loss; this is used in conjunction with extraction time $h$ to relate all situations to the same time base (when $n > h$ we use overall base $n$, and vice versa). However, the model is to some extent 'buffered' against relative differences in $t$, $h$, and $n$ because of the means of determining $l_{crit}$—the stake played for.

The effects of different levels of RHP disparity between extremes of the population are difficult to estimate accurately from the present model because of the assumption that the value of search time can be estimated directly from the proportion of resources held by lower RHP individuals. Clearly, however, the less the extent of RHP disparity across the competitor population, the closer the $c_{abs}$ values for combatants and the closer the $c_{crit}$ values (because there will be less search time disparity if the value of search time is only weakly influenced by RHP disparity). From the results obtained below, this might be expected to result in a wider 'escalation range' (see below).

Most animal conflict involves relatively little escalation and much conventional display, implying that the cost of damage and the degree of RHP disparity are both higher rather than lower, so that withdrawal is commonly the favourable strategy. An experimental computation was investigated using $t = 0\cdot1, h = 1, n = 10$, $G = 1$; $c_{crit}$ calculated after (10a). This ranking should give a moderately high incentive for withdrawal (search time short, damage prolonged). Because $h = 10t$ and the number of competitors is assumed to be considerably in excess of the number of resources, then $s$ might be expected to be fairly high and was taken as $0\cdot7$. RHP was arranged so that $r = \sigma$ (standard deviation) and where the mean $\mu = 2\sigma$. Hence $r_x = 1$ for a low ranking individual $-1\sigma$ below the mean, and $r_x = 3$ for a higher ranker $+1\sigma$ above the mean. A combat between two such individuals gives $c_{abs}$ values of $0\cdot25$ and $0\cdot75$ respectively; a moderate disparity. We shall consider the optimum strategy (escalate or withdraw) of a low, medium, and a high ranker ($r_x = 1, 2, 3$ respectively) in combats with a range of opponents within $\pm 2\sigma$ of the mean, and also consider the optimum strategy for the opponent.

Results are shown in Fig. 1. Each combatant shows a relatively narrow 'escalation range' (range of RHP opponents for which both individuals show a positive value for $c_{abs} - c_{crit}$). Outside this range the favourable strategy is withdrawal for the lower RHP individual and generally escalation for the higher one. How-

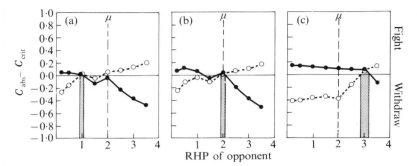

Fig. 1. Outcome of aggressive encounters of various combinations of RHP combatants using the model described in section 4 with $t = 0.1$, $h = 1.0$, $n = 10$, $G = 1$, $s = 0.7$, $r = \sigma$, $\mu = 2\sigma$. • and solid lines = $c_{abs} - c_{crit}$ for combatants of $r_z = 1$ (a), $r_z = 2 = \mu$ (b), $r_z = 3$ (c); ○ and dotted lines = $c_{abs} - c_{crit}$ for range of opponents between $r_z = 0.2$ to $3.5$ (abscissa) fighting each of these three combatants. In a fight the individual with the highest score for $c_{abs} - c_{crit}$ stays in possession of the resource and the loser withdraws, to determine the winner. Shaded zones = escalation ranges (ranges of opponent RHPs which will result in escalation with each of the three combatants considered).

ever, both the low and medium rankers show withdrawal as the favourable strategy for *both* combatants when the opponent has $r_x = 1.5$. Because $F(r_x)$ is sigmoidal with the maximum gradient at the mean, a unit drop in $r_x$ affects an average RHP individual more than an extreme one (the number of takeable resources falls more). Hence the relation between $l$ and $c_{crit2}$ is steeper the closer to the mean RHP. In Fig. 1(b) the $r_x = 1.5$ opponent has a higher fitness budget ($l_{crit}$) than the $r_x = 2$ combatant, even accounting for the difference in $c_{abs}$; however, $c_{crit} > c_{abs}$ for both individuals. For Fig. 1(a) though the $l_{crit}$ for $r_x = 1.5$ is higher than that for $r_x = 1$, its $c_{abs}$ value is still too low to allow escalation. These effects probably arise because for a short range of RHP close to the mean, the $l_{crit}$ is greater for the theoretical bout 2 than for bout 1 because the RHP disparity does not fully offset the effect of the steeper $F(r_x)$ gradient. Note, however, that the higher RHP individual still 'wins' in such cases (has less motivation to withdraw and can afford to be more persistent in conventional display).

The model also indicates a wider escalation range with the high ranking combatant (Fig. 1(c)). Because of the steeper gradient in $F(r_x)$ at the mean one might expect a narrower escalation range there than at the extremes. However, the escalation range of the low ranker is even smaller than that of the average one. This is probably because $c_{abs}$ disparity increases towards $r_x = 0$; obviously the disparity is greater between $r_x = 0{\cdot}3$ and $0{\cdot}5$ than between $r_x = 3{\cdot}3$ and $3{\cdot}5$.

## 5. Discussion

In summary, the main predictions of the type of model developed in section 4 are that there should be an escalation range of closely matched combatants and that on either side of the range for a given individual, the higher ranking opponent should usually be prepared to escalate and the lower one to withdraw. Much fighting follows this pattern. Size, strength, weaponry, and experience all seem involved in RHP. There are innumerable examples where the outcome of disputes depends to a large extent on the relative size of the opponents. For instance, large individuals dominate over small ones in green sunfish (Greenberg 1947; Hale 1956), crayfish (Bovbjerg 1953, 1956), mice (Ginsberg and Allee 1942), New Forest ponies (Tyler 1972) and a host of other species. Matched individuals often show the greatest tendency for escalation; mirror images are often very effective stimuli (e.g. Figler 1972). Though pushing contests are apparently commonplace as estimates of relative strengths, there are several examples where visual cues or physiological ones are used as indicators of weaker individuals. Chickens which are moulting (and hence likely to be weaker) are usually submissive (Collias 1943); crayfish avoid combat until the cuticle has hardened after moulting—newly-moulted specimens are less mobile and sustain greater damage and risk as a result of fights (Bovbjerg 1953). Antler and horn size appear to be judged directly in many deer and sheep and fights occur only between closely matched combatants (see Eibl-Eibesfeldt 1970); there seems little doubt from the literature that assessment of RHP is occurring in most cases of animal combat. To avoid any implications of teleology, it must be stated that 'assessment' in this context means only that the individual responds differentially to opponents on a basis of their RHP relative to its own; the only assessment of what is the appropriate response is the unconscious one performed by selection.

It is interesting that during a display, selection should mainly favour presenting an opponent with a maximal impression of one's RHP. Until a 'strategic decision' is reached, no information should be displayed to an opponent concerning withdrawal intentions, since there is the possibility that the opponent may withdraw first (see Maynard Smith 1974). In *Betta splendens* various display components increase in parallel for several minutes, and an outcome is not predictable until one individual finally gives up (Simpson 1968). For a given action of the opponents, there may be an optimal retaliatory action. For food fights of blue tits Stokes (1962) has shown that correlations between display components and the subsequent outcome (attack, escape, stay) are sometimes significant, but not generally high. Hinde (1972) has argued that this may be interpreted on the view that the next action of the displaying individual is not predetermined but dependent on the behaviour of the opponent.

Prior conditioning and experience can also be very important in determining the outcome of aggressive disputes. In some cases this is related to the holder: attacker imbalances mentioned in section 3, where the holder has a higher fitness budget than the attacker for reasons other than mere RHP disparity. However in some cases it seems likely that successful fighting experience markedly increases the readiness for escalation (e.g. in mice and rats, Scott and Fredericson 1951); an effect explicable in terms of experience increasing RHP.

Males are usually dominant over females. This often relates to RHP disparity because males are bigger; in some instances however secondary sexual characters are used as signals (e.g. comb size is a determinant of dominance in chickens (Collias 1943)). It seems possible that because of sexual selection male fitness may be increased by adopting a more dangerous strategy if this gives an overall increase insemination rate. Thus males of the same RHP as females may have a higher fitness budget for fighting over, say, food—because being in peak condition may affect male fitness more because of intra-sexual competition (see Trivers 1973 for a similar argument concerning male mortality). It is interesting in this context that females with young often (but not always) increase markedly in rank. They may have a higher fitness budget in such circumstances.

Prevention of damage during retreat is a common adaptation. Fish colour changes which accompany submission and retreat can often be explained on a basis of crypsis. A trapped retreater (presumably $c_{abs} < c_{crit}$) when faced with a potential escalator (presumably $c_{abs} > c_{crit}$) is often very frantic in its attempts to escape (e.g. Sabine 1949).

Much of RHP disparity must be environmental, due to experience, nutrition effects during development, accidental damage, etc. In insects adult size variation is very largely environmental in origin; if size is important in combat the main selective force acts on choice of oviposition site by the female (Prof. H. E. Hinton, personal communication). Selection will favour RHP increase until this is countered by opposing pressures; for sexual selection the selection coefficient of a competitively advantageous character actually accelerates as the character spreads through the population (Charlesworth and Charlesworth, in press).

If RHP variance is small, holder:attacker imbalances may be the main factors determining the outcome of aggressive disputes, and vice versa when RHP disparity is large and holder:attacker imbalances small. It is interesting that *within* social groups RHP disparity seems to be the main determinant of aggression and dominance rank. This fits the predictions because it is unlikely that (possibly food-fighting apart) holder:attacker imbalances will be of major importance within groups. The opposite may prevail for many *between* groups (or single individual) territorial situations; probably mainly because of the considerable reinvestment imbalances of the type discussed in case 1(b) and case 2, section 3. A fascinating effect is predictable here. Where the holder:attacker imbalances are high, the RHP disparities small, and the value of search time high (e.g. if alternative territories are relatively abundant), it seems possible that RHP assessment need not occur before withdrawal is favourable; mere signs that the territory is occupied may be enough to favour retreat for further searching. This may well be the explanation of territory-marking scents, songs and visual cues which appear to give little indication of RHP. Baker (1972) gives an excellent discussion and evidence for this sort of effect in the territorial behaviour of male nymphalid butterflies. For instance, in *Aglais urticae* searching males become less reluctant to share occupied territories as the afternoon wears on (and the chance of finding an unoccupied territory becomes reduced).

The models developed in the present paper have many inadequancies. In section 4 we have not considered how RHP will modify tenure-time of a resource because of its relation with the chances of take-over. Nor do we have an accurate assessment of the value of search time; nor is the exact relationship between $t$, $h$ and $s$ properly explored. Though the mere loss of RHP alone is adequate to explain much of observed behaviour, it would be interesting to examine the interaction of RHP loss with other possible sources of fitness loss through damage. A much more

## 292　Co-operation and disruption

rigorous analysis, though very complex, might allow a more exact set of predictions and enable quantitative consideration of real data.

Perhaps the implications of assessment strategy for human aggression are better left for a future occasion.

## References

Baker, R. R. (1972). *J. Anim. Ecol.* 41, 453.
Bovbjerg, R. V. (1953). *Physiol. Zoöl.* 26, 173.
—— (1956). *Physiol, Zoöl.* 29, 127.
Charlesworth, B. and Charlesworth, D. (in press).
Collias, N. E. (1943). *Amer. Nat.* 77, 519.
Crane, J. (1966). *Phil. Trans. R. Soc. Ser. B,* 251, 459.
Darwin, C. (1871). *Sexual selection and the descent of man.* Murray, London.
—— (1872). *The expression of emotions in man and animals.* Murray, London.
Eibl-Eibesfeldt, I. (1970). *Ethology, the biology of behaviour.* Holt, Rinehart and Winston, New York.
Ewer, R. F. (1968). *The ethology of mammals.* Logos Press, London.
Figler, M. H. (1972). *Behaviour* 42, 63.
Ginsberg, G. and Allee, W. C. (1942). *Physiol. Zoöl.* 15, 485.
Greenberg, B. (1947). *Physiol. Zoöl.* 20, 267.
Hale, E. B. (1956). *Physiol. Zoöl.* 29, 107.
Hamilton, W. D. (1964). *J. theoret. Biol.* 7, 17.
Hinde, R. A. (1972). *Social behaviour and its development in subhuman primates.* Condon Lectures. University of Oregon Press.
Jacobs, M. E. (1955). *Ecology* 36, 566.
Kastle, W. (1963). *Z. Tierpsychol.* 22, 751.
Leuthold, W. (1966). *Behaviour* 27, 214.
Maynard Smith, J. (1972). *On evolution.* Edinburgh University Press.
—— (1974). *J. theoret. Biol.* 47, 209.
—— and Price G. R. (1973). *Nature, Lond.* 246, 15.
Parker, G. A. (1970a). *Biol. Rev.* 45, 525.
—— (1970b). *Behaviour* 37, 113.
—— (1974) *Behaviour* 48, 157.
—— Hayhurst G. R. G., and Bradley, J. S. (1974). *Z. Tierpsychol.* (in press).
Sabine, W. S. (1949). *Physiol. Zoöl.* 22, 64.
Scott, J. P. and Frederickson, E. (1951). *Physiol. Zoöl.* 24, 273.
Simpson, M. J. A. (1968). *Anim. Behav. Monogr.* 1, 1.
Stokes, A. W. (1962). *Behaviour* 19, 118.
Trivers, R. L. (1973). In *Sexual selection and the descent of man,* 1871–1971. B. (Campbell, ed.). Aldine, Chicago.
Tyler, S. J. (1972). *Anim. Behav. Monogr.* 2, 121.
Williams, G. C. (1971). *Group selection.* Aldine, Chicago.
Wynne-Edwards, V. C. (1962). *Animal dispersion in relation to social behaviour.* Oliver and Boyd, London.

# Evolutionary rules and primate societies (extract)

T. H. CLUTTON-BROCK AND PAUL H. HARVEY†

### Disruptive behaviour

Disruptive or selfish actions diminish the fitness of the recipient. Interactions involving disruption are likely to occur where two or more individuals compete for limited resources and can, for our purposes, be regarded as one form of ecological competition (see Nicholson 1955; Brown 1964; Miller 1967; Geist 1974). Perhaps because disruption poses a less obvious problem to evolutionary theory than beneficence, it has attracted less attention.

*Aggression*

However agression is defined, it usually falls within the category of disruptive behaviour. The functions of aggression in animal societies are frequently viewed in terms which rely on group selection. For example, Lorenz (1966, p. 22) argues that it fulfils a 'species preserving function' (see also Scott 1962, 1974; Washburn and Hamburg 1968). Similar examples are widespread in recent literature on primate behaviour. Sorenson (1974, p. 14, quoting George, 1966) argues that, to have survival value, aggression must 'bring order into the social circle'. And in the same volume, Nagel and Kummer (1974, p. 175) suggest that its function is to 'maintain social structure and to relate it to ecological resources'.

There is no need to view the function of aggression in this way. The sifaka (*Propithecus verreauxi*) which fights to gain access to females (Richard 1974*b*) or the mangabey (*Cercocebus albigena*) which successfully disputes a clumped food source by threatening its neighbour (Chalmers, 1968) is likely to increase its own fitness. Behavioural traits which enable individuals to do this can spread through the population even if they diminish average reproductive success (see Hamilton 1971*b*). For example, in several primate species, males may disrupt each other's reproductive activities. In hanuman langurs (*Presbytis entellus*), males which occupy a troop

†Reprinted with permission from P. P. G. Bateson and R. A. Hinde (eds.) *Growing points in ethology*. Copyright © 1976 by Cambridge University Press.

after successfully ejecting the previous male(s) may attack and kill infants (Sugiyama 1965a, b; Mohnot 1971; Hrdy 1974). Similar behaviour may occur in black-and-white colobus (*Colobus polykomos*) and hamadryas baboons (*Papio hamadryas*) (J. F. Oates personal communication; Kummer, Gotz, and Angst 1974). By infanticide, such males remove potential competitors of their own offspring and ensure that the females of the group come quickly into oestrus. In addition, males of some species attack other males who are copulating (Gouzoules 1974) and sexual behaviour in subordinates may be inhibited by the presence of a dominant animal (Perachio, Alexander, and Marr 1973). Gouzoules (1974) has documented a number of cases of harassment in primate societies. His suggestion that the function of harassment may be to direct the copulating male's aggression away from his consort is difficult to state in terms of individual advantage. It seems more likely that, by harassing copulating animals, males may increase their own chances of fertilizing the females, though there is no available evidence on this point.

This approach to aggressive relationships requires us to explain why animals are not more aggressive. It is evident that aggression incurs costs as well as benefits (Tinbergen 1951; Hutchinson and MacArthur 1959; Ripley 1961; Geist 1974; E. O. Wilson 1975). Theoretical models of contest strategies (Maynard Smith and Price 1973; Parker 1974; Gale and Eaves 1975) show that neither the most persistent fighting strategy nor the most savage one is necessarily most advantageous because such strategies are likely to escalate costs to both contestants. This does not explain why, in contests where both animals have committed themselves and one has won, the winner does not kill the loser to avoid further competition. One possibility is that, even here, an attempt to kill or severely wound by the winner might lead to a violent counter-attack by the loser (Geist 1974). For example, when rhesus monkeys have been defeated they may passively accept incisor bites but will counter-attack viciously if the winners use their canines (Bernstein and Gordon 1974). In such situations, the cost of escalating the contest may exceed the advantages of removing a potential competitor. Alternatively, such advantages may fail to outweigh the time and energy which would be needed to kill the rival. Finally, in those species where winners are inhibited from killing losers this could be because they might be closely related and the act would diminish the winner's inclusive fitness.

We must not expect the relationship between temporary variation in resource availability and the frequency (or intensity) of aggression to be a simple one. Indeed, there is already some evidence that it is not. In several species, aggressive interactions are common during periods of intermediate food availability. When food is abundant, the frequency of aggression drops (Kruuk 1972; E. O. Wilson 1975) presumably because the benefit of winning contests does not justify the expenditure of time, energy or risk to further reproductive potential involved in fighting. At periods of very low food-availability, the frequency of aggression is also reduced (e.g. Hall 1963; Southwick 1967; Loy 1970), presumably because time/energy budgets are so finely adjusted that all forms of expenditure must be minimized. Similar rules are likely to apply to all species though we should expect to find inter-specific differences in the form of the relationship.

*Sexual dimorphism and aggression*

The theory of sexual selection states that the sex with lower reproductive potential will be competed for by the other sex (Fisher 1958; Trivers 1972). In most animals, the reproductive potential of males is considerably higher than that of females and their reproductive success usually varies more widely (Trivers 1972). Consequently, we should expect selection to favour greater development, in males, of traits which enhance successful competition for mates. Many sex differences in body size, weapon development (e.g. antlers, horns, canines) and pelage (Goss-Custard *et al.* 1972; Schaller 1972; Geist 1974) can thus be explained in terms of individual advantage and explanations relying on group advantage (e.g. DeVore and Washburn 1963; Coelho 1974) are unnecessary. Aggressiveness can be seen as a functionally similar trait since it may assist a male to compete successfully for females. In the great majority of primate species, males appear to be more aggressive than females (see Crook 1972; Chalmers 1973; Holloway 1974) and are generally dominant to them. (There is no obvious explanation why in some species, including ring-tailed lemurs (*Lemur catta*) (Jolly 1966), *Indri* (J. Pollock personal communication) and talapoins (*Miopithecus talapoin*) (Wolfheim and Rowell 1972) females are dominant to males.)

Convincing evidence that this view of the evolution of male characteristics is correct comes from studies of polyandrous bird species where females compete for access to males. Here, one would predict that females would possess most of the traits which usually typify males: they should be larger and more aggressive, initiating courtship and taking primary responsibility for defence of the territory. This is the case in most of these species (Jenni 1974).

The same approach helps us to understand the evolution of certain interspecific differences (Trivers 1972; Alexander 1974). We should expect to find the most marked sex differences in those societies where breeding competition is most intense. Thus, sex differences should be greatest in strongly polygamous societies and least in truly monogamous ones. By and large, this is true for primates as well as for other mammalian groups (e.g. ungulates; Jarman 1974). Species such as patas monkeys (*Erythrocebus patas*) or gelada baboons (*Theropithecus gelada*) which show the most unequal socionomic sex ratios* are characterized by the greatest sex differences in body size. In contrast, monogamous species such as gibbons or titis show the least. We would also predict that differences in aggressiveness between males and females should be most marked in the most polygynous societies, although suitable comparative data are rare (Nagel and Kummer 1974). For the same reasons, males of strongly polygamous species should be more likely to fight dangerously and to escalate conflicts than males of monogamous species.

The situation is probably further complicated by differences in aggressiveness between species which are not the product of sexual selection. For example, differences in aggressiveness between marmot species appear to control differences in the timing and dispersal in the young and may have developed for this reason (Barash 1974). Similarly, inter-specific differences in life span may affect aggressiveness: the longer the life span, the less willing individuals should be to risk reproductive potential for immediate access to resources (Geist 1974). Consequently, aggression may be commoner and more intense in short-lived species than in animals with long life spans. Finally, the nature of food supplies may be important. Where they are evenly dispersed, highly transient or super-abundant, the advantages of aggression may be minimal (Brown 1964; Geist 1974).

*Age differences in aggression*

Age differences are likely to affect the advantage of aggression to the individual (Parker 1974). Two processes will be involved. First, as an individual matures the chances of successfully winning dis-

* i.e. the ratio of adult males to adult females in reproductive groups.

putes are likely to increase initially and then decrease as it passes its prime. Secondly, older individuals should be prepared to invest more in these aggressive encounters than younger ones because their reproductive potential is lower (Trivers 1972). Age changes in aggressiveness in primate societies have not yet been well documented. However, there is evidence that juveniles and adolescents relatively rarely show aggression (e.g. Chalmers 1973) compared with adults. In addition, detailed studies of hunting forays in chimpanzees (Wrangham 1975) suggest that older individuals may be more prepared to contest access to supplies of meat.

## Kinship and aggression

Kin relationships, too, may modify aggression (see Hamilton 1971). One might predict that the amount of tolerance extended to different relatives would correlate with their degree of genetic similarity. This appears to be the case among Japanese macaques where genealogical relationships extending over several generations are known in a number of troops. To examine the extent of tolerance between different relatives, Yamada (1963) scattered wheat over ten areas of approximately one square metre each and recorded the frequency with which animals fed together at the same area. Co-feeding was limited by the tendency of dominant animals to threaten subordinates feeding at the same site. Consequently the frequency which individuals co-fed with different subordinates provided an approximate measure of the degree of tolerance extended to them (though there is an obvious danger that co-feeding frequency was also affected by the likelihood of subordinates approaching and feeding with dominant relatives). From Yamada's data, it is possible to calculate the average frequency with which (a) mothers tolerated children, (b) sibs tolerated other sibs, (c) grandmothers tolerated grandchildren, (d) aunts tolerated nieces and nephews, (e) mature females tolerated unrelated individuals. (In each case the first mentioned was dominant to the second.)

Table 1 compares the average frequency with which different relatives tolerated each other with their estimated degree of genealogical relationship. As can be seen, there is a close association between the two measures.

Kin selection may help to explain the function of unsolicited aggression (see p. 299) in animal societies. It is not uncommon for dominant individuals to threaten or attack animals which are neither initially close to them nor are competing for any obvious

Table 1. Average frequency of feeding tolerance (see text) shown by Japanese macaques to subordinate individuals compared with their degree of genealogical relationship (data from Yamada, 1963)

| Dominant | Mother | Sib | Grandmother | Aunt | Mature female |
|---|---|---|---|---|---|
| Subordinate | Child | Sib | Grandchild | Niece or nephew | Unrelated |
| Degree of relationship | 0·5 | 0·25– 0·5 | 0·25 | 0·125– 0·25 | 0 |
| Average frequency of observations where co-feeding occurred | 18·3 | 12·6 | 2·7 | 3·3 | 1·1 |
| Number of pairs of individuals involved | 16 | 14 | 4 | 6 | 154 |

resources (Hall and Mayer 1967; Richard 1974a, b). By doing this, they may possibly be depressing the latter's status and thus increasing their own relative rank or that of their offspring. Similarly, females may attempt to prevent other females from breeding. In several species of Old World monkeys, females harass copulating couples (see above) while, in wild dogs, breeding females may even kill litters born to other group members (van Lawick and van Lawick-Goodall 1970).

Inter-specific differences in the genealogical composition of social groups may help to explain some differences in the distribution of aggressive encounters. For example one would predict that in monogamous societies or in species living in single male troops where sibs have 50% of their genotype in common, they would be more tolerant of each other than in societies where promiscuous mating occurs (where they are likely to share only 25%). However, the relevant evidence is lacking in mammals.

*Parental manipulation*

In general, we should not expect parents to be aggressive to their own offspring (see above). However, in some situations, this may increase their inclusive fitness and it may even be to the advantage of the parent to kill one or more of its offspring (Alexander 1974; Trivers 1974). For example, some birds preferentially feed older offspring (e.g. Lockie 1955) with the probable result that the youngest fail to survive and may even be eaten by their sibs when food supplies are short. And in groups of Japanese and rhesus macaques, female offspring of the same mother tend to rank immediately below her, their rank within the matriline being inversely related to their age (Yamada 1963; Koyama 1967; Sade 1967). This may partly occur as a result of the mother's support for her younger offspring in agonistic interactions (see Koyama 1967, 1970). By doing this, she may help to protect her younger daughters from harmful competition with her older offspring, thereby increasing her inclusive fitness. In this case, we should predict that orphans would be out-ranked by their older sibs, though the available evidence for this is ambiguous (Sade 1967; Missakian 1972).

## Spiteful behaviour

So far, we have only discussed disruptive actions which are to the benefit of the initiator. In certain cases, disruptive behaviour may evolve which either does not benefit the individual's inclusive fitness or even diminishes it (Hamilton 1970, 1971). These are cases where an individual 'spitefully' disrupts other members of the population thus decreasing their fitness relative to its own. Hamilton (1970) argues that such behaviour is unlikely to evolve for three reasons: that all spiteful actions incur appreciable costs; that animals will not be able to differentiate between individuals which are of less than average relatedness to them and those which are more closely related, and may therefore damage their own kin; and that the trait is only likely to spread in small populations which it may help to extinguish. But these conditions are met in higher mammals. In species where marked dominance hierarchies exist, disruptive actions directed by dominants at subordinates may cost the dominant little. Also there is evidence that individuals may be able to distinguish their degree of relatedness to other animals quite nicely (see above). Finally, if one supposes that spiteful aggression is only shown in certain contexts (for example at high population densities) there is no reason to think that it would extinguish the lineage in which it arose. While it therefore seems possible that spite may evolve, it will be extremely difficult to distinguish from selfish disruption. For example, even where apparently unsolicited aggression occurs, it is possible that it increases the initiator's inclusive fitness (see above).

One possible example of spite might be called 'punishment'. Where a dominant individual's access to resources is jeopardised by the behaviour of a subordinate, it may be to the former's advantage to punish the latter, reducing its fitness enough to make further attempts unprofitable. Hyenas (*Crocuta crocuta*) that intrude into their neighbours' ranges may be attacked and killed even after they have submitted and are attempting to escape (Kruuk 1972). By occasionally killing such intruders, clans must reduce the potential advantages of poaching as well as removing competitors. And in Japanese macaques adults may punish individuals which attack other troop members (Kawamura 1967). Punishment is only likely to occur between individuals where there is an initial asymmetry in contesting ability, so that the cost of punishing is experienced or observed by several individuals, thus increasing the benefits of punishing. The reverse of punishment, where a dominant rewards a subordinate for behaviour which increases the former's fitness, constitutes one form of reciprocal altruism.

Similar behaviour may occur which does not involve transactions in fitness. Among most higher animals, the behaviour of individuals in social interactions is probably extensively conditioned by the consequences they experience as a result of each others' actions (see Hinde and Stevenson-Hinde 1976). We should expect the reactions which different individuals evoke to be adapted (so that social actions received by an individual which increase his/her fitness will be positively reinforcing and those that diminish it will be negatively reinforcing). However, it is difficult to believe that reactions could be equally well adapted to all contexts and it is possible that individuals may learn to take advantage of each other's learning systems. For example, if B behaves towards A in such a way as to increase A's fitness, A's most advantageous manoeuvre might be to reward B in 'artificial currency', trading a reinforcement which does not increase, or even detracts from, the latter's fitness. Similarly, it may be less costly to punish subordinates with behaviour which they find uncomfortable than actually to diminish their fitness. Clearly, this will not be a stable situation, for selection is likely to increase individuals' abilities to distinguish between 'real' and 'artificial' rewards or punishments. However, we might expect to find animals attempting to pull off such confidence tricks, particularly in social contests which are highly variable. As Humphrey (1976) has pointed out, selection operating through transactions of this kind would be likely to favour the evolution of intelligence.

## Functional aspects of social structure

### Social dominance

The concept of social dominance has been extensively employed since it was first described (Schjelderup-Ebbe 1922). Recently, several reviews have rightly condemned its uncritical use and have stressed that relationships between individuals cannot be described adequately in terms of linear hierarchies (Rowell 1974; Syme 1974). Since at least one reviewer (Gartlan 1968) has argued that dominance hierarchies may represent a pathological response to abnormal evironments, it is necessary to justify our acceptance of the concept before discussing its function.

Dominance has been defined in many ways (Rowell 1974). In functional terms, its central statement is that particular individuals in social groups have regular priority of access to resources (or of avoidance where noxious stimuli are involved) in competitive situations. If it is to be a useful explanatory concept or 'intervening variable' (Hinde 1974) the access of individuals to different resources and in different contexts should be correlated (Richards 1974). In its most developed form, dominance rank is 'transitive' (i.e. if male A has priority over male B he will also have precedence over all individuals over which B takes precedence) and is related to behaviour in agonistic and affiliative interactions as well as in asocial contexts (Bartlett and Meier 1971; Richards 1972).

Gartlan (1968) offers two important criticisms of dominance. First, he points out that access to different resources may be poorly correlated (see also Syme 1974). In particular, there is evidence that the frequency with which individuals copulate is often poorly correlated with their rank in other situations (Kummer 1957; Southwick, Begg and Siddiqi, 1965; Jay 1975; Jolly 1966; Bygott 1974). Secondly, he stresses that linear hierarchies occur in caged groups or in populations where density is unnaturally high and are usually associated with frequent aggression. The inference here is that hierarchies seldom occur in natural populations under normal conditions.

Neither of these two points requires us to reject the concept though both emphasize its complexity. In fact, evidence from a wide variety of species shows that access to many different resources is correlated in many species, though not in all (Bernstein 1969, 1970; Richards 1974; Syme 1974), and that agonistic dominance is generally associated with breeding success (see refs. in Wynne-Edwards 1962; Loy 1971; Suarez and Ackerman 1971; E. O. Wilson 1975). In addition, some degree of hierarchial ranking is present in most social species (Carpenter 1954; Washburn, Jay and Lancaster 1965; Mazur 1973) and in field studies where no hierarchy is reported (e.g. Carpenter 1934; Rowell 1966, 1967;

Neville 1972) it is usually difficult to be certain that this is not a product of inadequate sampling. However, there is evidence that the predictive value of ranking is lower in some species than in others (see below). The significance of these differences becomes clearer when one considers the adaptive significance of dominance.

Three common explanations of the function of dominance are that it reduces aggression (Collias 1953; Scott 1962; Poirier 1974), that it restricts the access of the most 'expendable' individuals to limiting resources (Chance and Jolly 1970; Kummer 1971) and that it helps to regulate population density (Wynne-Edwards 1962). Such arguments rely on group selection, and have arisen because a hierarchy has been regarded as a behavioural trait, whereas it is no more than 'the statistical consequence of a compromise made by each individual in its competition for food, mates and other resources. *Each compromise is adaptive but not the statistical summation*' (Williams 1966, p. 218).

The evolution of hierarchies is not difficult to explain. Individuals will benefit by gaining access to food and mates in competitive situations. Whether it will be to their advantage to contest access will depend on the potential benefit of acquiring the resource and on the potential costs of the interaction. Where possible benefits are high and costs low, individuals are likely to contest access and where benefits are low and costs high they will not do so (see Parker, 1974). Hierarchies occur because competitive ability inevitably varies between individuals and because less successful animals learn not to contest access to encounters where they are unlikely to win thus saving time and energy. They will be transitive if individuals' abilities to compete successfully are independent of the identity of their opponents (though not obviously, of the competitive ability of the latter). Animals may initiate dominance interactions outside the immediate context of competitive encounters (e.g. Struhsaker 1967a, b; Saayman 1972; Bygott 1974) if this enhances their ability to compete successfully in future.

Thus dominance may be analogous to territoriality (Wilson 1971). Both traits enable some individuals to obtain a larger-than-average share of available resources by disrupting competitors. The switch from territoriality at low population density to dominance behaviour at high density in many species (e.g. Archer, 1970), suggests that the two traits may be causally as well as functionally related.

Rowell (1974) proposes a different view of dominance, stressing two points. First, since the majority of approach–retreat interactions are initiated by subordinates, we should regard the hierarchy as being maintained by the subordinates rather than the dominants. Secondly, ranking may develop through observational learning in competitive situations. Thus if two strange monkeys are repeatedly presented with a raisin, which one of them consistently snatches, the other will learn that it is unlikely to get the reward and will eventually give up trying.

Neither point constitutes a serious criticism of the view of dominance suggested here. Even if the subordinates initiate the majority of interactions, they do not necessarily control the relationship. For example, monkey B may retreat from monkey A in 99 cases in the absence of any action or display on A's part. On the hundredth occasion he fails to retreat, and is attacked and chased away by A. As a result he is likely to initiate subsequent retreats without waiting for A to respond. For every retreat initiated by A, the observer will score a far greater number which are initiated by B, yet it is A's behaviour which largely controls the relationship.

Moreover, while observational learning may well be involved in the ontogeny of dominance hierarchies, it is only likely to be important in cases where it provides a reliable indicator of the observing animal's chances of acquiring the resource. Thus if the raisin, in Rowell's example, was an oestrus female which, by chance, immediately submitted to the blandishments of the first monkey, we should not expect the second to fail to contest such situations in future. Natural selection would be most unlikely to encourage such a fatalistic attitude.

Both costs and benefits of contesting another individual's priority in competitive dyadic encounters (as well as the costs and benefits of submitting) will be affected by the same factors that influence the advantages and disadvantages of aggressive behaviour generally. If our view of dominance is correct, it should be possible to understand many of the currently unexplained complexities of hierarchies (which so often confuse attempts to generalize about dominance, e.g. Gartlan 1964) in terms of variation in the costs and benefits of contesting and submitting (see Maynard Smith 1974*b*; Parker 1974).

This appears to be the case. The following subsections discuss the possible explanation of variation in interaction patterns associated with four different variables.

*Absolute value of resource.* Where the values of individual resource units are extremely low, it may not be advantageous to contest access to them. This may help to explain the apparent absence of hierarchical food access in some herbivorous and folivorous animals (see Geist 1974). Conversely, when the value of resources is extremely high, contesting will almost always be advantageous. In a number of species which live in individual territories or harem groups, males which do not gain a harem or a territory usually fail to breed (Kummer 1968; Watson and Moss 1972). In several such species, males usually fail to form stable hierarchies when caged together (e.g. Zuckerman 1932; Scott 1966; von Holst 1974).

*Inter-individual differences in resource value.* Where the costs or benefits of exploiting different resources vary between individuals, individual access to different resources should also be expected to vary. Cases where access to different kinds of food are not closely correlated (e.g. Harding 1973; Dunbar and Crook 1975; Wrangham 1975) may be a product of differences in dietetic needs and in food values: rhesus monkeys raised on low-protein diets outrank individuals raised on high-protein diets on food access but not on avoidance competition (Wise and Zimmerman 1973). It is, however, very difficult to explain the high frequency of copulations achieved by subordinates in many cases (e.g. Bygott 1974; Enomoto 1974) and the tolerance of breeding activity in subordinates by dominant animals belonging to the same group (e.g. Hanby, Robertson and Phoenix 1971; Saayman 1971). It is clearly important that studies should investigate the number of successful fertilizations achieved by high and low ranked individuals: in a troop of Japanese macaques Hanby *et al.* found that although some of the most sexually active males were low in rank, dominant animals ejaculated more frequently. Where tolerance of 'genuine' breeding actually exists, the usual degree of relatedness between dominants and subordinates must be determined.

*Changes in resource value.* In situations where either costs or benefits to individuals of acquiring resources vary through time, we should expect their willingness to contest access to do so too (Parker 1974). Where individuals have recently held a particular kind of resource, the advantage of maintaining access to it is likely to diminish. This may help to explain (i) why the longer initially dominant individuals have fed at a food source, the more likely they are to be displaced or to distribute a part of their food to other individuals when challenged (see Wrangham 1975); (ii) why deprivation, which is likely to affect individuals differentially, may produce changes in rank (Boelkins 1967; Castell and Heinrich 1971).

Alternatively, previous access to resources may increase their value (Parker 1974). In experiments with caged hamadryas baboons, Kummer *et al.* (1974) introduced a strange female to two strange males. In such cases, the two males would contest access to the female, and the winner then formed a social bond with her. However, if one of the males was allowed prior access to the female and the other was able to observe the two inter-acting before being introduced into the same enclosure, he 'respected' the bond and did not attempt to acquire the female even if he was dominant to the first male. Where the second male was much higher in dominance rank than the first, his inhibition sometimes disappeared and he attacked the first male. One possible explanation is that, in a natural situation, a female is likely to escape from an individual who defeats her established male (perhaps because the former may kill her infant). In this case, the value of the female to the holding male is greater than to the attacker and the latter might be expected to contest the situation only when he could do so at low cost. This inhibition may be absent in males which greatly out-rank the holders because the rank difference enables them to acquire the female at little potential cost. It is significant to note that when Kummer presented food (a resource whose value would not differ between the holder and the contestant in the same way as a female) to the animals in the same way, males were not inhibited from taking it away from its previous possessor.

Changes in resource value may be responsible for the appearance of dominance hierarchies only at certain times of year. For example, dominance rank is only apparent in male squirrel monkeys (*Saimiri sciureus*) during the mating season (Baldwin 1968, 1971). Changes in rank are also common at the onset of mating seasons, as in *Propithecus verreauxi* (Richard 1974a). Young males may be particularly likely to challenge at this time of year because the difference in fighting ability between them and older animals will have steadily decreased over the past year as they have grown. The onset of the breeding season will produce a sudden increase in the benefits of dominance and may conse-quently make challenging advantageous. We would predict that seasonal changes in rank stability would be most marked in species showing a well defined mating season.

Age changes, too, may affect the value of resources (Trivers 1972) and might be expected to influence the distribution of contests. In many long-lived primates, males show a relatively short period of reproductive activity (Kummer 1968; Rowell 1974). When individuals reach this period of their life, they should become more likely to contest encounters owing to the greater benefits that are possible. As individuals age, both fighting ability and potential loss in encounters will decline (see p. 297). Situations may occur where it is to an old male's advantage to contest access to valuable resources but not to do so for less valuable ones. This may help to explain cases where old males who have fallen in rank are still responsible for a large proportion of copulations (e.g. Hall and DeVore 1965; Saayman 1971). However, this phenomenon may also occur if females tend to select males with whom they have mated previously.

*Predictability of contest.* Aggressive interactions and unstable dominance relationships will be more likely to occur in situations where neither of the contestants can judge the outcome of the encounter (Parker 1974; Maynard Smith 1974). This may help to explain why aggessive interactions are relatively common (i) in newly formed groups (e.g. Bernstein 1969), (ii) between members of established groups and strangers (e.g. gorillas, A. H. Harcourt personal communication), (iii) between members of the same group who have not met for a while (e.g. Vessey 1971; Bygott 1974), (iv) between rank neighbours within established groups (Mazur 1973) and (v) in species living in very large groups where the same individuals are unlikely to encounter each other frequently.

## References

Alexander, R. D. (1974). The evolution of social behavior. *Ann. Rev. Ecol. Syst.* 5, 325–83.

Archer, J. (1970). Effects of population density on behaviour in rodents. In *Social behaviour in birds and mammals,* ed. J. H. Crook, pp. 169–210. Academic Press, London.

Baldwin, J. D. (1968). The social behaviour of adult male squirrel monkeys (*Saimiri sciureus*) in a seminatural environment. *Folia primatologica* 9, 281–314.

—— (1971). The social organization of a semifree-ranging troop of squirrel monkeys (*Saimiri sciureus*). *Folia primatologica* 14, 23–50.

Barash, D. P. (1974). The evolution of marmot societies: a general theory. *Science, Washington* 185, 415–20.

Bartlett, D. P. and Meier, G. W. (1971). Dominance status and certain operants in a communal colony of rhesus macaques. *Primates* 12, 209–19.

Bernstein, I. S. (1969). Stability of the status hierarchy in a pigtail monkey group (*Macaca nemestrina*). *Anim. Behav.* 17, 452–8.
—— (1970). Primate status hierarchies. In *Primate Behavior*, vol.1, ed. L. A. Rosenblum, pp. 71–109. Academic Press, New York.
—— and Gordon, T. P. (1974). The function of aggression in primate societies. *Amer. Sci.* 62, 304–11.
Boelkins, R. C. (1967). Determination of dominance hierarchies in monkeys. *Psychoanalytic Science* 7, 317–8.
Brown, J. L. (1964). The evolution of diversity in avian territorial systems. *Wilson Bull.* 6, 160–9.
Bygott, D. (1974). Agonistic behaviour in wild male chimpanzees. Ph.D. thesis, University of Cambridge.
Carpenter, C. R. (1934). A field study of the behaviour and social relations of howling monkeys. *Comparative Psychology Monographs* 10, no. 2.
—— (1954). Tentative generalisations on the grouping behaviour of non-human primates. *Human Biology* 26, 269–76.
Castell, R. and Heinrich, B. (1971). Rank order in a captive female squirrel monkey colony. *Folia primatologica* 14, 182–9.
Chalmers, N. R. (1968). The social behaviour of free living mangabeys in Uganda. *Folia primatologica* 8, 263–81.
—— (1973). Differences in behaviour between some arboreal and terrestrial species of African monkeys. In *Comparative ecology and behaviour of primates,* ed. R. P. Michael and J. H. Crook, pp. 69–100. Academic Press, London.
Chance, M. and Jolly, C. (1970). *Social groups of monkeys, apes and men.* Cape, London.
Coelho, A. M. (1974). Socio-bioenergetics and sexual dimorphism in primates. *Primates* 15, 263–9.
Collias, N. E. (1953). Social behavior in animals. *Ecology* 34, 810–11.
Crook, J. H. (1972). Sexual selection, dimorphism and social organization in the primates. In *Sexual selection and the descent of man 1871–1971,* ed. B. Campbell, pp. 231–81. Aldine-Atherton, Chicago.
DeVore, I. and Washburn, S. (1963). Baboon ecology and human evolution. In *African ecology and human evolution,* ed. F. C. Howell and F. Boulière, pp. 335–367. Wenner-Gren Foundation, New York.
Dunbar, R. I. M. and Crook, J. H. (1975). Aggression and dominance in the weaver bird, *Quelea quelea. Anim. Behav.* 23, 450–59.
Enomoto, T. (1974). The sexual behaviour of Japanese monkeys. *J. hum. Evol.* 3, 351–72.
Fisher, R. A. (1958). *The genetical theory of natural selection.* Dover, New York.
Gale, H. S. and Eaves, L. J. (1975). Logic of animal conflict. *Nature, Lond.* 254, 463–4.
Gartlan, J. S. (1964). Dominance in East Africa monkeys. *Proceedings of the East African Academy* 2, 75–9.
—— (1968). Structure and function in primate society. *Folia primatologica* 8, 89–120.
Geist, V. (1974). On the relationship of social evolution and ecology in ungulates. *Amer. Zool.* 14, 205–20.
George, J. (1966). Why do animals fight? *Audubon* 68, 18–20.
Goss-Custard, J. D., Dunbar, R. I. M. and Aldrich-Blake, F. P. G. (1972). Survival, mating and rearing strategies in the evolution of primate social structure. *Folia primatologica* 17, 1–19.
Gouzoules, H. (1974). Harassment of sexual behaviour in the stumptail macaque, *Macaca arctoides. Folia primatologica* 22, 208–17.
Hall, K. R. L. (1963). Variations in the ecology of the chacma baboon. *Symposia of the zoological Society of London* 10, 1–28.
—— and DeVore, I. (1965). Baboon social behaviour. In *Primate Behaviour,* ed. I. DeVore, pp. 53–110. Holt, Rinehart and Winston, New York.
—— and Mayer, B. (1967). Social interaction in a group of captive patas monkeys, *Erythrocebus patas. Folia primatologica* 5, 213–236.

Hamilton, W. D. (1970). Selfish and spiteful behaviour in an evolutionary model. *Nature, Lond.* 228, 1218–20.
—— (1971). Selection of selfish and altruistic behavior in some extreme models. In *Man and beast: comparative social behavior*, ed. J. F. Eisenberg and W. S. Dillion, pp. 55–91. Smithsonian Institution Press, Washington, D.C.
Hanby, J. P., Robertson, L. T. and Phoenix, C. H. (1971). The sexual behaviour of a confined troop of Japanese macaques (*Macaca fuscata*). *Folio primatologica* 16, 123–43.
Harding, R. S. O. (1973). Predation by a troop of olive baboons (*Papio anubis*). *American Journal of Physical Anthropology*. 38, 589–91.
Hinde, R. A. (1974). *Biological bases of human social behaviour*. McGraw-Hill, New York.
Holloway, R. L. (1974) ed. *Primate aggression, territoriality and xenophobia*. Academic Press, New York.
Holst, A. von (1974). Social stress in the tree-shrew: its causes and physiological and ethological consequences. In *Prosimian biology*, ed. R. D. Martin, G. A. Doyle and A. C. Walker, pp. 389–411. Duckworth, London.
Hrdy, S. B. (1974). Male–male competition and infanticide among the langurs (*Presbytis entellus*) of Abu' Rajasthan. *Folia primatologica* 22, 19–58.
Hutchinson, G. E. and MacArthur, R. H. (1959). On the theoretical significance of aggressive neglect in interspecific competition. *Amer. Natur.* 93, 133–4.
Jarman, P. J. (1974). The social organisation of antelope in relation to their ecology. *Behaviour* 48, 215–67.
Jay, P. (1965). The common langur of North India. In *Primate Behavior*, ed. I DeVore. Holt, Rinehart and Winston, New York.
Jenni, D. A. (1974). Evolution of polyandry in birds. *Amer. Zol.* 14, 129–44.
Jolly, A. (1966). *Lemur Behavior*. Chicago University Press, Chicago.
Kawamura, S. (1967). Aggression as studied in troops of Japanese monkeys. *Brain Function* 5, 195–223.
Koyama, N. (1967). On dominance rank and kinship of a wild Japanese monkey troop in Arashiyama. *Primates* 8, 189–216.
—— (1970). Changes in dominance rank and division of a wild Japanese monkey troop in Arashiyama. *Primates* 11, 335–90.
Kruuk, H. (1972). *The Spotted Hyena*. Chicago University Press, Chicago.
Kummer, H. (1957). Soziales Verhatten einer Mantelpavian-Gruppe. *Beiheft Schweiz Zeitschrift für Psychologie* 33, 1–91.
—— (1968). *Social organisation of hamadryas baboons*. Chicago University Press, Chicago.
—— (1971). *Primate societies: group techniques of ecological adaptation*. Aldine-Atherton, Chicago.
—— Gotz, W. and Angst, W. (1974). Triadic differentiation: an inhibitory process protecting pair bonds in baboons. *Behaviour* 49, 62–87.
Lawick, H. van and Lawick-Goodall, J. van (1970). *Innocent Killers*. Collins, London.
Lockie, J. D. (1955). The breeding and feeding of jackdaws and rooks. *Ibis*, 97, 341–69.
Lorenz, K. (1966). *On Aggression*. Methuen, London.
Loy, J. (1970). Behavioural responses of free-ranging rhesus monkeys to food shortage. *American Journal of Physical Anthropology*, 33, 263–71.
—— (1971). Estrous behaviour of free-ranging rhesus monkeys (*Macaca mulata*). *Primates* 12, 1–31.
Maynard Smith, J. (1974). The theory of games and the evolution of animal conflict. *J. Theoret. Biol.* 47, 209–21.
—— and Price, G. R. (1973). The logic of animal conflict. *Nature, Lond.* 246, 15–18.
Mazur, A. (1973). A cross-species comparison of status in small established groups. *American sociological Review* 28, 513–30.
Miller, R. S. (1967). Pattern and process in competition. *Advances in ecological Research* 4, 1–74.

Missakian, E. A. (1972). Genealogical and cross-genealogical dominance relations in a group of free-ranging rhesus monkeys (*Macacca mulatta*) on Cayo Santiago. *Primates* 13, 169—80.

Mohnot, S. M. (1971). Some aspects of social changes and infant killing in the hanuman langur, *Presbytis entellus* (Primates: Cercopithecidae) in Western India. *Mammalia* 35, 175—98.

Nagel, U. and Kummer, H. (1974). Variation in cercopithecoid aggressive behavior. In *Primate aggression, territoriality and xenophobia,* ed. R. L. Holloway, pp. 159—84. Academic Press, New York.

Neville, M. K. (1972). Social relations within troops of red howler monkeys (*Alouatta seniculus*). *Folia primatologica* 18, 47—77.

Nicholson, A. J. (1955). An outline of the dynamics of animal populations. *Australian J. Zool.* 2, 9—65.

Parker, G. A. (1974). Assessment strategy and the evolution of fighting behaviour. *J. theoret. Biol.* 47, 223—43.

Perachio, A. A., Alexander, M. and Marr, L. (1973). Hormonal and social factors affecting evoked sexual behaviour in rhesus monkeys. *American Journal of physical Anthropology* 38, 227—32.

Poirier, F. E. (1974). Colobine aggression: a review. In *Primate Aggression, Territoriality and Xenophobia,* ed. R. L. Holloway, pp. 123—58. Academic Press, New York.

Richard, A. (1974*a*). Intra-specific variation in the social organization and ecology of *Prophithecus verreauxi. Folia primatologica* 22, 178—207.

—— (1974*b*). Patterns of mating in *Propithecus verreauxi.* In *Prosimian biology,* ed. R. D. Martin, G. A. Doyle and A. C. Walker, pp. 49—74. Duckworth, London.

Richards, S. M. (1972). Tests for behavioural characteristics in rhesus monkeys. Ph.D. thesis, University of Cambridge.

—— (1974). The concept of dominance and methods of assessment. *Anim. Behav.* 22, 914—30.

Ripley, S. D. (1961). Aggressive neglect as a factor in inter-specific competitions in birds. *Auk* 78, 366—71.

Rowell, T. E. (1966). Forest living baboons in Uganda. *J. Zool.* 147, 344—64.

—— (1967). A quantitative comparison of the behaviour of a wild and caged baboon troop. *Anim. Behav.* 15, 499—509.

—— (1974). The concept of social dominance. *Behavioural Biology* 11, 131—54.

Saayman, G. S. (1971). Behaviour of the adult males in a troop of free-ranging chacma baboons (*Papio ursinus*). *Folia primatologica* 15, 36—57.

—— (1972). Aggressive behaviour in free-ranging chacma baboons. (*Papio ursinus*). *Journal of Behavioural Science* 1, 77—83.

Sade, D. S. (1967). Determinants of dominance in a group of free-ranging rhesus monkeys. In *Social Communication among Primates,* ed. S. A. Altmann, pp. 99—114. Chicago university Press, Chicago.

Schaller, G. B. (1972). *The Serengeti Lion.* Chicago University Press, Chicago.

Schjelderup-Ebbe, Th. (1922). *Beiträge zur Biologie und Social-und Individual Psychologie bei Gallus domesticus.* Muller, Greifswald, Germany.

Scott, J. P. (1962). Hostility and aggression in animals. In *Roots of behaviour,* ed. E. L. Bliss, pp. 167—78. Harper and Row, New York.

—— (1974). Agnostic behavior of primates: a comparative perspective. In *Primate aggression, territoriality and xenophobia,* ed. R. L. Holloway, pp. 417—34. Academic Press, New York.

Sorenson, M. W. (1974). A review of aggressive behaviour in the tree shrews. In *Primate aggression, territoriality and xenophobia,* ed. R. L. Holloway, pp. 13—30. Academic Press, New York.

Southwick, C. H. (1967). An experimental study of intragroup agonistic behaviour in rhesus monkeys (*Macaca mulatta*). *Behaviour* 28, 182—209.

—— Begg, H. A. and Siddiqi, M. R. (1965). Rhesus monkeys in North India. In *Primate Behavior,* ed. I. DeVore, pp. 111—159. Holt, Rinehart and Winston, New York.

Struhsaker, T. T. (1967*a*) Behavior of vervet monkeys (*Cercopithecus aethiops*). *University of California Publications in Zoology* 82, 1–64.

—— (1967*b*). Social structure among vervet monkeys (*Cercopithecus aethipos*). *Behaviour* 29, 83–121.

Suarez, B. and Ackerman, D. R. (1971). Social dominance and reproductive behavior in male rhesus monkeys. *American Journal of physical Anthropology* 35, 219–22.

Sugiyama, Y. (1965*a*). Behavioural development and social structure in two troops of hanuman langurs (*Presbytis entellus*). *Primates* 6, 213–47.

—— (1965*b*). On the social change of hanuman langurs (*Presbytis entellus*) in their natural conditions. *Primates* 6, 381–418.

Syme, G. J. (1974). Competitive orders as measures of social dominance. *Animal Behaviour,* 22, 931–40.

Tinbergen, N. (1951). *The Study of Instinct.* Oxford University Press, London.

Trivers, R. L. (1972). Parental investment and sexual selection. In *Sexual Selection and the Descent of Man, 1871–1971,* ed. B. Campbell, pp. 136–79. Aldine-Atherton, Chicago.

—— (1974). Parent-offspring conflict. *Amer. Zool.* 14, 249–64.

Vessey, S. H. (1971). Free-ranging rhesus monkeys: behavioural effects of removal, separation and reintroduction of group members. *Behaviour* 40, 216–27.

Washburn, S. L., Jay, P. C. and Lancaster, J. B. (1965). Field studies of old world monkeys and apes. *Science, Washington,* 150, 1541–47.

Watson, A. and Moss, R. (1972). A current model of population dynamics in red grouse. *Proceedings of the 15th international Ornithology Conference,* pp. 134–49. E. J. Brill, Netherlands.

Williams, G. C. (1966). *Adaptation and Natural Selection.* Princeton University Press, Princeton.

Wilson, E. O. (1971). Competitive and aggressive behaviour. In *Man and Beast: Comparative Social Behavior,* ed. J. F. Eisenberg and W. S. Dillion, pp. 182–217. Smithsonian Institution Press, Washington, D.C.

—— (1975). *Sociobiology.* Harvard University Press, Cambridge, Mass.

Wise, L. A. and Zimmerman, R. R. (1973). The effect of protein deprivation on dominance measured by shock and avoidance competition and food competition. *Behavioural Biology* 9, 317–29.

Wolfheim, J. H. and Rowell, T. E. (1972). Communication among captive talapoin monkeys. *Folia primatologica* 18, 224–255.

Wrangham, R. W. (1975). The behavioural ecology of chimpanzees in Gombe National Park, Tanzania. Ph.D. thesis, University of Cambridge.

Wynne-Edwards, V. C. (1962). *Animal dispersion in relation to social behaviour.* Oliver and Boyd, Edinburgh.

Yamada, M. (1963). A study of blood-relationship in the natural society of the Japanese macaque. *Primates* 4, 43–65.

Zuckerman, S. (1932). *The social life of monkeys and apes.* Routledge, London.

# Section 4

# Comparative social behaviour

# Section 4

# Comparative social behaviour

## Introduction

The second half of E. O. Wilson's *Sociobiology, the new synthesis* surveys current knowledge of comparative social behaviour in insects, fish, reptiles, amphibia, birds, and mammals. We cannot attempt similar coverage here though the following reviews may prove useful: Brattstrom 1974 (reptiles); Barlow 1974 (reef fish), Wiley 1974 (grouse); Goss-Custard 1970 (wading birds); Newton 1972 (finches); Kaufman 1974 (marsupials); French, Stoddart, and Bobek 1975 (rodents); Estes 1974; Jarman 1974 (ungulates); Kruuk 1972, 1975 (carnivores); Jolly 1972; Eisenberg, Muckenhirn, and Rudran 1972 (primates).

Instead, we have included three papers representing different approaches to understanding species differences in social behaviour. Wilson's lucid review of the social systems of ants (1975) provides a broad coverage but has the disadvantage that it does not show to what extent current evolutionary theory fails to explain observed differences in behaviour. This is a common failing in reviews of social systems, which frequently base arguments concerning relationships between social behaviour and ecology on selected examples—a technique criticized by Wilson himself (1975). One alternative is to examine the frequency with which particular traits appear in different groups of species. This kind of evidence was extensively used by Lack (1968); we produce an excerpt from his chapter on the adaptive significance of the pair bond and its relation to diet in birds (p. 328). Finally, it is sometimes possible to adopt a more formal statistical approach. For example, in many birds and mammals, breeding animals defend large territories which supply the bulk of their food supply and that of their mates and offspring as well (Hinde 1956; Tinbergen 1957; Stokes 1974). If the primary function of defending such territories is to secure a reliable supply of food, we should predict that territory size would be positively related to the weight of the individual or group occupying it and that herbivores should have relatively smaller territories than carnivores on the grounds that their food supplies generally show a higher density. Relationships between territory size, body weight, and diet type have been formally tested for small mammals (McNab 1963), birds (Schoener 1968), and primates (Milton and May 1976) and confirm both these predictions. With some misgivings, we have included our own recent survey of primate social organization as the final paper in this selection. This investigates the relationships between ecological factors and various aspects of morphology and behaviour: body weight, group size, activity budgets, home range size, day range length, population density, breeding sex ratio, and sexual dimorphism in body weight. Perhaps too clearly, it demonstrates the problems of this approach: the necessity for reliance on heterogeneous data, the problem of dependence on groups of closely related species, and the difficulty in distinguishing between causal relationships and non-causal associations. Nevertheless, the results show that, where relationships can be firmly predicted on the basis of evolutionary or energetic theory, significant correlations can usually be demonstrated though the strength of these varies widely.

# The Ants

E. O. WILSON†

Ants are in every sense of the word the dominant social insects. They are geographically the most widely distributed of the major eusocial groups, ranging over virtually all the land outside the polar regions. They are also numerically the most abundant. At any given moment there are at least $10^{15}$ ants on the earth, if we assume that C. B. Williams (1964) is correct in estimating a total of $10^{18}$ individual insects—and take $0 \cdot 1$ per cent as a conservative estimate of the proportion made up of ants. The ants contain a greater number of known genera and species than all eusocial groups combined.

The reason for the success of these insects is a matter for conjecture. Surely it has something to do with the innovation as far back as the mid-Cretaceous period 100 million years ago, of a wingless worker caste able to forage deeply into soil and plant crevices. It must also stem partly from the fact that primitive ants began as predators on other arthropods and were not bound, as were the termites, to a cellulose diet and to the restricted nesting sites that place colonies within reach of sources of cellulose. Finally, the success of ants might be explained in part by the ability of all the primitive species and most of their descendants to nest in the soil and leaf mold, a location that gave them an initial advantage in the exploitation of these most energy-rich terrestrial microhabitats. And perhaps the behavioral adaptation was made possible in turn by the origin of the metapleural gland, the acid secretion of which inhibits growth of microorganisms. It may be significant that the metapleural gland (or its vestige) is the one diagnostic anatomical trait that distinguishes all ants from the remainder of the Hymenoptera.

The 'bulldog ants' of the genus *Myrmecia* are important in several respects for the study of sociobiology. They are among the largest ants, workers ranging in various species from 10 to 30 milli-metres in length, but are nevertheless easy to culture in the lab-oratory. They are, next to *Nothomyrmecia* and *Amblyopone,* the most primitive of the living ants. The first encounter with foraging *Myrmecia* workers in the field in Australia is a memorable experi-ence for an entomologist. One gains the strange impression of a wingless wasp just on its way to becoming an ant: 'In their incess-ant restless activity, in their extreme agility and rapidity of motion, in their keen vision and predominant dependence on that sense, in their aggressiveness and proneness to use the powerful sting upon slight provocation, the workers of many species of *Myrmecia* and *Promyrmecia* show more striking superficial resemblance to certain of the Myrmosidae or Mutillidae than they do to higher ants' (Haskins and Haskins 1950).

Yet so little do ants vary in the broad features of their societal life cycles that *Myrmecia* can be taken as an adequate paradigm for most of this group of insects. The colonies are moderate in size, containing from a few hundred to somewhat over a thousand workers. They capture a wide variety of living insect prey, which they cut up and feed directly to the larvae. The ants are formidable predators, being able to haul down and to paralyze honeybee workers. They also collect nectar from flowers and extrafloral nectaries, which appears to be the main article in their diets when the nest is without larvae. In most species the queens are winged when they emerge from the pupae, whereas the workers are smaller and wingless—the universal condition of ants. Intermediates between the two castes normally occur in some species, and occa-sionally the usual queen caste has been replaced either by ergato-gynes with reduced thoraces and no wings or by mixtures of ergatogynes and short-winged queens. However, these exceptions represent secondary evolutionary derivations and not the primitive states left over from the ancestral wasps. In some of the larger species, such as *M. gulosa,* the worker caste is differentiated into two overlapping subcastes. The workers do most or all of the foraging, while the smaller ones devote themselves principally to brood care.

Many species of *Myrmecia* engage in a spectacular mass nuptial flight. The winged queens and males fly from the nests and gather in swarms on hilltops or other prominant landmarks. As the females fly within reach they are mobbed by males, who form solid balls around them in violent attempts to copulate. After being inseminated, the queen sheds her wings, excavates a well-formed cell in the soil beneath a log or stone, and commences rearing the first brood of workers. In 1925 John Clark made the discovery, later confirmed and extended by Wheeler (1933) and Haskins and Haskins (1950), that the queens do not follow the typical 'claustral' pattern of colony founding seen in higher ants. That is, they do not remain in the initial cell and nourish the young entirely from their own metabolized fat bodies and alary muscle tissues. Instead, they periodically emerge from the cells through an easily opened exit shaft and forage in the open for insect prey. This 'partially claustral' mode of colony foundation, which is now known to be shared with most of the Ponerinae, is regarded as a holdover from a more primitive form of progressive provisioning practice by the nonsocial tiphiid wasp ancestors. More recently C. P. Haskins (1970) has shown that *Myrmecia* queens also use nutrients metabolized from their own tissues to

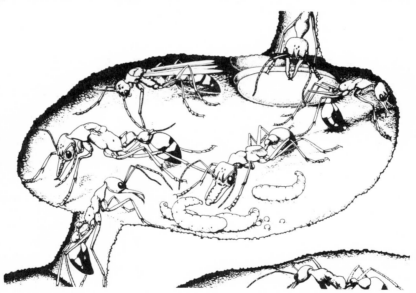

Fig. 1. A view inside the earthen nest of a colony of primitive Australian bull-dog ants (*Myrmecia gulosa*). To the extreme left is the mother queen, distinguished by her larger size, heavier thorax, and three ocelli in the center of her head. Behind her is a winged male, who is her son. The other adults are workers, all daughters of the queen. To the right, one lays a trophic egg while another offers one of its trophic eggs to a wormlike larva. Spherical queen-laid eggs, which will be permitted to hatch into larvae, are scattered singly over the nest floor. To the rear of the chamber are three cocoons containing pupae of the ants. (Drawing by Sarah Landry; from Wilson 1971a.)

help raise the brood. The important difference is that they do not depend upon it.

A typical colony of *Mymecia* is depicted in Fig. 1. As Haskins has stressed, bulldog ants display a mosaic of primitive and advanced traits in their social biology. In Table 1 I have classified many of the recorded traits according to this simple dichotomy. It must be added at once that this effort at a synthesis is no more than a set of phylogenetic hypotheses. The 'higher ants' with which *Myrmecia* is compared are all the living subfamilies except Myrmeciinae and Ponerinae. The last two subfamilies, which are the most primitive living subfamilies of the myrmecioid and poneroid complexes, respectively, share some (but not all) of the primitive traits listed for *Myrmecia*. In sum, the behavior of

Table 1. Behavioral and other traits of *Myrmecia*. (Modified from Wilson 1971; based on data of C. P. Haskins.)

*Primitive traits*
   1. Multiple queens occur in many nests
   2. The eggs are spherical and lie apart from one another on the nest floor
   3. The larvae are fed directly with fresh insect fragments
   4. The larvae are able to crawl short distances unaided
   5. The adults are highly nectarivorous and collect insects mainly as food for the larvae
   6. Transport of one adult by another is rare, awkward in execution, and not accompanied by tonic immobility on the part of the transportee
   7. There is neither recruitment among workers to food sources nor any other apparent form of co-operation during foraging
   8. Alarm communication is slow and inefficient; the nature of the signal is still unknown
   9. Colony founding is only partially claustral
  10. When deprived of workers, nest queens can revert to colony-founding behavior, including foraging above ground

*Advanced traits also found in higher ants*
   1. The queen and sterile worker castes are very distinct from each other, and intermediates are rare
   2. Worker polymorphism occurs in many species, manifested as the co-existence of two well-defined worker subcastes
   3. The colonies are moderately large and the nests regular and fairly elaborate in construction
   4. Regurgitation occurs both among adults and between adults and larvae
   5. Adults groom one another as well as the brood
   6. Trophic eggs are laid by the workers and fed to other workers and the queen
   7. The workers cover the larvae with soil just prior to pupation, thus aiding them in spinning cocoons; and they assist the newly enclosed adults in emerging from the cocoons
   8. Nest odors exist and territorial behavior among colonies is well developed
   9. Workers respond to the odor of oleic acid and possibly other decomposition products in dead nestmates by carrying the corpses out of the nest.

*Myrmecia* is well advanced into the eusocial level in most essential features, yet marked by a residue of primitive traits which gives us an indistinct and tantalizing view of what the ancestral Mesozoic ants must have been like.

The adaptive radiation of ants from something like the *Myrmecia* prototype has been extraordinarily full. Food specialization in many species is extreme, exemplified by the species of the ponerine genus *Leptogenys* that prey only on isopod crustaceans; by certain *Amblyopone* that feed exclusively on centipedes (Wilson 1958a; Gotwald and Lévieux 1972); by species of the ponerine genera *Discothyrea* and *Proceratium* that feed only on arthropod eggs, especially those of spiders (Brown 1957); by certain members of the myrmicine tribe Dacetini that prey only on springtails (Brown and Wilson 1959); and by ponerines in the genus *Simopelta* and in the tribe Cerapachyini, all of which, so far as we know, prey exclusively on other ants (Wilson 1958b; Gotwald and Brown 1966). The majority of ant groups exhibit a high degree of variability in prey choice, while a few have come to subsist on seeds. Still others rely primarily or exclusively on the anal 'honeydew' excretions of aphids, mealybugs, and other homopterous insects reared in their nests. Unquestionably the most remarkable group of all are the fungus-growing ants of the myrmicine tribe Attini. The 11 genera and 200 attine species are limited entirely to the New World. They are extremely successful in the tropics—in Brazil *Atta* is the most destructive insect pest of agriculture—and a few species range as far north as New Jersey in the United States (Weber 1972). These ants rear specialized symbiotic yeasts or fungi on organic material that they gather and carry into their nests. The substratum varies according to species: in *Cyphomrymex rimosus,* for example, it is chiefly or entirely caterpillar feces; in *Myrmicocrypta buenzlii,* dead vegetable matter and insect corpses; and in the famous leaf-cutter ants in *Atta* and *Acromyrmex,* fresh leaves, stems, and flowers. The art of gardening has been highly developed in these ants, and has even been extended to the 'manuring' of the fungi with faecal droplets rich in chitinases and proteases (Martin and Martin 1971; Martin *et al.* 1973).

Social parasitism attains its most advanced development in ants. A finely graded series of stages in the evolution of the phenomenon is displayed by various species up to and including degenerate forms of slavery in which the slave-maker workers are capable only of conducting raids and are totally dependent for minute-to-minute care on their slave workers. Other evolutionary lines lead to total inquilinism, in which the worker caste is lost.

Nesting habits have been no less diversified. A few ant species, such as members of the genus *Atta* and the extreme desert dwellers *Monomorium salomonis* and *Myrmecocystus melliger*, excavate deep galleries and shafts down into the soil, sometimes to depths of 6 meters or more (Creighton and Crandall 1954). In contrast, some arboreal members of the subfamily Pseudomyrmecinae and dolichoderine genus *Azteca* are limited to cavities of one or a very few species of plants. Some of the host plants in turn are highly specialized to house and nourish ant colonies. Experiments have shown that these plants are probably unable to survive without their insect guests (Janzen 1967, 1968, 1972). The tiny myrmicine *Cardiocondyla wroughtoni* sometimes nest in cavities left in dead leaves by leaf-mining caterpillars, while a few formicine species, *Oecophylla longinoda* and *O. smaragdina, Camponotus senex,* and certain species of *Polyrhachis*, have evolved the habit of using silk drawn from their own larvae to construct tentlike arboreal nests.

In certain respects the army ants constitute one of the most advanced grades of social evolution within the insects. A colony on the march presents one of the great spectacles of nature. Wheeler expressed it in the following way in *Ants: Their Structure, Development and Behavior* (1910): 'The driver and legionary ants are the Huns and Tartars of the insect world. Their vast armies of blind but exquisitely co-operating and highly polymorphic workers, filled with an insatiable carnivorous appetite and a longing for perennial migrations, accompanied by a motley host of weird myrmecophilous camp-followers and concealing the nuptials of their strange, fertile castes, and the rearing of their young, in inaccessible penetralia of the soil—all suggest to the observer who first comes upon these insects in some tropical thicket, the existence of a subtle, relentless and uncanny agency, directing and permeating all their activities.'

The years since Wheeler's characterization have seen the mystery largely solved. It was T. C. Schneirla (1933–1971) who, by conducting patient field and laboratory studies over virtually his entire career, first unraveled the complex behavior and life cycles of *Eciton, Neivamyrmex,* and other New World species. His results have been confirmed and greatly extended by others, especially Rettenmeyer (1963). Meanwhile, the essential features of the life cycle of the African driver ant (*Dorylus*) have been worked out by Raignier and van Boven (1955) and Raignier (1972).

Let us turn to *Eciton burchelli,* a big, conspicuous swarm

raider found in humid lowland forests from southern Mexico to Brazil. A day in the life of an *E. burchelli* colony begins at dawn, as the first light suffuses the heavily shaded forest floor. At this moment the colony is in bivouac, meaning that it is temporarily camped in a more or less exposed position. The sites most favored for bivouacs are the spaces beneath the buttresses of forest trees and beneath fallen tree trunks, or any sheltered spot along the trunks and main branches of standing trees to a height of 20 meters or more above the ground. Most of the shelter for the queen and immature forms is provided by the bodies of the workers themselves. As they gather to form the bivouac, they link their legs and bodies together with their strong tarsal claws, forming chains and nets of their own bodies that accumulate layer upon interlocking layer until finally the entire worker forces constitutes a solid cylindrical or ellipsoidal mass up to a meter across. For this reason Schneirla and others have spoken of the ant swarm itself as the 'bivouac'. Between 150,000 and 700,000 workers are present. Toward the centre of the mass are found thousands of immature forms, a single mother queen, and, for a brief interval in the dry season, a thousand or so males and several virgin queens. The dark-brown conglomerate exudes a musky, somewhat fetid odor.

When the light level around the ants exceeds 0·5 foot candle, the bivouac begins to dissolve. The chains and clusters break up and tumble down into a churning mass on the ground. As the pressure builds, the mass flows outward in all directions. Then a raiding column emerges along the path of least resistance and grows away from the bivouac at a rate of 20 meters an hour. No leaders take command of the raiding column. Instead, workers finding themselves in the van press forward for a few centimeters and then wheel back into the throng behind them, to be supplanted immediately by others who extend the march a little farther. As the workers run on to new ground, they lay down small quantities of chemical trail substance from the tips of their abdomens, guiding others forward. A loose organization emerges in the columns, based on behavioral differences among the castes. The smaller and medium-sized workers race along the chemical trails and extend them at the points, while the larger, clumsier soldiers, unable to keep a secure footing among their nestmates, travel for the most part on either side. The location of the *Eciton* soldiers misled early observers into believing that they are the leaders. As Thomas Belt put it, 'Here and there one of the light-colored officers moves backwards and forwards directing the columns.' Actually the soldiers, with their large heads and exceptionally long, sickle-shaped mandibles, have relatively little control over their nestmates and serve instead almost exclusively as a defense force. The smaller workers, bearing shorter, clamp-shaded mandibles, are the generalists. They capture and transport the prey, choose the bivouac sites, and care for the brood and queen.

At the height of their raids the *Ection burchelli* workers spread out into a fan-shaded swarm with a broad front. Dendritic columns, splitting up and recombining again like braided ropes, extended from the swarm back to the bivouac site where the queen and immature forms remain sequestered in safety. The moving front of workers flushes a great harvest of prey: tarantulas, scorpions, beetles, roaches, grasshoppers, wasps, ants, and many others. Most are pulled down, stung to death, cut into pieces, and quickly transported to the rear. Even some snakes, lizards, and nestling birds fall victim.

As one might expect, the *burchelli* colonies have a profound effect on the animal life of those particular parts of the forest over which the swarms pass. E. C. Williams (1941), for example, recorded a sharp depletion of the arthropods at those spots on the forest floor where a swarm had struck the previous day. But the total effect on the forest at large may not be very significant. On Barro Colorado Island, which has an area of approximately 16 square kilometers, there exist only about 50 *burchelli* colonies at any one time. Since each colony travels at most 100 to 200 meters in a day (and not at all on about half the days), the collective population of *burchelli* colonies raids only a minute fraction of the island's surface in the course of one day, or even in the course of a single week.

Even so, it is a fact that the food supply is quickly and drastically reduced in the immediate vicinity of each colony. Early writers jumped to the seemingly reasonable conclusion that army ant colonies change their bivouac sites whenever the food supply is exhausted. At an early stage in his work, Schneirla (1933, 1938) discovered that the emigrations are subject to an endogenous, precisely rhythmic control unconnected to the immediate food supply. He proceeded to demonstrate that each *Eciton* colony alternates between a *statary phase,* in which it remains at the bivouac site for as long as two to three weeks, and a *nomadic phase*, in which it moves to a new bivouac site at the close of each day, also for a period of two to three weeks. The basic *Eciton* cycle is summarized in Fig. 2. Its key feature is the correlation

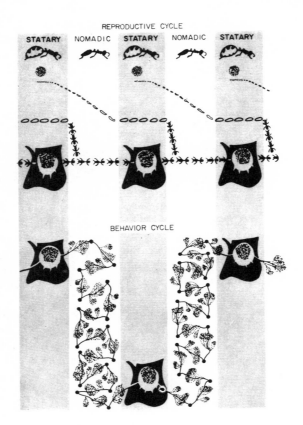

REPRODUCTIVE CYCLE

| STATARY | NOMADIC | STATARY | NOMADIC | STATARY |

BEHAVIOR CYCLE

Fig. 2. The monthly colony cycle of the army ant *Eciton burchelli*. The alternation of the statary and nomadic phases consists of distinct but tightly synchronized reproductive and behavior cycles. During the statary phase the queen, shown at the top, lays a large batch of eggs in a brief span of time; the eggs hatch into larvae; the pupae derived from the previous batch of eggs develop into adults; and, as indicated in the lower diagram, the colony remains in one bivouac site. During the nomadic phase, the larvae complete their development; the new workers emerge from their cocoons; and, as indicated in the lower diagram, the ants change their nest sites after the completion of each day's swarm raid. (Redrawn from Schneirla and Piel 1948.)

between the *reproductive cycle* , in which broods of workers are reared in periodic batches, and the *behavior cycle*, consisting of the alternation of the statary and nomadic phases. The single most important feature of *Eciton* biology to bear in mind in trying to grasp this rather complex relation is the remarkable degree to which development is synchronized within each successive brood.

The ovaries of the queen begin to develop rapidly when the colony enters the statary phase, and within a week her abdomen is greatly swollen by 55,000 to 66,000 eggs. Then, in a burst of prodigious labor lasting for several days in the middle of the statary period, the queen lays from 100,000 to 300,000 eggs. By the end of the third and final week of the statary period, larvae hatch, again all within a few days of one another. A few days later the 'callow' workers (so called because they are at first weak and lightly pigmented) emerge from the cocoons. The sudden appearance of tens of thousands of new adult workers has a galvanic effect on their older sisters. The general level of activity increases, the size and intensity of the swarm raids grow, and the colony starts emigrating at the end of each day's marauding. In short, the colony enters the nomadic phase. The nomadic phase itself continues as long as the brood initiated during the previous statary period remains in the larval stage. As soon as the larvae pupate, however, the intensity of the raids diminishes, the emigrations cease, and the colony (by definition) passes into the next statary phase.

The activity cycle of *Eciton* colonies is truly endogenous. It is not linked to any known astronomical rhythm or weather event. It continues at an even tempo month after month, in both wet and dry seasons throughout the entire year. Propelled by the daily emigrations of the nomadic phase, the colony drifts perpetually back and forth over the forest floor. The results of experiments performed by Schneirla indicate that the phases of the activity cycle are determined by the stages of development of the brood and their effect on worker behavior. When he deprived *Eciton* colonies in the early nomadic phase of their callow workers, they lapsed into the relatively lethargic state characteristic of the statary phase, and emigrations ceased. Nomadic behavior was not resumed until the larvae present at the start of the experiments had grown much larger and more active. In order to test the role of larvae in the activation of the workers, Schneirla divided colony fragments into two parts of equal size, one part with larvae and the other without. Those workers left with larvae showed much greater continuous activity. The nature of the stimuli inducing the activity, whether chemical, tactile, or whatever, remains to be determined.

In his interpretive writings, Schneirla typically failed to distinguish between proximate and ultimate causation. After demonstrating the endogenous nature of the cycle, and its control by synchronous brood development, he dismissed the role of food depletion. The emigrations, he repeatedly asserted, are caused by the appearance of callow workers and older larvae; they are not caused by food shortage. He overlooked the fuller evolutionary explanation combining the two causations: that the adaptive significance of the emigrations is to take the huge colonies to new food supplies at regular intervals, and that in the course of evolution the emergence of callows has come to be employed as the timing signal. To state this another way, if there is a selective advantage for colonies to move frequently to new feeding sites (and all the evidence from the *Eciton* studies suggests that this is so), then workers' behavior would tend to evolve in such a way as to synchronize the emigrations precisely with the presence of the life stages that cause the greatest food shortage. The internalization of the proximate cause of emigration does not alter the nature of the ultimate cause of emigration, which seems almost certainly to be the chronic depletion of food sources.

In 1958 I traced the probable evolutionary steps leading to army-ant behavior by comparing the behavior of group-raiding ants in the subfamily Dorylinae (the advanced army ants) and in the related subfamily Ponerinae. It has been stated repeatedly by previous entomologists that compact armies of ants are more efficient at flushing and capturing prey than are assemblages of foragers acting independently. This observation is certainly correct, but it is not the whole story. Another, primary function of group raiding becomes clear when the prey preferences of the group-raiding ponerines and dorylines are compared with those of ponerines that forage in solitary fashion. Most nonlegionary ponerine species for which the food habits are known take living prey of approximately the same size as their worker caste or smaller. As a rule they must depend on proportionately small animals that can be captured and retrieved by lone foraging workers. Group-raiding ants, on the other hand, feed on large arthropods or the brood of other social insects, prey not normally accessible to ants foraging solitarily. Thus, the species of *Onychomyrmex* and the *Leptogenys diminuta* group specialize on large arthropods; those of *Eciton* and *Dorylus* prey on a wide variety of arthropods that include social wasps and other ants; species of *Simopelta* and the Cerapachyini specialize on other ants; and *Megaponera foetans* and certain other large African and South African Ponerini prey on termites.

From this generalization and a close comparison of species it has been relatively easy to reconstruct the steps in evolution leading to the full-blown legionary behavior of the Dorylinae.

1.   Group raiding was developed to allow specialized feeding on large arthropods and other social insects. Group raiding without frequent changing of nests sites might occur in *Cerapachys* and related genera, but if so, this probably represents a short-lived stage, which soon gives way to the next step.

2.   Nomadism was either developed at the same time as group-raiding behavior, or it was added shortly afterward. The reason for this new adaptation was that large arthropods and social insects are more widely dispersed than other types of prey, and the group-predatory colony must constantly shift its foraging area to tap new food sources. With the acquisition of both group-raiding and nomadic behavior, the species is now truly 'legionary', that is, an army ant in the functional sense. Most of the group-raiding ponerines have evidently reached this adaptive level. Colony size in these species is on the average larger than in related, nonlegionary species, but it does not approach that attained by *Eciton* and *Dorylus*.

3.   As group raiding became more efficient, still larger colony size became possible. This stage has been attained by many of the Dorylinae, including the species of *Aenictus* and *Neivamyrmex* and at least a few members of *Eciton*.

4.   The diet was expanded secondarily to include other smaller and nonsocial arthropods and even small vertebrates and vegetable matter; concurrently, the colony size became extremely large. This is the stage reached by the driver ants of Africa and tropical Asia (*Dorylus*), the species of *Labidus,* and *Eciton burchelli,* most or all of which also utilize the technique of swarm raiding as opposed to column raiding.

The Dorylinae, then, constitutes either a phyletic group of species or a conglomerate of two or more convergent phyletic groups that have triumphed as legionary ants over all their competitors. They not only outnumber other kinds of legionary ants in both species and colonies, but they tend to exclude them altogether. Cerapachyines, for example, are relatively scarce throughout the continental tropics wherever dorylines abound, but they are much more common in remote places not yet reached by dorylines—for example, Madagascar, Fiji, New Caledonia, and most of Australia.

# References

Brown, W. L. (1957). Predation of arthropod eggs by the ant genera *Proceratium* and *Discothyrea*. *Psyche, Cambridge* **64**(3), 115.

— — and Wilson, E. O. (1956). The evolution of the dacetine ants. *Quarterly Review of Biology* **34**, 278–94.

Creighton, W. S., and Crandall, R. H. (1954). New data on the habits of *Myrmecocystus melliger* Forel. *Biological Review, City College of New York* **16**(1), 2–6.

Gotwald, W. H., and Brown, W. L. (1966). The ant genus *Simopelta* (Hymenoptera: Formicidae). *Psyche, Cambridge* **73**(4), 261–77.

— — and Lévieux, J. (1972). Taxonomy and biology of a new West African ant belonging to the genus *Amblyopone* (Hymenoptera: Formicidae). *Annals of the Entomological Society of America* **65**(2), 383–96.

Haskell, P. T. (1970). The hungry locust. *Science Journal* (January), pp. 61–7.

Haskins, C. P. and Haskins, Edna F. (1950). Notes on the biology and social behavior of the archaic ponerine ants of the genera *Myrmecia* and *Promyrmecia*. *Annals of the Entomological Society of America* **43**(4), 461–91.

Janzen, D. H. (1967). Interaction of the bull's-born acacia (*Acacia cornigera* L.) with an ant inhabitant (*Pseudomyrmex ferruginea* F. Smith) in eastern Mexico. *Kansas University Science Bulletin* **47**(6), 315–558.

— — (1969). Allelopathy by myrmecophytes: the ant *Azteca* as an allelopathic agent of *Cecropia*. *Ecology*, **50**(1), 147–53.

— — (1972). Protection of *Barteria* (Passifloraceae) by *Pachysima* ants (Pseudomyrmecinae) in a Nigerian rain forest. *Ecology* **53**(5), 884–92.

Martin, M. M., Gieselmann M. J. and Martin J. S. (1973). Rectal enzymes of attine ants, α-amylase and chitinase. *Journal of Insect Physiology* **19**(7), 1409–16.

— — and Martin J. S. (1971). The presence of protease activity in the rectal fluid of primitive attine ants. *Journal of Insect Physiology* **17**(10), 1897–1906.

Raignier, A. (1972). Sur L'origine des nouvelle sociétés des fourmis voyageuses africaines (Hyménoptères Formicidae, Dorylinae). *Insectes Sociaux* **19**(3), 153–70.

— — and Van Boven J. (1955). Etude taxonomique, biologique et biométrique des *Dorylus* du sous-genre *Anomma* (Hymenoptera Formicidae). *Annales du Musée Royal du Congo Belge, Tervuren* (Belgium), n.s. 4 (Sciences Zoologiques) **2**, 1–359.

Rettenmeyer, C. W. (1963). Behavioral studies of army ants. *Kansas University Science Bulletin* **44**(9), 281–465.

Schneirla, T. C. (1933). Studies on army ants in Panama. *Journal of Comparative Psychology* **15**(2), 267–99.

— — (1938). A theory of army-ant behavior based upon the analysis of activities in a representative species. *Journal of Comparative Psychology* **25**(1), 51–90.

— —and Piel, G. (1948). The army ant. *Scientific American*, **178**(6) June, 16–23.

Weber, N. A. (1972). *Gardening ants: the attines.* Memoirs of the American Philosophical Society no. 92. American Philosophical Society, Philadelphia. xx + 146 pp.

Wheeler, W. M. (1933). *Colony-founding among ants, with an account of some primitive Australian species.* Harvard University Press, Cambridge. x + 179 pp.

Williams, C. B. (1964). *Patterns in the balance of nature and related problems in quantitative biology.* Academic Press, New York. vii + 324 pp.

Williams, E. C. (1941). An ecological study of the floor fauna of the Panama rain forest. *Bulletin of the Chicago Academy of Science* **6**(4), 63–124.

Wilson, E. O. (1958a). The beginnings of nomadic and group-predatory behavior in the ponerine ants. *Evolution* **12**(1), 24–31.

— — (1958b). Observations on the behavior of the cerapachyine ants. *Insectes Sociaux* **5**(1), 129–40.

— — (1971). *The insect societies.* Belknap Press of Harvard University Press, Cambridge. x + 548 pp.

# The significance of the pair-bond, and sexual selection in birds (extract)

D. LACK†

In reviewing the literature on the pair-bond in birds, there is danger that the attention inevitably given to unusual pairing habits, brilliant plumage or bizarre displays may mislead the reader as to their prevalence. Similarly, if a bird could read out daily news-papers, it might get an exaggerated idea of the frequency of divorce or of glamorous screen stars in our midst. Monogamy, in birds and men, tends to be dressed in drab colours, but it is far more frequent than the exotic alternatives. As shown in Table 1 over nine-tenths of all birds are monogamous, in nidicolous species, many of which do not feed their young, nearly four-fifths. More-over, all except two of the subfamilies with unusual pairing habits consist of very few species, the exceptions being the humming-birds, in which most of the few species so far studied are promis-cuous, and the pheasants Phasianinae, many of which have harem polygyny. The above proportions are unlikely to be changed by further research (unless many more hummingbirds should prove to be monogamous than is at present supposed).

Since, in the species which fed their young, clutch-size has been evolved in relation to the number of young which they can feed, and since more food can be brought by two parents than one, then provided that the sex ratio is fairly near to equality, each adult will tend to leave more offspring if it has one mate and both help to feed the brood than if it tries to have more than one mate and only one parent feeds the young. So far as is known in birds, the sex ratio tends to be about equal at hatching, and there are very few species in which the adults of one sex are as much as twice as com-mon as those of the other. Moreover, except in the polygynous and promiscuous species discussed later males tend to be rather commoner than females. In general, therefore, natural selection may be expected to favour monogamy in species which feed their young.

† Reprinted with permission from the publisher from D. Lack: *Ecological Adaptations for Breeding in Birds.* Copyright © 1968 by Methuen Publishing Co. Ltd.

## Table 1. Relation of diet to pair bond

| Pair-bond | Percentage of nidicolous species of each type | | |
|---|---|---|---|
| | Monogamous | Polygynous | Promiscuous |
| *Main diet* | | | |
| Land animals | 42 | 0 | 0 |
| Land animals and plants | 27 | x | 4 |
| Aquatic animals | 5 | 0 | 0 |
| Seeds | 11 | 1 | 0 |
| Fruit | 8 | 0 | 2 |
| *All nidicolous* | 93 | 1 | 6 |

| | Percentage of nidifugous species of each type | | | | |
|---|---|---|---|---|---|
| | Monogamous | Polygynous | Promiscuous | Polyandrous | Polygynous and Polyandrous |
| Land animals | 12 | 0 | 0 | 0 | 0 |
| Land animals and plants | 4 | 0 | x | 2 | 0 |
| Aquatic animals | 36 | 0 | x | x | 0 |
| Aquatic plants | 14 | 0 | x | 0 | 0 |
| Land plants | 17 | 8 | 2 | 0 | 4 |
| *All nidifugous* | 83 | 8 | 2 | 2 | 4 |
| *All birds* | 92 | 2 | 6 | 0·4 | — |

Notes. The number of species in each group has been assessed on the assumption that those whose habits are unknown are similar to their nearest relatives. The percentages are calculated separately for the two halves of the table, x means less than 0·5 per cent. In the percentages for all birds, those species of Tinamidae which are both polygynous and polyandrous have been scored as ½ in each category.

Birds are termed monogamous if a male and female stay together for one brood. In some species, the male and female usually stay together, or re-mate with each other after separation in winter, for as many seasons as both remain alive. At least in one species of gull, the pairs which do this raise, on the average more offspring than those which change mates. The same probably applies to the other long-lived birds which usually pair for life. But in birds with a high annual mortality, say of around 50%, as in many passerine species, such re-mating is rare.

Given that the sex ratio is fairly near to equality, monogamy will be advantageous not only to the species which fed their young, but to all in which the co-operation of two parents results in more young being raised than by a single unaided adult. In geese, for instance, the male guards the sitting female and in these and many other nidifugous birds both parents escort and protect their young, though they do not feed them. Migratory geese even take their young with them to the winter-quarters, and they pair for life.

It is more surprising to find that monogamy is the rule in many ducks and anicoline waders, in which only one parent incubates and cares for the young and in certain parasitic cuckoos, in which neither parent does so. Presumably, even in such species, a male tends to leave more offspring if mated with only one female than with several. This will be helped if the sex ratio is nearly equal so that it is difficult to obtain more than one mate. Moreover, in the parasitic cuckoos concerned, the male regularly feeds the female, which perhaps enables her to lay more eggs than would otherwise be possible. In various arctic wading birds, monogamy may be advantageous in allowing the pair to arrive mated on the breeding grounds, and so to lay their eggs without delay, which could be important in the short summer. The same might apply to northern ducks, and in the latter, probably owing to the extra mortality through predation during incubation, females are scarcer than males. Under these circumstances, a drake will perhaps raise more offspring if it secures a mate in the winter and thereafter drives away other drakes, than if it tries to copulate with several ducks and, with the intense competiton, perhaps succeeds with none.

**The exceptions to monogamy**

The figures in Table 1 show that about 2% of the species of birds are polygynous, 0·4% of the species are polyandrous, and 6% of the species are promiscuous. (The percentage of promiscuous species might be too high if it is wrong to assume that almost all hummingbirds are promiscuous.) In the polygynous and promiscuous species, only the female incubates and raises the young, while in the polyandrous species only the male does so, but there are many monogamous species in which the female alone, or less commonly the male alone, raises the family, so this habit has not necessarily led to the unoccupied sex having more than one mate.

Only in certain nidifugous species do the young feed for themselves from hatching, and probably it is owing to the reduced need for parental care that polygyny is much commoner in nidifugous than nidicolous species, while polyandry is confined to nidifugous species. However, many other nidifugous species are monogamous. Promiscuity is proportionately about as common in nidicolous as nidifugous species.

Unusual pairing habits seem linked indirectly with diet, though the link is less certain than in the case of nesting dispersion. In both nidicolous and nidifugous birds, nearly all the species which eat animal food are monogamous, while nearly all the nidicolous polygynous species eat many seeds, nearly all the nidicolous promiscuous species eat much fruit or nectar, and nearly all the nidifugous species which are either polygynous or promiscuous eat much plant food. Snow (1962) for the manakins Pipridae, and Crook (1962, 1964) for the ploceine weavers, independently suggested that fruit or seeds respectively are much easier to obtain for the young than are insects, so that, in such species, there is less need for the cock to feed the young, and his emancipation from family duties may have been an important predisposing factor favouring promiscuity or polygyny. (That a diet of seed is linked with polygyny and a diet of fruit with promiscuity might perhaps be related in some way with the associated difference in habitat, as grass seed is found mainly in open country and fruits in evergreen forest.) However, diet, is, at most, only a subsidiary cause of unusual pairing habits, since many other seed-eating, fruit-eating and nectar-eating birds are monogamous, and so are some of the vegetarian nidifugous species.*

## Polygyny

Polygyny, with a pair-bond between one male and several females, has been evolved about eight times in birds, but it is hard to be sure exactly how often because it is found in certain related genera in which it may, or may not, be of independent origin. Successive polygyny is restricted to passerine birds and harem (simultaneous) polygyny to cursorial nidifugous species, with the possible addition of the parasitic viduines.

Successive polygyny, in which each male usually has two to four females in its territory is found in two genera of Ploceidae which nest in colonies, *Ploceus* and *Bubalornis*, and one genus of Ploceidae and two of Icteridae which nest in grouped territories in tall grass, *Euplectes, Agelaius* and *Xanthocephalus*. In certain other species, some males are monogamous and others are regularly bigamous, including various solitary and colonial species of

* During one of the perennial discussions on divorce in the daily press, a correspondent suggested that mankind ought to be monogamous because this is nature's rule, particularly in birds. Apart from the mistaken notion that what happens in nature is necessarily the best guide to what ought to happen in man, the present survey would suggest that irregular mating habits are permissible in vegetarians.

Icteridae, four species of wrens Troglodytidae, two flycatchers in the genus *Ficedula,* the Penduline Tit *Remiz pendulinus* and the pyrrhuloxiine finch *Spiza americana.* In many other passerine species, bigamy occurs but is very rare, showing that it is selected against. This is probably because it is normally advantageous for the male to help in feeding the brood. In *Ficedula hypoleuca,* for instance, more young are raised if the male assists, as in a monogamous pair, than if he does not help, as often happens if it is his second mate (von Haartman 1956).

If polygyny is to succeed, food must be sufficiently plentiful to enable the female to raise a brood unaided. That is presumably why the insectivorous forest species of Ploceinae are monogamous, whereas those which eat seeds in savanna or grassland are polygynous, as are a few savanna species which eat insects where they are plentiful beside water. Similarly the insectivorous woodland species of Icteridae are monogamous, but several species in grassland or reeds are polygynous. Again, the wren *T. troglodytes* is monogamous on bleak Scottish islands but may be bigamous in rich Dutch woods.

However, many other species with a rich food supply are monogamous, presumably because, if the male helps to feed the young, a larger brood can be raised. Hence some further factor is needed to account for the evolution of polygyny. The polygynous species do not at first sight appear to have any other ecological factor in common. Only some of them build elaborate nests, and only some of them breed in open (two-dimensional) habitats, while other species which have elaborate nests or which breed in open habitats are monogamous.

In the wren *Cistothorus palustris,* the Dickcissel *Spiza americana* and the flycatcher *Ficedula hypoleuca,* some females pair with already mated males even though other males are unmated, probably because the unmated males occupy less suitable nesting territories. This argument, due to Verner (1964), can be extended to the truly polygynous species, provided that their nesting places are so restricted, and their territories are so easily defended, that a proportion of the males can exclude all the rest, and in one polygynous species, *Agelaius phoeniceus*, Orians (1961) proved by shooting that the owners of territories in fact prevent many other males from breeding. Under these conditions, many females must join already mated males if they are to raise young.

Once polygyny has been evolved, there will be unusually strong competition for mates, hence the males will evolve brilliant plumage or displays, which may render them more liable to predation, so that the one-year-old males, which are unlikely to get mates, will not attempt to breed. While it has not been proved in any polygynous species that the mature males have a higher mortality than the females, this would explain why the brightly coloured males of the polygynous Ploceidae revert to cryptic plumage when not breeding; moreover, the mortality is higher in adult males than females in a promiscuous icterid. Because of these trends, once polygyny has been evolved, the proportion of females in the breeding population is likely to increase, and hence to strengthen selection in favour of polygyny. Even so, species in at least four genera of Ploceinae have secondarily evolved monogamy, presumably because food is too sparse for the female to raise her young unaided.

Outside the passerine birds, the only nidicolous species which seems often to be polygynous is the Bittern *Botaurus stellaris*. The factors involved are unknown. The male is unique among polygynous species, as is the male Great Snipe *Capella media* among promiscuous species, in being cryptically coloured, presumably because it is highly vulnerable to predators.

As regards nidifugous species, in the polygynous Phasianidae each male has a harem and the females disperse to nest and raise their families unaided. As the young feed for themselves, two parents are less necessary for raising them than in nidicolous birds. Even so, many other Phasianidae are monogamous and a few are promiscuous, and the factors favouring these different types of pair-bond are not known. In the polygynous and promiscuous species, the males are brightly coloured and start breeding when a year or more older than the females, presumably for the same reason as in the polygynous and promiscuous passerine species.

Polygyny takes a different form in various ratites, also in one anatid, as two or more females lay in a common nest. In most of the ratites concerned, the male raises the brood alone, and in some Tinamidae the females later move away to lay for another male. It is hard to see what advantage the latter system could have over monogamy, except that it substantially reduces the period for which the eggs are left uncovered during laying, when predation is perhaps highest. Once again, selection may reverse the trend towards polygyny, as one tinamou is monogamous. The sex ratio is not reliably known in any nidifugous polygynous or promiscuous species.

## Polyandry

Polyandry is ecologically equivalent to polygyny, the difference being that the female plays the dominant role in courtship and the male raises the family unaided. Polyandry is much rarer than polygyny, probably because it is rare in birds for the normal sex roles to be reversed, but this happens in many limicoline species, nearly all of which are monogamous. Hence while reversal of the sex roles is a prerequisite for polyandry, it does not necessarily lead to it. Why in any species a reversal of the sex roles should be advantageous is not known.

Polyandry has probably been evolved independently five times, in buttonquails Turnicidae, painted snipe Rostratulidae, various but not all lilytrotters Jacanidae, various but not all tinamous Tinamidae, and one rail Rallidae. All these species are nidifugous but, in contrast to the nidifugous polygynous and promiscuous species, most of them eat mainly animal food. There seems nothing peculiar in their ecology to suggest why they should have evolved polyandry. In most of them, each female lays clutches for a succession of males, but in tinamous several females lay a common clutch first for one male and then another, as just mentioned, and in the polyandrous rail, one female associates with several males which help to raise a common family.

## Promiscuity

Among nidicolous birds, promiscuity has been evolved in the hummingbirds, Trochilidae, which eat nectar and insects, the Indicatoridae, which are brood parasites, one seed-eating euplectine species, and five frugivorous families. Among the last, the manakins Pipridae are closely related to the cotingas Cotingidae, and the birds of paradise Paradisaeidae to the bowerbirds Ptilonorhynchidae, so the habit may have been evolved only twice independently in these four families, though each, as at present classified, also includes monogamous insectivorous species. Among nidifugous birds promiscuity has been evolved in many grouse Tetraonidae, a few pheasants Phasianidae, a few ducks Anatidae (twice independently), one bustard Otididae and several waders Scolopacidae (twice independently). In all, therefore, promiscuity has probably been evolved thirteen times in birds, more often than either polygyny or polyandry.

The evolution of promiscuity raises the same problems as those just discussed for polygyny. The males do not breed until a year later than the females, the habit is much commoner in plant-feeding than animal-feeding birds, and in nidifugous than nidicolous species, but most fruit or nectar-eating nidicolous species and various vegetarian nidifugous species are monogamous or poly-gynous. Hence there seems no clear link between diet or other ecological factors and unusual pairing habits in birds. In this respect the promise of Crook's study of the Ploceinae has not been fulfilled, in contrast to his findings on breeding dispersion.

### Lek displays

The frugivourous Icteridae, unlike the other promiscuous birds, breed in colonies. In the promiscuous Anatidae the males chase the females about. In the other promiscuous groups, the males display prominently on separate territories, where they are visited for copulation by the females. In many of them, the males are widely separated, though within earshot, in a few they are nearer, in a 'dispersed lek', and in yet others they assemble on communal display grounds, the leks or arenas, where each male has a small territory or court.

Communal lek display has been evolved separately nine times in birds, a striking instance of convergent evolution. It occurs in most species of manakins Pipridae, in at least one cotingid, the Cock-of-the-Rock *R. rupicola,* in two species of birds of paradise Paradisaeidae, in one ploceine weaver *Euplectes jacksoni,* in a few hummingbirds Trochilidae, in various grouse Tetraonidae, in the bustard *Otis tarda,* and in two waders the Ruff *Philomachus pugnax* and the Great Snipe *Capella media.* The male manakins and Cock-of-the-Rock have brilliant plumage and make loud noises by mechanical means, the male birds of paradise and many of the grouse have bright plumage and loud calls, the Ruff displays its bright plumage silently, and the male hummingbirds and Great Snipe have dull plumage but call repeatedly. Nearly all these species have exaggerated postures. Leks are conspicuous, as the males have to attract the females, but in all the species concerned, the behaviour is adapted to reduce the risk of predation so far as possible. The grouse and limicoline species display primarily in the early morning or late evening when, through lack of upcurrents, most raptorial birds are not hunting, and most of them use very open flat ground, where mammals cannot approach undetected. Further, many of the species that perform on the forest floor remove fallen leaves, so that snakes and other ground predators may find it hard to approach undetected, while the birds of paradise display in the tops of the trees (Wallace 1869). All of them also, tend to use traditional sites, which presumably persist as such because they have proved safe in the past. Further, just as an approaching predator is more likely to be noticed by one member of a feeding flock than by a bird feeding solitarily, so it is more likely to be noticed by one member of a displaying group than by a bird displaying solitarily. This may, indeed, be the main advantage of communal display (Koivisto 1965). In addition, it perhaps helps the females to find males in breeding condition, though Koivisto found no difference in the number of females attracted to large and small leks respectively.

Fig. 1. Lek display of Ruff *Philomachus pugnax*. The only species with marked individual variations in male plumage.

In the Pipridae, Cotingidae, Paradisaeidae, Trochilidae, Tetraonidae and Otididae, various other species are monogamous, and the males of these species tend to be dully coloured and to help in raising the family. Moreover the monogamous species of Pipridae and Cotingidae eat proportionately more insects than the promiscuous species, suggesting that the correlation of promiscuity with a vegetable diet is more than a coincidence. In these familes, the monogamous species probably represent the primitive condition.

The next stage in evolution is that found in a few Pipridae, many Cotingidae, most Paradisaeidae, most Trochilidae, and some Tetraonidae, in which the males display solitarily, and the females visit them solely for copulation. In at least one cotingid and several birds of paradise, such males display nearer to each other than would be expected by chance, and this leads on to the final stage, shown by the species in which the males display close together on leks. The first step away from monogamy, the emancipation of the male from family duties, may have been assisted by a vegetarian diet or by the possession of nidifugous young which feed for themselves. Once the male's sole function in breeding is to copulate with the female, the pair-bond is likely to break down, and once the male mates with several females, sexual selection will be intensified, leading to the evolution of brilliant plumage and displays, up to a limit set by predation. The final stage, from solitary to communal display, would follow if the males tend to display in earshot of each other, and if a closer grouping increases the chance, either of their detecting predators or of each male acquiring a mate.

A similar evolutionary sequence may have occurred in the Scolopacidae, in which there are (a) many monogamous species in which both sexes raise the brood, (b) a few in which each male displays solitarily, the sexes meet solely for copulation, and the female raises the family unaided and (c) a very few with lek displays.

The sequence must, however, have been different in the genus *Euplectes*, all except one species of which breed in grouped nesting territories and have successive polygyny. In various of these other species, the territories are extremely small, and perhaps it was through such an intermediate stage that the exceptional species *E. jacksoni* evolved a lek display, with the females nesting outside the display areas. The sequence must also have been different in the promiscuous icterids, since these are colonial, and in this family there are no promiscuous species which display solitarily or in leks (while those with successive polygyny appear to constitute a separate line of evolution). Promiscuity takes yet another form in ducks, in which it perhaps arose through an intensification of the habit, found in many more-or-less monogamous species, of other males pursuing and attempting to copulate with mated females. The puzzle in ducks is why most species have remained monogamous although the males take no part in rearing the family.

## Pair formation

The Reverend J. M. McWilliam, Scotland's 'minister of ornithology', said in a lecture on birds that if he took his wife to Edinburgh and did not wish to lose her, he had two alternatives, (i) to have a fixed address to which they could both return, and (ii) to follow her about continuously. Similarly the pair-formation of birds is of two main types (Lack 1940). In one, the male (or in a few species the female) isolates itself in a territory, and in the other the pairs form in the winter flocks and the male and female keep continuously near each other. The first method is found in territorial passerine, limicoline and many other birds, and also in many colonial species in which each male defends a small nesting territory. The second method is found in cardueline finches, some tits, other limicoline birds, various Laridae, most Anatidae, various Phasianidae, and others. In these latter groups, it is evidently advantageous for the birds to remain in flocks until breeding commences, presumably because this assists either the finding of food or the avoidance of enemies. Both types of pair-formation are found in monogamous species. In many polygynous and promiscuous species, the males are likewise isolated in territories, but the polygynous pheasants and the promiscuous ducks perhaps mate when in flocks.

## Secondary sexual characters

Clearly, it is nowadays more meaningful to speak of the functions of secondary sexual characters in birds than of the theory of sexual selection. That, as pointed out by Wallace, many dull-coloured species have elaborate displays is not really surprising, since they are chiefly birds such as warblers Sylviinae and waders Scolopacidae, in which there is evidently need for cryptic colouring in both sexes; and many such species have striking songs or calls. It is also not surprising that the most elaborate male plumage and displays occur in promiscuous and polygynous species, because in these a successful male will acquire several mates, and hence there will be unusually strong selection for those characters which enable it to attract females. There is the further need, particularly among related species breeding in the same area, that females should recognize males of their own species. For this there is strong selection as, in general, hybrids survive or breed less successfully than either parent species. Hence in many species the males in breeding plumage look distinctive, even though their females and juveniles may look rather alike. Moreover where there is strong selection for cryptic, and hence similar, plumage in both sexes, as in the warblers, the males have usually evolved distinctive songs instead (Mayr 1940). Specific recognition may be particularly necessary in birds which meet solely for copulation, or in which casual matings are frequent. Probably it is for this reason that strikingly distinctive male plumage has been evolved by many species of Paradisaeidae, Trochilidae, Phasianidae and Anatidae. It is only curious that, despite this, hybrids are more frequent in these families than any others (Sibley 1957).

Various further puzzles remain. One is the tendency for essentially 'male' characters to be transferred to the female, in which they appear to be functionless (Winterbottom 1929). In the genus *Passer*, for instance, the female is much duller than the male in the House Sparrow *P. domesticus*, but has plumage of typical male pattern in the otherwise very similar Tree Sparrow *P. montanus*. Many other examples could be cited from birds, and also from mammals, in which the females of various species have horns, while the nipple of the human male provides an example of transfer in the opposite direction. Another tendency is for the males of land birds to lose their distinctive secondary plumage on oceanic islands (Lack 1947). This is perhaps due to the absence of related species, as the female then has no difficulty in recognizing a male of her own species. However, distinctive male plumage has also been partly lost in the geospizine finches of the Galapagos, where there are several species on each island (and if it were lost before their divergence, one might have expected it to be reacquired).

### Differences in size between the sexes

Amadon (1959) postulated that in most birds the male is slightly larger than the female because the female has evolved the size best suited to the ecology, including feeding habits, of the species, while the male is slightly stronger to help it in excluding rival males and in obtaining a mate. In various polygynous and promiscuous species, however, the male is not slightly but much larger than the female, nearly twice as large in the Turkey *Meleagris gallopavo* and over twice as large in the Capercaillie *Tetrao urogallus*, in contrast, for instance to the monogamous Willow Grouse *L. lagopus*, in which the male is only 13% larger than the female. The male is also much larger than the female in the polygynous Phasianinae and the promiscuous Icteridae, Anatidae, and Great Bustard *Otis tarda*. But Selander (1966) doubted whether larger size had been evolved in such species because it helps the male to acquire more mates, because the male is not appreciably larger than the female in the promiscuous Pipridae, Paradisaeidae or Trochilidae. In my view, however, the association between a large male and promiscuous or polygynous habits is too frequent to be due to chance, and I would suppose that, in these other groups, either the pairing habits differ, so that large size is not particularly advantageous to the male, or that any advantage in large size for mating is offset by a disadvantage in regard to feeding. The biggest difference between the sexes in birds occurs in the Australian passerine species *Cinclorhamphus cruralis* ($\delta$ wing 65—68 mm, $\mathcal{P}$ wing 28—32 mm) (Amandon 1959); the habits of this species are not properly known, but it is probably not monogamous (H. J. Frith, personal communication).

There are also some birds in which the female is larger than the male, notably those which prey on mammals or birds, such as the Falconidae, Accipitridae and Strigidae, an extreme case being the hawk *Accipiter fasciatus* ($\delta$ wing 269 mm, $\mathcal{P}$ wing 477 mm) (Amadon 1959). As these three families are not closely related to each other, their similarity in this respect is attributable to convergent evolution, and here the size difference is presumably adapted in some way to their feeding habits. This view is reinforced by the fact that in the carrion-feeding Falconiformes, notably the Cathartidae, Aegypiinae and Daptriinae, which likewise are not closely related to each other, the two sexes are very similar in size. Other groups in which the female is larger than the male are the frigate birds Fregatidae, the skuas Stercorariidae and the coucals *Centropus*, together with many but not all limicoline birds.

Where the two sexes of one species differ in size, it was formerly assumed that this must be related in some way to sexual behaviour. But there is nothing in the sexual behaviour of raptorial birds to suggest why this should be so. In another group, the Anatidae, a sexual difference in size is correlated with the type of nesting site, the female being proportionately smaller than the male in hole-nesting than open-nesting species, presumably because the smallest holes into which the females can squeeze are the safest (Bergmann 1965). A 21% difference in the size of the beaks of the two sexes in the woodpecker *Melanerpes striatus* in Dominica is associated with a difference in feeding habits, and both sexes benefit, as it reduces the potential competition between them for food (Selander

1966). The latter author postulated that the same factor explains the difference in size of the two sexes in various passionate birds, especially on islands where related species are absent, and also in various limicoline birds and grebes. Probably the size difference in hawks is similarly of advantage in reducing competition for food, as in several of the species concerned the two sexes catch prey of different size. That the male is the smaller and catches smaller prey is perhaps because he catches most of the prey when the nestlings are small and are still brooded by the female. Hence some of the size differences between the sexes have been evolved in relation to sexual behaviour and others in relation to ecology.

### References

Amadon, D. (1964). The evolution of low reproductive rates in birds. *Evolution* **18**, 105–10.

Bergman, G. (1965). Der sexuelle Grössendimorphismus der Anatiden als Anpassung an das Höhlenbrüten. *Commentationes Biological Soc. Sci. Fenn.* **28**, 1–10.

Crook, J. H. (1962). The adaptive significance of pair formation types in weaver birds. *Symp. zool. Soc. Lond.* **8**, 57–70.

–– (1964). The evolution of social organisation and visual communication in the weaver birds (*Ploceinae*). *Behav. Suppl.* **10**.

Haartman, L. V. (1956). Der Trauerfliegenschnapper. III Die Nahrungsbiologie. *Acta Zool. Fenn.* **83**, esp. p. 37.

Koivisto, I. (1965). Behaviour of the Black Grouse, *Lyrurus tetrix* (L.) during the spring display. *Finnish Game Res.* **26**, 60.

Lack, D. (1940). Pair-formation in birds. *Condor* **42**, 269–86.

–– (1974). *Darwin's Finches.* Cambridge University Press.

Mayre, E. (1940). Speciation phenomena in birds. *Amer. Nat.* **74**, 249–78.

Orians, G. H. (1961). The ecology of blackbird (*Agelaius*) social systems. *Ecol. Monog.* **31**, 285–312.

Selander, R. K. (1966). Sexual dimorphism and differential niche utilization in birds. *Condor* **68**, 113–51.

Sibley, C. G. (1957). The evolution and taxonomic significance of sexual dimorphism and hybridization in birds. *Condor* **59**, 166–91.

Snow, D. W. (1962). A field study of the Black and White Manakin, *Manacus manacus,* in Trinidad. *Zoologica* **47**, 65–104.

Verner, J. (1964). Evolution of polygamy in the Long-billed Marsh Wren. *Evol.* **18**, 252–61.

Wallace, A. R. (1869). *The Malay Archipelago.* (p. 354 of edition of 1902). Constable, London.

Winterbottom, J. M. (1929). Studies in sexual phenomena. VII. The transference of male secondary sexual display characters to the female. *J. Genet.* **21**, 367–87.

# Primate ecology and social organization

T. H. CLUTTON-BROCK AND PAUL H. HARVEY †

## Introduction

Despite wide intraspecific variation, different primate species show characteristic modal patterns of social organization. Differences between species are often extremely large: breeding group size varies from two animals (as in *Loris, Galago, Pongo*) to more than 70 (*Miopithecus*); adult sex ratio in breeding groups from parity (*Lemur, Propithecus, Hylobates*) to more than seven females per male (*Erythrocebus patas*); home range size from less than 1 ha (*Lepilemur, Callicebus*) to more than 50 km² (*Erythrocebus patas*).

Early attempts to investigate the functions of these differences relied on comparisons of social behaviour between species allocated to gross ecological categories (Crook and Gartlan 1966; Denham 1971; Jolly 1972; Eisenberg, Muckenhirn, and Rudran 1972) and made little attempt to examine differences within as well as between ecological groups. Subsequent critiques (Struhsaker 1969; Clutton-Brock 1974b; Altmann 1974; E. O. Wilson 1975) pointed out that social behaviour varies widely between species allocated to the same ecological category and challenged the view that interspecific differences were closely related to ecological variation at this level. Instead, they stressed that differences in social organization were frequently associated with relatively fine differences in ecology, such as the proportion of mature foliage in the diet (see Hladik and Hladik 1972; Clutton-Brock 1974b; Sussman in press). They also emphasized the point that close associations between behavioural and ecological variables might not be expected since different species have probably evolved different adaptations to ecologically similar situations.

Systematic investigation of relationships between ecological and behavioural differences is necessary to show at what level the two are related. Two reviews have recently attempted this approach with primates. Milton and May (1976) investigated some of the factors affecting interspecific differences in home range size (see

† Reprinted from *J. Zool., Lond.* (1977) **183**.

p. 359) while Jorde and Spuhler (1974) examined interrelationships between a large number of behavioural and ecological variables. Both studies used considerably fewer species than the sample used in this analysis.

In this paper, we attempt to describe both associations between gross ecological categories and behavioural differences, and relationships between different behavioural variables. Our sample includes a total of 100 primate species (see Table 1).

Table 1. Species included in the analysis and sources used for calculating species averages. Napier and Napier (1967) was also used in virtually all cases and provided the majority of body weights. Where no references are shown, only the latter source was used. Genera are arranged in taxonomic order, species within genera in alphabetical order.

Lemuridae
*Lemur catta* (Ringtailed lemur)
Jolly (1966)
Jolly (pers. comm.)
Sussman and Richard (1974)
*Lemur fulvus* (Ruffed lemur)
Sussman and Richard (1974)
Sussman (1975)
Sussman (in press)
*Lemur macaco* (Brown or black lemur)
Petter (1962)
Jolly (1972)
*Lemur mongoz* (Mongoz lemur)
Tattersall and Sussman (1975)
*Lemur rubriventer*
Petter (1962)
Jolly (1972)
*Lemur variegatus* (Variegated lemur)
Petter (1962)
Jolly (1972)
*Hapalemur griseus* (Gentle lemur)
Pollock (pers. comm.)

*Lepilemur mustelinus* (Sportive lemur)
Charles-Dominique and Hladik (1971)
Hladik and Charles-Dominique (1974)
*Cheirogaleus* spp. (Dwarf lemurs)
Petter (1962)
Jolly (1972)
*Microcebus murinus* (Mouse lemur)
Charles-Dominique and Martin (1972)
*Phaner furcifer* (Forked lemur)
Jolly (1972)
*Indri indri* (Indri)
Pollock (in press)
Pollock (pers. comm.)
*Avahi laniger* (Woolly lemur)
Petter (1962)
Jolly (1972)
*Propithecus diadema* (Diademed sifaka)
Pollock (pers. comm.)

*Propithecus verreauxi* (White sifaka)
Sussman and Richard (1974)
Jolly (pers. comm.)
Jolly (1966)
Sussman and Richard (1974)
Richard (in press)
*Daubentonia madagascariensis* (Aye-aye)
Petter (1962)
Jolly (1972)
Lorisidae
*Loris tardigradus* (Slender loris)
Jolly (1972)
Petter and Hladik (1970)
*Nycticebus coucang* (Slow loris)
Jolly (1972)
*Arctocebus calabarensis* (Golden potto)
Jewell and Oates (1969)
Charles-Dominique (1974a)
Jolly (1972)
*Perodicticus potto* (Potto)
Charles-Dominique (1974a)
Charles-Dominique (1974b)

*Galago alleni* (Allen's
bushbaby)
Charles-Dominique
(1974a)
*Galago crassicaudatus*
(Greater bushbaby)
Bearder and Doyle
(1974)
*Galago demidovii* (Dwarf
bushbaby)
Charles-Dominique and
Martin (1972)
Charles-Dominique
(1974a)
*Galago (Euoticus)
elegantulus* (Needle-
clawed bushbaby)
Jewell and Oates (1969)
Charles-Dominique and
Martin (1972)
Charles-Dominique
(1974a)
*Galago senegalensis* (Lesser
bushbaby)
Bearder and Doyle
(1974)

Callithricidae
*Callithrix jaccus* (Common
marmoset)
*Leontideus rosalia* (Golden
lion tamarin)
Coimbra-Filho and
Mittermeier (1973)
*Saguinus midas* (Red-
handed tamarin)
Thorington (1968a)
*Saguinus geoffroyi* (Cotton
top)
Moynihan (1970)

Cebidae
*Cebus apella* (Blackcapped
capuchin)
Schultz (1956)
Jolly (1972)
*Cebus capucinus* (White
throated capuchin)
Schultz (1956)
Oppenheimer (1969a)
Oppenheimer (1969b)
Jolly (1972)
Eisenberg and
Thorington (1973)

*Cebus nigrivittatus* (Weeper
capuchin)
Fooden (1964)
Oppenheimer and
Oppenheimer (1973)
*Saimiri oerstdii* (Red-
backed squirrel
monkey)
Schultz (1956)
Baldwin and Baldwin
(1972)
*Saimiri sciureus* (Squirrel
monkey)
Fooden (1964)
Thorington (1967)
Thorington (1968b)
*Alouatta caraya* (Black
howler)
Pope (1966)
Neville (1972)
Jolly (1972)
*Alouatta (palliata) villosa*
(Mantled howler)
Chivers (1969)
Eisenberg and Thoring-
ton (1973)
Jolly (1972)
*Alouatta seniculus* (Red
howler)
Fooden (1964)
Neville (1972)
*Aotus trivirgatus* (Night
monkey)
Moynihan (1964)
Jolly (1972)
*Callicebus moloch* (Dusky
titi)
Mason (1968)
*Ateles belzebuth* (Colom-
bian or long-haired
spider monkey)
Klein and Klein (in press)
*Ateles geoffroyi* (Black-
headed spider monkey)
Eisenberg and Kuhn
(1966)
Jolly (1972)
*Ateles paniscus* (Black
spider monkey)
Durham (1971)
*Lagothrix lagothricha*
(Humboldt's woolly
monkey)
Kavanagh and Dresdale
(1975)

*Pithecia pithecia* (Pale-
headed saki)

Cerocopithecinae
*Cercopithecus aethiops*
(Vervet monkey)
Hall and Gartlan (1965)
Hill (1966)
Struhsaker (1967a,b)
Jolly (1972)
*Cercopithecus ascanius*
(Redtailed monkey)
Haddow (1952)
Struhsaker (1975)
*Cercopithecus campbelli*
(Campbell's monkey)
Bouliére, Hunkeler and
Bertrand (1970)
*Cercopithecus cephus*
(Moustached monkey)
Struhsaker (1969)
Gautier (1971)
*Cercopithecus l'hoesti*
(Hoest's monkey)
Struhsaker (1969)
Clutton-Brock (unpub-
lished)
*Cercopithecus mitis*
(Blue monkey)
Hill (1966)
Aldrich-Blake (1970a)
Kingdon (1971)
DeVos and Omar (1971)
Struhsaker (1975)
*Cercopithecus mona* (Mona
monkey)
Struhsaker (1969)
Kingdon (1971)
*Cercopithecus neglectus*
(de Brazza monkey)
Gautier (1971)
Jolly (1972)
*Cercopithecus nictitans*
(Spotnosed monkey)
Struhsaker (1969)
Gautier (1971)
*Cercopithecus pogonias*
(Crowned monkey)
Struhsaker (1969)
Gautier (1971)

*Miopithecus talapoin*
(Talapoin)
Gautier-Hion (1970,
1971)
Gautier (1971)
Rowell (1973)
*Erythrocebus patas* (Patas
monkey)
Hall (1966)
Jolly (1972)
*Cercocebus albigena* (Gray-
cheeked mangabey)
Chalmers (1968, pers.
comm.)
Waser and Floody (1974)
Waser (in press)
Struhsaker (1975)
*Cercocebus galeritus*
(Agile mangabey)
Homewood (pers.
comm.)
*Cercocebus torquatus*
(Whitecollared manga-
bey)
Jones and Sabater Pi
(1968)
*Cynopithecus niger*
(Celebes black ape)
*Macaca fascicularis* (Crab-
eating macaque)
Rodman (1973)
*Macaca fuscata* (Japanese
macaque)
*Macaca maurus* (Celebes
macaque)
*Macaca mulatta* (Rhesus
macaque)
Southwick, Beg and
Siddiqi (1965)
Neville (1968)
Lindberg (1971, in press)
*Macaca nemestrina* (Pig-
tailed macaque)
Bernstein (1967)
Rodman (1973)
Jolly (pers. comm.)
*Macaca radiata* (Bonnet
macaque)
Nolte (1955)
Simonds (1965)
Rahaman and
Parthasarathy (1969)
Sugiyama (1971)
*Macaca sinica* (Toque
macaque)

*Macaca sylvanus* (Barbary
macaque)
Deag and Crook (1971)
*Papio anubis* (Olive
baboon)
Devore and Hall (1965)
Rowell (1966, 1968)
Altmann and Altmann
(1970)
*Panio cynocephalus*
(Yellow baboon)
Altmann and Altmann
(1970)
Jolly (1972)
Maples (1972)
*Papio hamadryas*
(Hamadryas baboon)
Kummer (1968)
Altmann and Altmann
(1970)
*Papio papio* (Guinea
baboon)
Dunbar and Nathan
(1972)
*Papio ursinus* (Chacma
baboon)
Hall (1962)
Altmann and Altmann
(1970)
Stolz and Keith (1970)
*Mandillus leucophaeus*
(Drill)
Struhsaker (1969)
Gartlan (1970)
*Theropithecus gelada*
(Gelada baboon)
Crook (1966)
Dunbar and Dunbar
(1975)

Colobinae
*Colobus angolensis*
(Angolan black and
white colobus)
Groves (1973)
*Colobus badius* (Red
colobus)
Clutton-Brock (1972)
Clutton-Brock (1974*b*)
Clutton-Brock (1975*a,b*)
Struhsaker (1975)

*Colobus guereza* (Black
and white colobus)
Kuhn (1964)
Marler (1969)
Clutton-Brock (1972)
Dunbar and Dunbar
(1974)
Oates (in press)
*Colobus verus* (Olive
colobus)
Hill (1952)
Booth (1957)
Jolly (1972)
*Presbytis aygula* (Sunda
Island langur)
Rodman (1973)
Wilson and Wilson
(1975)
*Presbytis cristatus* (Silver
langur)
Bernstein (1968)
*Presbytis entellus*
(Hanuman langur)
Sugiyama (1966)
Yoshiba (1968)
Jay (1972)
*Presbytis johnii* (Nilgiri
langur)
Poirier (1968)
Poirier (1969)
Horwich (1972)
Jolly (1972)
Tanaka (1965)
*Presbytis melalophos*
(Banded langur)
Chivers (1972*a*)
*Presbytis obscurus* (Dusky
langur)
Chivers (1972*a*)
Mukherjee and Saha
(1974)
*Presbytis rubicundus*
(Maroon langur)
Wilson and Wilson
(1975)
*Presbytis senex* (Purple
faced langur)
Eisenberg, Muckenhirn,
and Rudran (1972)
Hladik and Hladik (1972)
*Pygathrix nemaeus* (Douc
langur)
Pfeiffer (1969)

| | | |
|---|---|---|
| *Nasalis larvatus* (Proboscis monkey) | *Hylobates hoolock* (Hoolock gibbon) | Pongidae |
| Kawabe and Mano (1972) | Tenaza and Hamilton (1971) | *Pongo pygmaeus* (Orang-utan) |
| D. MacDonald (pers. comm.) | *Hylobates klossii* (Kloss' gibbon) | Rodman (1973*a,b*, in press) |
| | Tenaza and Hamilton (1971) | Mackinnon (1974) |
| Hylobatidae | Tenaza (1975) | Galdikas-Brindanour (in press) |
| *Hylobates agilis* (Dark handed gibbon) | *Hylobates lar* (White-handed gibbon) | *Pan troglodytes* (Chimpanzee) |
| Tenaza and Hamilton (1971) | Tenaza and Hamilton (1971) | van Lawick-Goodall (1968) |
| Chivers (1974) | Chivers (1972*a,b*, 1974) | Wrangham (1975) |
| Tenaza (1975) | Ellefson (1974) | Struhsaker (1975) |
| *Hylobates concolor* (Black gibbon) | *Hylobates moloch* (Silvery gibbon) | *Gorilla gorilla beringei* (Mountain gorilla) |
| Schultz (1956) | Tenaza and Hamilton (1971) | Schaller (1965) |
| Tenaza and Hamilton (1971) | | Fossey (1974) |
| Tenaza (1975) | *Symphalangus syndactylus* (Siamang) | Fossey and Harcourt (in press) |
| | Chivers (1972*a,b*) | |
| | Chivers (1974) | |
| | Chivers, Raemakers and Aldrich-Blake (1975) | |

Two fundamental problems face an analysis of this kind. First, what level of evidence should be required for inclusion? If the results of detailed and extensive ecological studies only are included, sample size will be small and is likely to be biased by representatives of particular habitat types and taxonomic groups. Conversely, if all available estimates are included, some are likely to be wrong (see Aldrich-Blake 1970*b*) and others to be non-comparable due to methodological differences. Although their inclusion should not lead to consistent bias, the amount of random error introduced may be sufficient to obscure important relationships. In this analysis, we have tried to maximize the number of species included, while restricting ourselves to information collected in biological field studies.

A second problem is how to construct realistic species-averages in cases where two or more studies have produced different results. This becomes acute with variables such as population density, which can show interpopulation differences to two orders of magnitude. Wherever possible, we have taken the median of such estimates in order to minimize the effects of extreme values.

## Methods

All the primate species for which we could find data were classified as:

(1)    primarily nocturnal or primarily diurnal;
(2)    primarily arboreal or primarily terrestrial;
(3)    primarily insectivorous, frugivorous or folivorous.

Species usually classified as omnivorous (e.g. *Papio* and *Macaca*) were placed in the frugivorous category on the grounds that fruit and seeds constitute the bulk of their diet. Neither *Theropithecus gelada* nor *Presbytis entellus* could be easily allocated to a dietetic group. Eventually, both were classified as folivores on the grounds that a larger proportion of their diet (or feeding time) consists of foliage than of any other food category (Dunbar in press; Hladik and Hladik 1972).

This classification gave 12 possible combinations which, for convenience, we shall refer to as 'ecological categories', of which only 7 were represented in the sample:

nocturnal, arboreal, insectivorous;
nocturnal, arboreal, frugivorous;
nocturnal, arboreal, folivorous;
diurnal, arboreal, frugivorous;
diurnal, arboreal, folivorous;
diurnal, terrestrial, frugivorous;
diurnal, terrestrial, folivorous.

Species were also classified for convenience into eight taxonomic groups; Lemuridae, Lorisidae, Callithricidae, Cebidae, Cercopithecinae, Colobinae, Hylobatidae, and Pongidae.

Average figures for each species were then calculated for the following variables;

(1)    Body weight: average of adult male and female weight (kg).
(2)    Sexual dimorphism: male weight divided by female weight.
(3)    Feeding group size: the number of individuals usually found feeding together.
(4)    Breeding group size: the number of individuals forming a reproductive unit (e.g. the harem groups of *Papio hamadryas* or *Theropithecus gelada* rather than the bands or herds).
(5)    Population group size: the number of animals which regularly associate together and share a common home range. (This is the same as the feeding group or the breeding group

in most species, but not in all (e.g. *Ateles, Pan, Theropithecus*).)

(6) Socionomic sex ratio: the number of adult females per adult male in breeding groups.

(7) Breeding group type: whether typical breeding groups contain only one adult male (single-male) or more than one (multi-male).

(8) Feeding group weight: the total weight (kg) of the typical feeding group, calculated on the assumption that 50% of the group was made up of juveniles and adolescents with an average body weight of one-half the average adult body weight (1). Calculations of group weight allowed for biased sex ratios among adults and sex differences in body weight.

(9) Population group weight: the total weight (kg) of the population group, calculated in the same way as feeding group weight.

(10) Home range size: the total area (km²) used by a population group. Where two figures for home range size were quoted, one including and one excluding small lacunae within the perimeter of the range (e.g. Struhsaker 1975), we have taken the larger figure on the grounds that the majority of studies have included such lacunae. Figures are not corrected for range overlap between neighbouring troops (see Hladik 1975) since the necessary quantitative data are not available in most studies.

(11) Day range length: the daily path length of the feeding group (km).

(12) Population density: estimates of the number of animals per km².

(13) Biomass: weight per unit area (kg/km²) calculated by dividing feeding group weight (8) by feeding group size (3) and multiplying by population density (12).

In calculating values for individual species we used medians wherever possible. In a number of cases, body weights were quoted as ranges and we were forced to take the average of the two extremes (see Napier and Napier 1967).

The references which we used in the calculation of species averages are shown in Table 1. In the statistical analysis of the results we have (generally) used linear regression (Sokal and Rohlf 1969) though in some cases sample sizes were too small to allow statistical testing. Because of the nature of the data, we have sometimes suggested the presence of trends which cannot be supported by statistical evidence. Where we have done so, we generally also refer to paired comparisons between related species.

Where we discuss theoretical predictions about behaviour we use the convenient shorthand of describing what animals should do to maximize their fitness. This should not be taken to imply any form of conscious decision on the part of the animal.

## Results and discussion

### Taxonomy and behaviour

Within many genera, variation in social organization is slight. This is the case in *Galago, Alouatta, Ateles, Macaca, Cercopithecus* and, particularly, *Hylobates*. Moreover, within at least three of the families (*Lorisidae, Callithricidae,* and *Hylobatidae*) there is also relatively little varation.

Neither families nor genera are evenly distributed across ecological categories. Thus, all insectivores belong to the Lorisidae and all but four terrestrial species (*Lemur catta, Presbytis entellus, Pan,* and *Gorilla*) to the Cercopithecidae. Of this last category, more than 50% are members of the two genera, *Macaca* and *Papio*.

Together, these two points raise one theoretical problem and one methodological one. If relationships between social organization and ecology are confounded with taxonomic variation, how can we base functional arguments on simple associations between the first two? For example, if only macaques and baboons (*Macaca* and *Papio*) are terrestrial, can we tell whether the large group size typical of these species is an adaptation to terrestriality or a phylogenetic characteristic of these genera?

The problem is illusory, for there is no reason to suppose that differences in behaviour at a generic level are less likely to be adaptive than differences between species. Evidence that a trait is similar throughout a particular genus (despite interspecific differences in ecology) does not indicate that it is non-adaptive—only that phylogenetic inertia (*sensu* Wilson 1975) may have been involved.

The association between taxonomy and behaviour raises a more pressing methodological problem. If phylogenetic inertia is strong, the potential adaptations which related species may evolve will be similarly constrained, with the effect that species cannot be regarded as independent of each other. For example, monogamous breeding seems to be an extremely stable characteristic and in those primate genera where it is found, it apparently occurs in all congeneric species. Moreover, facultative polygyny is not known to occur in any of these species. It seems likely that the switch from monogamy to polygyny is not easily achieved and that phylogenetic inertia is probably important. To include, within diurnal, arboreal, frugivores six (almost) identical gibbon species

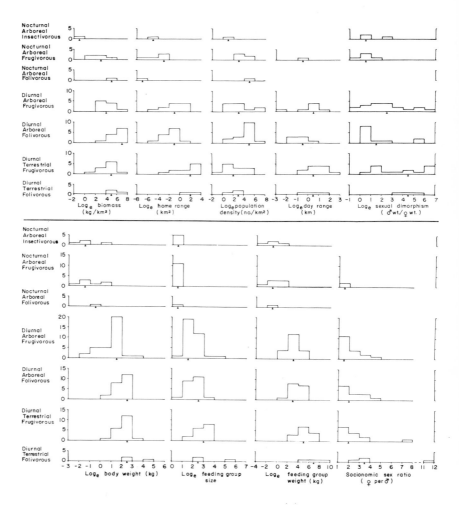

Fig. 1. Distribution of variables across ecological categories. Number of species shown on ordinate. ▲ indicates median value for each category.

would severely bias this ecological category and would inflate the number of statistically independent points.

There is no simple answer to this problem. A multiway analysis of variance among taxonomic and ecological categories is impractical owing to the unequal distribution of species across categories. When describing the distribution across ecological categories we have ignored the problem and these histograms (Fig. 1) show the frequency of species, irrespective of taxonomic affinity. This was done because our primary aim in these plots was to describe the distribution of particular variables. In the analysis of the relationships between variables, where we have been more concerned with formal statistical evidence, we have taken the mean of all congeneric species belonging to the same ecological category but have included congenerics belonging to different categories as separate points. This is only a partial solution since there is evidently some degree of similarity between species belonging to the same family. However, a nested analysis of variance revealed significant differences among genera within families for seven out of eight behavioural or ecological variables, but significant additional variation among families for only two (Table 2).

Table 2. Nested analysis of variance (two-level with unequal sample sizes) for eight variables (a) among genera within families, (b) among families. Variables were normalized using logarithmic transformations. ns = not significant.

| Variable | Among genera within families | Among families |
|---|---|---|
| Body weight | $P < 0.001$ | ns |
| Sexual dimorphism | ns | $P < 0.001$ |
| Home range | $P < 0.001$ | ns |
| Feeding group size | $P < 0.001$ | $P < 0.05$ |
| Population density | $P < 0.001$ | ns |
| Day range | $P < 0.01$ | ns |
| Feeding group weight | $P < 0.001$ | ns |
| Population group weight | $P < 0.001$ | ns |

*Body weight*

Body weight differs significantly between ecological categories (Table 3). All nocturnal primates are small (Fig. 1). Among diurnal primates, there is a tendency for terrestrial species to be heavier than arboreal ones and, within the latter two categories, for

Table 3. One-way analyses of variance for eight variables across ecological categories (all variables normalized using log transformation)

| Variable | Degrees of freedom | $F$ | Significance |
|---|---|---|---|
| Body weight | 6,84 | 22·06 | $P < 0.001$ |
| Feeding group size | 6,86 | 22·93 | $P < 0.001$ |
| Feeding group weight | 6,58 | 33·76 | $P < 0.001$ |
| Home range | 6,36 | 4·95 | $P < 0.001$ |
| Day range | 4,23 | 3·31 | $P < 0.05$ |
| Population density | 6,42 | 2·49 | $P < 0.05$ |
| Biomass | 6,39 | 8·79 | $P < 0.001$ |
| Sexual dimorphism | 5,49 | 4·57 | $P < 0.005$ |

folivores to be heavier than frugivores. Thus among pairs of related species, the siamang (*Symphalangus syndactylus*) is larger than the more frugivorous gibbons (*Hylobates* spp.) and the gorilla (*Gorilla gorilla*) is larger than the chimpanzee (*Pan troglodytes*). Similar associations between diet and body size are found among antelope: browsing species which live on a (relatively) nutritious diet are usually smaller than grazing species (Jarman 1974).

Most reviews of primate ecology have considered body weight as an independent variable, though it is clear that species differences are adaptive (Geist 1974*a, b*; Schoener 1969). In mammals, where differences in body size are functionally related to a wide array of life history parameters (including litter size, size of young at birth, growth rate, reproduction frequency, and lifespan), it is a difficult topic to consider.

Small body size may be associated with nocturnal activity among primates because most nocturnal species occupy the middle and lower strata of forest or woodland (and consequently need to be able to travel on relatively fine twigs) and because they rely on crypsis rather than on flight or defence to avoid predation (e.g. Charles-Dominique 1971). The same relationship between body size and crypsis occurs among the ungulates (Eisenberg and Lockhart 1972; Eisenberg and McKay 1974; Geist 1974*b*; Estes 1974): species inhabiting dense cover where movement is easily audible and rapid escape is difficult tend to be small and to rely on crypsis for protection.

There are two possible explanations of increased size in terrestrial species. It may represent an adaptation to increased predator pressure, permitting individuals to escape or defend themselves more efficiently. An alternative hypothesis is that constraints on body size imposed on arboreal species by the necessity of retaining access to food supplies located on terminal twigs are removed in terrestrial species, most of which live in savannah country where large size is advantageous because it permits increased day range length. This could explain why body size does not increase greatly in most of the terrestrial species living in woodland such as *Lemur catta* or *Cercopithecus aethiops*. However, it fails to explain why *Gorilla* is so large and one might, too, have predicted an appreciably larger body size for *Papio hamadryas* and a smaller one for *Mandrillus*. Arguments relating body size to diet are complex and are discussed in more detail elsewhere (Clutton-Brock and Harvey 1977).

### Group size and weight

Feeding group size differs significantly between categories. All nocturnal species, irrespective of diet type, feed alone or in pairs. Among the diurnal species there is a wide variation within the four categories but some tendency for terrestrial species to feed in larger groups than arboreal ones. There is little average difference in group size between folivores and frugivores (the apparent difference between arboreal frugivores and arboreal folivores is largely a result of the former category being weighted by the inclusion of several species of *Hylobates* and *Cercopithecus*). When related species are compared there is a tendency for the more frugivorous species to live in larger troops than the more folivorous ones. Thus *Lemur catta, Colobus badius,* and *Presbytis entellus* respectively feed in larger groups than *Lemur fulvus, Colobus guereza,* and *Presbytis senex*. Larger species tend to live in larger groups and (see Fig. 2) group size and body weight are significantly associated ($F_{1,39} = 18.33, P < 0.001$).

Feeding group weight also shows marked differences between categories with rather less variation within them. Again, values for terrestrial species tend to be greater than for arboreal ones. Though the heaviest groups occur among the folivores (e.g. *Colobus badius, Theropithecus,* and *Gorilla*) there is little average difference between folivores and frugivores.

Population and breeding group sizes and weights are the same, in the great majority of species, as feeding group sizes and weights show nearly identical distribution.

A wide variety of functions of grouping have been suggested. They fall into four main categories: those related to (1) detection,

Fig. 2. Feeding group size plotted against body weight (kg) for different genera. ★ nocturnal, arboreal insectivores; △ nocturnal, arboreal frugivores; ▲ nocturnal, arboreal folivores; ○ diurnal, arboreal frugivores; □ diurnal, terrestrial frugivores; ● diurnal, arboreal folivores; ■ diurnal, terrestrial folivores.

avoidance, or defence against predators (Patterson 1965; Hamilton 1971; Lack 1968; Crook 1970; Vine 1971; Treisman 1975a, b); (2) food finding, handling, defence, or exploitation (Krebs, Mac-

Roberts, and Cullen 1972; E. O. Wilson 1975; Kruuk 1972; Cody 1971; Alexander 1974; Clutton-Brock 1974a, 1975b); (3) direct reproductive advantages including regular access to the opposite sex (Brown 1975; Wrangham 1975); (4) facilitation of learning processes (particularly for juveniles and adolescents) resulting from the presence of several older animals (Harcourt personal communciation).

None of the theories concerning the functions of grouping easily explains the distribution of group size across primate species. Although terrestrial species tend to show larger group size than arboreal ones (as might be expected if grouping represents an adaptation to predators), there are also marked differences between species whose liability to predation is probably similar (*Colobus guereza* and *Colobus badius*; *Cercopithecus* spp. and *Miopithecus talapoin*). Nor is there any evidence that group size is closely associated with the relative difficulty with which food supplies can be located. In several species where feeding group size varies throughout the year (including *Cercopithecus mitis*, *Ateles belzebuth*, *Theropithecus gelada*, and *Pan troglodytes*), groups tend to be largest when food is most abundant and smallest when it is scarce (Aldrich-Blake 1970a; Klein and Klein in press; Crook 1966; Wrangham 1975). In fact, most primate groups living in relatively small home ranges appear to know the distribution of their food supplies extremely well (e.g. Wrangham 1975) and locating food sources may pose few problems. This may contrast both with bird species (Thompson, Vertinsky, and Krebs 1974), where food supplies are often cryptic and where individuals spend a considerable proportion of their time searching for food, and with primate species living in large home ranges where food supplies tend to be widely and sparsely dispersed (Milton and May 1976): it is significant that yellow baboons tend to form files when travelling, but spread out into ranks when they begin to feed (Altmann and Altmann 1970).

To understand the distribution of grouping it may be more useful to assume that it has multiple advantages and to consider possible constraints on group size. Grouping is likely to have several disadvantages (Jarman 1974; Alexander 1974). The two most obvious are that it

(1)    makes individuals more conspicuous to predators (although see Vine 1971; Treisman 1975a, b). This is supported by evidence that cryptic species tend to be solitary (Geist

1974*a*; Estes 1974; Jarman 1974) and that, in social species, individuals which cannot escape from predators (e.g. sick individuals or females about to give birth) also become solitary (Jarman 1974; Clutton-Brock and Guinness 1975).

(2)    decreases feeding rate either through mutual interference at food sources (e.g. Goss-Custard 1970) or through indirect competition for food supplies (Jarman 1974). Evidence for the latter effect is rare in primates but intraspecific comparisons suggest that it may be important. Mangabeys living in large groups move further each day than those living in small groups (Waser in press) while the proportion of time spent feeding declines with increasing group size in chimpanzees (Wrangham 1975).

In the nocturnal Lemuridae and Lorisidae, which rely on crypsis to avoid predation, small group size may represent an adaptation for avoiding predators. Reduction of feeding interference may also be involved since the diet of most of the species contains a considerable proportion of insects which can be easily distributed. However, this cannot be the only reason since nocturnal folivores (e.g. *Lepilemur mustelinus*) are also solitary (Eisenberg, Muckenhirn, and Rudran 1972). In several other mammalian taxa, nocturnal species are usually solitary or occur in small groups (e.g. Jarman 1974; Kaufmann 1974).

Increased group size in terrestrial versus arboreal species may well represent an alternative adaptation to increased predator pressure, as Crook and Gartlan (1966) originally suggested. In the relatively open environments inhabited by most terrestrial species, crypsis would be impractical and in other mammalian groups, too, large group size is typical of open-country species (Jarman 1974; Geist 1974*a*, *b*; Estes 1974; Kaufmann 1974). A contributory factor here may be the tendency for the food supplies of savannah-dwelling primates (all of which are terrestrial) to be aggregated in large clumps (see Altmann 1974), thus allowing many individuals to feed simultaneously at the same food source. The association between food source size and group size found within several species (e.g. *Theropithecus gelada* (Crook 1966)) supports the suggestion that grouping may be limited by the size of food sources.

There are at least two reasons why, in pairs of related species, those which feed more on mature leaves should live in smaller groups. First, large group size will increase the distance which each individual will have to move per day to collect food.

Species living on energetically poor food supplies may be unable to afford the extra costs which large group size imposes (see Smith 1968). Alternatively, the difference in group size may be related to a difference in the distribution of food supplies (Crook 1972). Evidence from a number of studies shows that 'specialized' folivores feed on a small number of food species throughout the year and that, even when numerous foods are available, they tend to exploit only a few of them (Charles-Dominique and Hladik 1971; Hladik and Hladik 1972; Clutton-Brock 1975a; Oates in press). In contrast, species feeding more on young foliage, flowers, and fruit utilize a wide range of species and select for dietetic variety each day (Clutton-Brock in press). If their food supplies are clumped, group size may be able to increase without affecting day range length (see above) whereas this would not be the case in specialized folivores feeding on a less varied diet. Secondly, the difference in dietetic diversity may permit specialized folivores (but not species living on a more diverse diet) to exist throughout the year in home ranges small enough to be defended (Aldrich-Blake 1970b) and reduced group size may represent an adaptation to territorial defence (see Goss-Custard, Dunbar, and Aldrich-Blake 1972; Crook 1972; Clutton-Brock 1974b).

Body weight and group size are probably associated because similar energetic constraints apply to both variables. Associations between body size and group size are found in bovids (Jarman 1974; Estes 1974), cervids (personal observation), carnivores (Kruuk 1975) and some marsupials (Kaufmann 1974).

At least four primate species live in population groups which regularly break up into smaller feeding units: *Ateles belzebuth, Ateles geoffroyi, Papio hamadryas,* and *Pan troglodytes* and some evidence indicates that several other species, including *Mandrillus leucophaeus* and *Nasalis larvatus* may also live in unstable groups. The most likely explanation of group splitting is that it occurs where food is widely distributed in small clumps (Altmann 1974), so that the costs of feeding competition are high. This is supported by the fact that all species which live in unstable groups of this kind (with the possible exception of *Nasalis*) are frugivores and that, in some other species, population groups may divide into smaller feeding parties when food becomes scarce (e.g. *Cercopithecus mitis, Cercocebus galeritis*). A parallel situation is found both among the grazing ungulates (Jarman 1974; Geist 1974a) in kangaroos (Kaufmann 1974) and in some carnivores (Schaller 1972; Kruuk 1972).

*Home range size*

Home range differs significantly between groups (see Table 3). It is consistently smaller in nocturnal species than in diurnal ones. Terrestrial species tend to have larger home ranges than arboreal ones (as might be expected both because their movements are restricted to two dimensions and because they mostly live in habitats where food is relatively sparse), and frugivores tend to have larger home ranges than folivores.

Comparisons between ecological categories are complicated by the close relationship between home range size and group weight (see Milton and May 1976). As Fig. 3 shows, there is a close association between population group weight and home range size both among frugivores ($F_{1,15} = 26.49$, $P < 0.001$, $b = 0.99$) and folivores ($F_{1,9} = 66.89$, $P < 0.001$, $b = 1.15$). In contrast to the dietetic groups of birds compared by Schoener (1968), there is no difference in slope between the two regressions ($F_{1,24} = 0.36$), but frugivores have consistently larger home ranges than folivores for a given population group weight and the difference in elevation is significant ($F_{1,25} = 9.12, P < 0.01$).

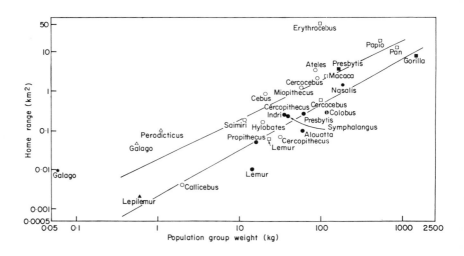

Fig. 3. Home range size (in km²) plotted against population group weight (kg) for different genera. ★ nocturnal, arboreal insectivores; △ nocturnal, arboreal frugivores; ▲ nocturnal, arboreal folivores; ○ diurnal, arboreal frugivores; □ diurnal, terrestrial frugiovres; ● diurnal, arboreal folivores; ■ diurnal, terrestial folivores. Separate regression lines shown for folivores and frugivores.

Differences in home range size relative to group weight between frugivores and folivores are supported by comparisons between pairs of related species. Thus the ratio of population group weight to home range area is smaller for *Lemur catta, Presbytis entellus, Hylobates lar*, and *Pan troglodytes* than (respectively) for *Lemur fulvus, Presbytis senex, Symphalangus syndactylus*, and *Gorilla gorilla*. Species which spend a considerable proportion of their time feeding on insects tend to occupy particularly large home ranges in relation to their population group weight (e.g. *Microcebus murinus, Galago demidovii, Callicebus torquatus* (Kinzey, Rosenberger, Heisler, Prowse, and Trilling in press; Kinzey in press), *Cebus capucinus, Cercocebus albigena*).

It is evident that differences in food density are important in determining home range size (McNab 1963; Schoener 1968; Altmann 1974). Both the relationship between population group weight and home range size and the occurrence of relatively larger home range in frugivores than folivores and in terrestrial species than arboreal species would be expected on this basis.

Similar relationships between body weight (in solitary or pair-living species) or group weight and home range or territory size have been found in several vertebrate groups including birds (Armstrong 1965; Schoener 1968), small mammals (McNab 1963), and smaller samples of primates (Jorde and Spuhler 1974; Milton and May 1976). The first three studies differed from this one in being primarily concerned with animals living either in pairs or in individual home ranges. Milton and May's study of 36 primate species covers many of those included in our analysis but differs in methodology: instead of relating group weight to home range size, they related individual bödy weight (usually the mean of adult male and female weights) to the ratio of home range size to group size ('individual home range allocation'). Their methodology has the disadvantage that it does not take into account either sex or age differences in body weight and must consistently overestimate the home range:body weight ratio. However, as they point out, such differences are unlikely to affect their general conclusions. A more serious objection is that they include as independent points not only closely related species from the same genus but also different studies of the same species, thus over-estimating the number of independent points.

The slope of home range or territory size on body or group weight can be compared in three groups of vertebrates:—*small mammals*: herbivorous species (e.g. *Microtus pennsylvanicus*,

*Pitmys pinetorum, Sylvilagus floridanus*) slope 0.65 (McNab 1963); omnivorous species (e.g. *Mus musculus, Peromyscus leucopus, Zapus hudsonicus*) slope 0.67 (calculated for the species assigned to McNab's 'hunter' category excluding the primarily insectivorous or carnivorous species *Didelphis virginiana, Sorex vagrans, Blarina brevicauda, Mustela rixosa, Procyon lotor, Vulpes fulva*); *primates*: frugivores slope 0.99 (this study); folivores slope 1.15 (this study); *predators*: predatory birds (e.g. *Buteo, Accipiter, Falco* spp.) slope 1.39; predatory mammals (e.g. *Mustela, Martes, Spilogale, Vulpes, Lynx, Blarina* spp.) slope 1.41 (Schoener 1968).†

A steeper slope is to be expected among predators than among omnivores, frugivores, or herbivores on energetic grounds. Larger predators can only economically feed on relatively large sized prey. Where prey biomass declines with (prey) body size, the size of the predator's home range should rise sharply with increasing body size (see Schoener 1968). Among omnivorous, frugivorous, herbivorous, and folivorous mammals, increased body size does not constrain the size of food which is selected to the same extent, with the effect that food density is less likely to decline with increasing (predator) body size. The same point probably explains why slopes do not differ between folivorous and frugivorous primates despite differences in elevation. There is no obvious explanation of the difference in slope between small mammals and primates.

The distribution of food supplies in time and space is clearly important, too. Primate species relying on food supplies which are heavily clumped and spatially variable occupy larger home ranges than related species living on less variable food supplies (Clutton-Brock 1975a, b; Hladik 1975) and a similar difference between home range or territory size and the dispersion of food supplies is found amoung ungulates (Jarman 1974; Geist 1974a) as well as in birds (Orians 1961; Schoener 1968). The relationship between food dispersion and home range size is a difficult one to identify since the situation is usually complicated by differences in group size and it is possible to argue that increased home range size represents an adaptation to increased group size, which has evolved for some other reason. However, the present evidence suggests that large group size may be functionally related to variable and unpredictable food supplies.

† We have not formally tested these differences in slope since there is little overlap between samples in individual or group weight.

The association between the density of food supplies and home range size is supported by evidence at an intraspecific level:

(1)  Populations living in areas of low food availability tend to have larger home ranges than those living in areas where food is more abundant (Yoshiba 1968; Schoener 1968; Clutton-Brock 1972).

(2)  Within populations, groups whose ranges include a large proportion of preferred habitat tend to have the smallest ranges (e.g. Struhsaker 1967*b*).

(3)  Large groups may occupy bigger home ranges than smaller groups (e.g. Gatinot 1975) though this is not always the case (e.g. Struhsaker 1967*b*).

(4)  Home ranges tend to be largest at those times of the year when food is least available (e.g. Clutton-Brock 1975*b*).

There is also evidence that interspecific competition affects home range size. In bird species, differences in average territory size of up to 12 times can occur between populations living in areas where different numbers of competing species are present (Cody 1974). Red colobus at Gombe and in Kibale forest, where heterospecific competitors are common, live in troops of about 50 individuals in ranges of the order of 50 to 100 ha (Clutton-Brock 1975*a, b*; Struhsaker 1975) whereas in Senegal, on the Tana River (Kenya), and in Zanzibar, where competitors are relatively scarce, they occur in somewhat smaller troops in ranges approximately one quarter as large (Gatinot 1975; Marsh personal communication; Kingston personal communication). Reduced home range size and high population density may be associated with low numbers of interspecific competitors in several other species (*Colobus guereza*: Schenkel and Schenkel-Hulliger 1967; Oates in press; *Cercocebus albigena*: Chalmers 1968; Waser in press). However, in most cases, it is impossible to be certain that differences in competition rather than in food density unrelated to competition are involved.

Finally, in primates, (e.g. Chivers 1969) as well as in birds (Huxley 1934; Schoener 1968; Maher 1970; Cody 1974) reduced home range size may occur in high density populations where intraspecific competition is intense.

## Territorial defence

There is no clear-cut distinction among primates between territorial and non-territorial species (see Bates 1970). Inter-group relationships grade from situations where (virtually) exclusive use of territories is maintained by frequent displays and interactions (e.g. *Callicebus moloch, Hylobates* spp.), through species where territories or ranges overlap widely but one or other group is consistently dominant in each overlap area (e.g. *Presbytis* spp.), to cases where extensive home range overlap exists and groups usually avoid each other (e.g. *Papio* spp.). As Deag's detailed study of Barbary macaques shows, even where no overtly territorial interactions occur, well-defined systems of approach/withdraw relationships may be present (Deag 1973).

As in other vertebrates (Hinde 1956; Tinbergen 1957; Lack 1968; Geist 1974a) the functions of territoriality are probably multiple. Defence of a food supply is particularly likely to be advantageous in relatively stable environments where competition is intense (Geist 1974a). Territoriality probably also assists individuals in maintaining exclusive breeding access to their mates and it may be for this reason that territorial behaviour is generally more strongly developed in males than females (Clutton-Brock and Harvey 1976).

The distribution of territorial behaviour among primates conforms closely to Brown's (1964) hypothesis concerning the constraints on territoriality. Where food supplies are relatively dense and evenly distributed, individuals tend to be dispersed singly or in small groups which defend feeding territories. In contrast, where food supplies are clumped, widely dispersed, and unpredictable, larger groups form which do not form territories and may or may not show hostility to neighbouring troops. A similar situation exists among ungulates (Geist 1974a, b) and in birds (Schoener 1968).

Though there are some exceptions, the extent of home range or territory overlap is roughly related to home range size among primates species. This is probably both because larger territories cannot be as efficiently defended as smaller ones, even if the number of defenders increases proportionately (Crook 1972), and because clumped and variable food supplies may be temporarily super-abundant, thus reducing the benefits of defending them (see Altmann 1974).

*Day range length*

Day range length also differs significantly between categories. Terrestrial species tend to have longer day ranges than arboreal ones and frugivores have longer ranges than folivores (see Fig. 4).

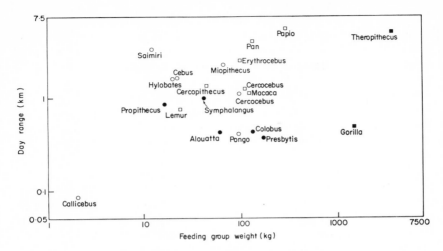

Fig. 4. Day range length (km) plotted against feeding group weight (kg) for different genera. ○ diurnal, arboreal frugivores; □ diurnal, terrestrial frugivores; ● diurnal, arboreal folivores; ■ diurnal, terrestrial folivores.

There is a significant association between feeding group weight and day range length in frugivores ($F_{1,12} = 7\cdot17, P < 0\cdot025$) though not in folivores ($F_{1,6} = 1\cdot54$). As in the case of home range size, elevation but not slope differs significantly between the two categories (for difference in elevation, $F_{1,18} = 6\cdot69, P < 0\cdot025$; for difference in slope, $F_{1,17} = 0\cdot80$).

As in the case of home range size, many differences in day range length are probably related to variation in food density. On energetic grounds, it would be predicted that day range length would be positively related to feeding group weight, that terrestrial species would have longer ranges than arboreal ones and that frugivores would have longer ranges than folivores (see Jorde and Spuhler 1974). The absence of any significant correlation between group weight and day range length in folivores is puzzling but may be the product of the small number of genera involved. It is also surprising that group weight is more closely related to variation in home range size than variation in day range length. One explanation is that differences in group dispersion affect day range length but not home range size.

The association between day range length and food density is supported by several interspecific comparisons. Day range length tends to be longer (a) in populations or groups where food density is particularly low (Struhsaker 1967$b$), (b) in larger than in smaller groups (e.g. Waser in press), and (c) at times of year when food availability is reduced (e.g. Clutton-Brock 1975$b$).

*Population density and biomass*

Population density differs significantly between the categories. Nocturnal species tend to show lower densities than diurnal ones and arboreal species higher densities than terrestrial ones. The highest densities are found among the arboreal folivores (e.g. *Lepilemur mustelinus, Lemur fulvus,* and *Colobus badius*).

There is a significant negative association between body weight and population density both in frugivores ($F_{1,14} = 7.36, P < 0.025$) and in folivores ($F_{1,19} = 19.65, P < 0.005$). The elevation of the two regression lines (see Fig. 5) differs significantly ($F_{1,24} = 8.10$, $P < 0.001$) but the slope does not ($F_{1,23} = 0.38$).

Biomass also differs between categories (see Table 3). There is some tendency for nocturnal species to have lower biomasses than diurnal ones and an evident association between biomass and diet. Folivorous species consistently show the highest biomasses and insectivores the least. Among the frugivorous species, those with a considerable proportion of insects in their diet (e.g. *Callicebus torquatus* (Kinzey *et al.* in press), *Cebus capucinus, Saimiri sciureus, Cercocebus albigena*) tend to have relatively low biomasses (Clutton-Brock in press). Relationships between ecology and biomass are discussed in more detail elsewhere (Clutton-Brock and Harvey in press).

These differences too, are probably related to variation in food density and both the (negative) relation between body size and population density and the observed differences in biomass between ecological groupings would be expected on this basis. The similarity in regression slope between dietetic groups suggests that variation in food density associated with increasing body size is similar in both dietetic groups.

*Socionomic sex ratio*

Socionomic sex ratio has a heavily skewed distribution which cannot be normalized for analysis of variance. The frequency of species with socionomic sex ratios above parity does not differ significantly between categories ($\chi^4_2 = 7.21, 0.2 > P > 0.1$). However, the frequency of single male troops, multimale troops, and solitary species shows marked variation between categories ($\chi^2_{12} =$

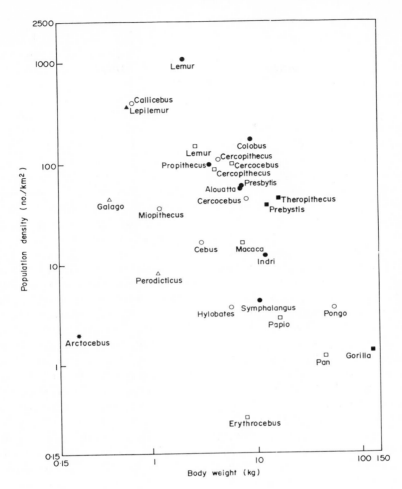

Fig. 5. Population density (no./km²) plotted against body weight (kg) for different genera. ★ nocturnal, arboreal insectivores; △ nocturnal, arboreal frugivores; ▲ nocturnal, arboreal folivores; ○ diurnal, arboreal frugivores; □ diurnal, terrestrial frugivores; ● diurnal, arboreal folivores; ■ diurnal, terrestrial folivores.

66·84, $P < 0·001$). Multimale troops are commoner in terrestrial species than in arboreal ones but show no obvious differences between diet types.

The distribution of socionomic sex ratio across primate species is difficult to understand. It is commonly argued that sex ratios biased towards females represent a more efficient use of resources than more equal ones, on the grounds that one male can fertilize several females (Gartlan and Brain 1968; Denham 1971; Kummer 1971; Estes 1974; Coelho 1974). However, this argument is difficult to state in terms of individual advantage (Altmann and Altmann 1970; Clutton-Brock and Harvey 1976).

A more useful approach is to consider the advantages and disadvantages of different (socionomic) sex ratios to individuals. It is helpful to break the problem down into three questions:

(1) *Why are some species monogamous and others polygynous?* At least 14 primate species are monogamous: *Indri indri, Callithrix jaccus, Saguinus midas, Saguinus geoffroyi, Leontideus rosalia, Callicebus moloch, Callicebus torquatus* (Kinzey *et al.* in press), all six species of *Hylobates,* and *Symphalangus syndactylus.*

The functions of monogamy are not obvious (Clutton-Brock and Harvey 1976). There is likely to be strong selection pressure for males to breed polygynously wherever possible, since by doing so individuals will increase their reproductive success and some form of polygyny probably represents the primitive condition among mammals (E. O. Wilson 1975).

Monogamy is likely to occur (1) where the male leaves more offspring by assisting the female to rear their joint offspring (Trivers 1972), or (2) where females mate selectively with monogamous males (see Verner 1964; Verner and Willson 1966, 1969) because the latter are likely to invest more time and energy assisting in rearing offspring than polygynous males.

For both reasons, monogamy is likely to evolve where male assistance is most necessary (Brown 1975). Among primates, males do not directly feed the young as in birds and male assistance takes the form of infant carrying (as in the Callithricidae, *Callicebus,* and *Symphalangus*) and territorial defence.

Assistance in carrying the infant will be necessary where more than one offspring is born simultaneously and this probably explains the evolution of monogamy in the Callithricidae where twins are usually born (Clutton-Brock and Harvey 1976).

Among those species where territorial defence is permitted by a relatively even distribution of food in time and space (see Brown 1964; Clutton-Brock 1975a, b), it is likely to be most advantageous where (a) intraspecific competition is most intense† (as in relatively

† This hypothesis is supported by intraspecific evidence from studies of howler monkeys and baboons. In the Barro Colorado howler population, high density is appar-

stable environments: Crook 1970, 1972; Geist 1974a) and (b) where the costs of feeding competition are high (e.g. where food is easily disturbed).

In both the latter situations, territoriality may evolve but not necessarily monogamy, for it may still be to the male's advantage to defend an area large enough for several females. Whether monogamy evolves will depend on the size of the area which the male can defend and on the number of females and juveniles which the area can supply.

Since intruders are usually located by sight and sound, there is probably a maximum size of territory which can be effectively defended (Crook 1970, 1972) and this may be largely independent of male body size. If there is strong selection for territoriality, monogamy is likely to develop where males cannot breed polygynously without exceeding this territory size limit. Thus, we might expect it to occur in territorial species living at relatively low population densities either because of large body size, specialized dietary habits, or high intensity of interspecific competition. Below the maximum limit for territory size, polygyny may develop but will be limited by (a) the relationship between increasing territory size and increasing costs of defence (usually to the male), (b) any deterioration in the efficiency of defence with increasing territory size which could cause females to mate selectively with males occupying smaller territories.

This hypothesis may help us to understand the distribution of monogamy. All species in which it occurs live in the kinds of environment where intraspecific competition is likely to be consistent and are territorial. Most occupy relatively large home ranges for arboreal species and occur at comparatively low population densities (*Indri indri*, 23 ha, $12 \cdot 5/\text{km}^2$; *Callicebus torquatus*, 20 ha, $13/\text{km}^2$ (Kinzey *et al.* in press); *Hylobates lar*, 40 ha, $3 \cdot 3/\text{km}^2$; *Hylobates agilis*, not known, $5 \cdot 1/\text{km}^2$; *Symphalangus syndactylus*, 23 ha, $4 \cdot 2/\text{km}^2$) either as a result of large body size (e.g. *Indri*, *Symphalangus*) or low density food supplies (*Callicebus torquatus*). Two species conflict with this hypothesis: *Callicebus moloch* (home range 0·4 ha, probable population density approximately $750/\text{km}^2$) and *Hylobates klossii* (home range 7 ha, density $30/\text{km}^2$). One

---

ently associated with a reduction in group and home range size and an increase in territorial behaviour (Chivers 1969). And in at least some high density baboon populations (Hamilton *et al.* 1975) territorial interactions occur between neighbouring troops whereas in those at lower population density (e.g. Altmann and Altmann 1970) troops tend to avoid each other.

possible explanation is that both the latter studies were carried out on isolated populations where many of the species' usual competitors were lacking and where, as a result, home range size may have been unusually small and population density unusually high (see p. 360). Further studies on *Callicebus* species are clearly needed.†

One prediction of this theory is that, within polygynous species, males which hold particularly large territories will tend to breed with a greater number of females than males which hold smaller or poorer territories (see Brown 1975). This is the case in several bird species (e.g. Zimmerman 1966, 1971) though there is no evidence available for primates.

(2) *Why do some species live in single-male troops and others in multi-male troops?* The increase in the frequency of multi-male troops among terrestrial species is usually taken to indicate that it represents an adaptation to predator pressure (Crook and Gartlan 1966; Crook 1970, 1972). However, this offers no explanation of why multi-male troops occur in some arboreal genera (e.g. *Lemur, Propithecus, Alouatta, Cercocebus, Colobus*) but not in others.

It may be useful to consider the costs and benefits to a dominant male in an (initially) single-male troop of either allowing an extra male to join the troop or of consistently driving him away. The costs of the first alternative will include:

(a)    reduction of breeding access
(b)    increased energetic expenditure and direct competition for food for all group members, particularly the original male's mates and offspring, owing to increased group size.

Benefits may include:

(a)    reciprocal assistance in defending the troop against other males (e.g., Bertram 1975) or against predators
(b)    advantages in terms of inclusive fitness if the second male is related (see Clutton-Brock and Harvey 1976).

† Since writing the above, further studies of *C. moloch* and *C. torquatus* have been completed. Although *C. moloch* has a consistently smaller range than *C. torquatus* (which is associated with a larger proportion of leaves in the diet) the home ranges estimated by Mason (1968) are unusually small (Kinzey personal communication). Average home range size for four groups observed over a period of a month in southern Peru was 4·2 ha with a population density of 24/km². The low population density relative to home range size probably indicates that annual home range size is appreciably larger than the monthly ranges measured.

Ejecting additional males will reduce the costs of sharing access to food and females but will involve others resulting from:

(a)    frequent agonistic interactions; those costs are likely to be highest in species where the disadvantages of being ejected are great (see Altmann 1974) and where the contestant should fight strongly to gain entry to the troop

(b)    the danger of solitary males joining co-operatively to eject the resident male (e.g. Sugiyama 1965).

It is difficult to explain the distribution of single-male and multi-male troops on any unitary hypothesis and different factors are probably involved in different species. Crook's (1972) suggestion that multi-male groups occur in savannah but not in rainforest or arid country species because of 'the relaxation of food poverty so that the broad dispersion of "surplus" males and their separate ranging from the reproductive unit is of no particular value to either sex' is unconvincing since there is little reason to suppose that populations of savannah species would be less likely to be food limited than those of forest-dwelling species. Moreover, it fails to explain the presence of multi-male troops in several rain-forest species (e.g. *Colobus badius, Colobus verus, Miopithecus talapoin, Alouatta* spp.). In general, we suspect that the energetic costs of allowing extra males to join the troops are less important than the reproductive ones.

(3) *What determines the socionomic sex ratio?* Imbalances in the socionomic sex ratio are presumably the product of at least four different mechanisms:

(a)    Differential maturation rate between males and females. This is sufficiently large to produce a significant prepon-derance of adult females in some species (see Altmann and Altmann 1970; Crook 1972). It is likely to be greatest in species where inter-male competition is most intense (Trivers 1972; E. O. Wilson 1975) and therefore might be expected to have the greatest effect in species where socio-nomic sex ratio is already most heavily skewed.

(b)    Differential mortality. In some species, inter-male com-petition is associated with increased mortality rates in males (Trivers 1972; Geist 1971, 1974*b*; Grubb 1974).

(c)    Differential migration rates. Ejection of adolescent males from breeding groups by mature animals (usually males) is common among primates and much of the interspecific variation in sex ratio is a product of differences in the tolerance shown by adult males to subordinates. An additional factor which may affect the benefits of migration is any tendency in females to breed selectively with polygynous males. This may occur either when the fitness of polygynous males tends to be higher than that of monogamous males (see E. O. Wilson 1975) or when the former occupy the richest or largest territories (Verner 1964; Verner and Willson 1966, 1969; Orians 1969; Brown 1975). There is, as yet, little evidence for the presence of mechanisms of this kind in primates but the relevant data are hard to collect and it is currently impossible to judge the importance of female selection. However, none of these four mechanisms offer an explanation of the distribution of socionomic sex ratios across primates, which remain an anomaly.

*Sexual dimorphism*

Sexual dimorphism in body weight shows less variation between categories than body weight and there is wide interspecific variation in all ecological categories. Sexual dimorphism is slightly less marked in nocturnal than diurnal species and in arboreal than terrestrial species. There is no obvious difference between folivores and frugivores.

The two most commonly suggested functions of sexual dimorphism in body weight are:

(1)   that it allows the sexes to exploit different niches (Selander 1966, 1972)
(2)   that it has arisen as a result of differential competition for mates (Darwin 1871; Trivers 1972).

Although sex differences in feeding behaviour are common among primates (Clutton-Brock in press) there is little evidence to suggest that they have evolved to minimize feeding competition between the sexes. There are no cases where obvious sex differences occur in feeding apparatus but not in other aspects of morphology (as in some bird species (Selander 1972)) and all differences observed so far can be satisfactorily explained as consequences of behavioural or morphological differences arising from differential breeding competition (Clutton-Brock in press).

There is reasonable evidence that sexual dimorphism in body

size is related to differences in breeding competition between the sexes. As would be expected (Trivers 1972), males are larger than females in virtually all primates and can vary widely in reproductive success. Secondly, there is some indication that dimorphism in body size is greatest in species where inter-male competition is most intense. On the assumption that inter-male competition would be strongest in species showing the most heavily biased socionomic sex ratios, we predicted a correlation between sexual dimorphism in body size and socionomic sex ratio. Monogamous species show the smallest degree of dimorphism and strongly polygynous species (e.g. *Erythrocebus patas*) the greatest (see Fig. 6) and both within particular taxonomic groups (Gautier-Hion 1975) and across the order as a whole, sexual dimorphism

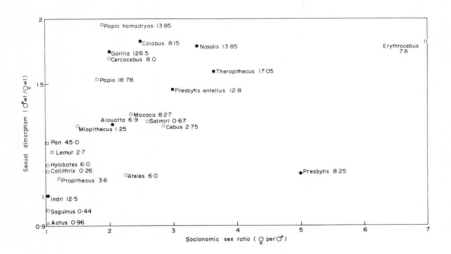

Fig. 6. Sexual dimorphism (male weight divided by female weight) plotted against socionomic sex ratio for different genera. The regression line used in text weighted for several values of $y$ per value of $x$ (Sokal and Rholf 1969). Body weight given after generic name △ nocturnal, arboreal frugivores; ○ diurnal, arboreal frugivores; □ diurnal, terrestrial frugivores; ● diurnal, arboreal folivores; ■ diurnal, terrestrial folivores.

is related to socionomic sex ratio. However, this trend is complicated by a positive association between sexual dimorphism and body weight. When dimorphism is regressed on both independent variables (breeding sex ratio and body weight) using partial regression analysis (Sokal and Rohlf 1969) it is found to be positively related to both of them ($t = 2\cdot144, P < 0\cdot05, df = 21; t = 2\cdot785, P < 0\cdot02, df = 21$).

Although dimorphism can be shown to be statistically related to breeding sex ratio, it is evident that a considerable proportion of the variance is not accounted for. There are several possible explanations of the lack of close relationship:

(1)   Our measures of sexual dimorphism and socionomic sex ratio are unreliable and have introduced considerable random error.
(2)   Socionomic sex ratio fails to provide a reliable index of inter-male competition. This is particularly likely in multi-male troops where a marked dominance hierarchy exists.
(3)   Ecological factors, including energetic limitations (which presumably limit the increase in male size) may be more constricting in some dietetic groups than in others. For example, the increased dimorphism found in terrestrial species (see Fig. 1) may be the product of relaxation of the constraints on increased body size in a situation where it is unlikely to restrict the male's access to food resources (see Jorde and Spuhler 1974).
(4)   Increased intensity of intraspecific competition (and hence increased sexual dimorphism) would be expected in those species (a) where breeding is markedly seasonal; (b) where the reproductive lifespan of the male is relatively short (Geist 1974a).
(5)   Particular taxonomic groups may have characteristic degrees of sexual dimorphism as a result of phylogenetic inertia. There is, for example, some indication that dimorphism tends to be low among the Lemuridae and high in the Cercopithec-idae (Crook 1972).

Similar relationships between body size and sexual dimorphism have been found in other vertebrate groups (e.g. Wiley 1974). One possible explanation is that, in some taxonomic groups, larger species tend to have fewer heterospecific competitors and that competitive release permits increased variance in body size (see Wilson 1975).

### Categorizing social organization

Although this analysis shows that it is possible to define ecological categories (see Table 4), four points should be stressed about categories of this kind. First, some categories are much more variable

Table 4. Rank order of log mean values for each variable in the seven ecological groups (median value for socionomic sex ratio). The number of species is shown in brackets in each cell.

| | Nocturnal Arboreal Insectivore | Nocturnal Arboreal Frugivore | Nocturnal Arboreal Folivore | Diurnal Arboreal Frugivore | Diurnal Arboreal Folivore | Diurnal Terrestrial Frugivore | Diurnal Terrestrial Folivore |
|---|---|---|---|---|---|---|---|
| Body weight | 1(4) | 2(7) | 3(1) | 4(34) | 5(22) | 6(20) | 7(3) |
| Feeding group size | 1·5(5) | 1·5(11) | 3(1) | 4(34) | 5(23) | 6(16) | 7(3) |
| Feeding group weight | 1(4) | 2(7) | 3(1) | 4(20) | 5(16) | 6(14) | 7(3) |
| Home range | 2(1) | 3(5) | 1(1) | 5(11) | 4(13) | 6(10) | 7(2) |
| Day range | — | 2(1) | — | 3(7) | 1(6) | 5(11) | 4(3) |
| Population density | 1(1) | 5(5) | 7(1) | 4(11) | 6(16) | 2(12) | 3(3) |
| Biomass | — | 2(5) | 5(1) | 4(10) | 7(12) | 3(11) | 6(3) |
| Sex ratio | — | 1(2) | — | 2(18) | 4(15) | 3(14) | 5(3) |
| Sex dimorphism | 2(3) | 1(5) | — | 4(18) | 3(12) | 5(14) | 6(3) |

than others. Thus, while the nocturnal species represent a moderately cohesive group within which variation in most characteristics is not large, the arboreal folivores and the arboreal frugivores show extensive variation in virtually all traits. Second, some traits show much more distinct differences between ecological categories than others (see Table 1). Third, although it may be possible to identify categories of social organization and to relate them to ecological variation, this does not necessarily indicate that they represent evolutionary sequences or grades (Huxley 1958). Finally, the results of this analysis indicate that examination of the distribution of individual variables (rather than of types of social organization (Crook, Ellis and Goss-Custard 1976)) is likely to lead to the construction of useful hypotheses concerning their functional significance.

## Summary

Estimates of body weight, sexual dimorphism, socionomic sex ratio, home range size, day range length, and population density of 100 primate species were extracted from the literature. Species were allocated to seven ecological categories:

nocturnal, arboreal, insectivorous; nocturnal, arboreal, frugivorous; nocturnal, arboreal, folivorous; diurnal, arboreal, frugivorous; diurnal arboreal, folivorous; diurnal, terrestrial frugivorous; diurnal, terrestrial, folivorous.

Congeneric species belonging to the same ecological category tend to show similar patterns of social organization. To avoid biasing by unequal representation of genera, averages were calculated for congeneric species belonging to the same ecological category.

Body weight is larger in diurnal species than in nocturnal, in terrestrial than in arboreal and in folivores than in frugivores. Nocturnality is probably associated with small size both because most nocturnal species occupy the middle and lower strata of the forest and need to be able to travel on relatively fine twigs and because they rely on crypsis rather than on flight or defence to avoid predation. Large body size may be advantageous in terrestrial species because it permits increased day range length and enhances the ability of individuals to defend themselves against predators. Increased body size in folivores over frugivores may be related to a relaxation of energetic constraints on body size in the former group.

None of the current theories concerning the function of grouping easily explains the distribution of group size in primates, and several different factors may be involved. Group size is small in all nocturnal species, perhaps because the same factors which limit body size (see above) also constrain group size. Large group size may occur in savannah-dwelling species both because food supplies are clumped and widely dispersed and as an anti-predator adaptation. The distribution of group size in arboreal folivores and frugivores is less easily understood. As among ungulates, group size is correlated with body weight, probably because similar constraints affect both variables.

Differences in home range size are closely related to variation in the density and distribution of food supplies. Home range size is related to group weight and is greater in frugivores than in folivores. Species feeding on clumped and unpredictable food supplies tend to live in large groups in large home ranges.

Day range length is less closely related to variation in food density, perhaps because differences in group dispersion may be involved. Day range length is significantly related to group weight among frugivores and is consistently longer in frugivores than folivores.

As would be predicted on energetic grounds, population density is negatively related to body weight and the highest population densities (and biomasses) occur among the arboreal folivores. There is no close relationship between body weight and biomass.

Monogamy occurs in territorial species living at relatively low density (though there are some exceptions). This may be because there is an upper limit on the size of the area which a male can defend efficiently, irrespective of body size. Monogamy may consequently evolve where there is strong selection for territoriality and where the defensible area cannot supply more than one female and her offspring. There is currently no satisfactory explanation of the distribution of one-male versus multi-male troops, nor of species differences in socionomic sex ratio.

Sexual dimorphism in body weight shows wide variation within most ecological categories. Sexual dimorphism tends to be largest in large species with heavily imbalanced sex ratios and smallest in monogamous species, probably because higher competition for breeding access in the former group selects for increased male body size. However, much variation in sexual dimorphism is still difficult to explain.

The study showed that most variables had associated distributions across ecological categories, providing some support for attempts to categorize social systems where this is necessary. However, it also demonstrated that investigation of quantitative relationships between variables is likely to yield more tangible results.

# References

Aldrich-Blake, F. P. G. (1970a). *The ecology and behaviour of the blue monkey, Cercopithicus mitis.* Ph.D. thesis, University of Bristol.

—— (1970b). Problems of social structure in forest monkeys. In *Social behaviour of birds and mammals* 79–101. (ed. J. H. Crook). Academic Press, London.

Alexander, R. D. (1974). The evolution of social behavior. *Ann. Rev. Ecol. Syst.* 5, 325–383.

Altmann, S. A. (1974). Baboons, space, time and energy. *Amer. Zool.* 14, 221–48.

—— and Altmann, J. (1970). *Baboon ecology.* Chicago University Press.

Armstrong, J. T. (1965). Breeding home range in the nighthawk and other birds; its evolutionary and ecological significance. *Ecology* 46, 619–29.

Baldwin, J. D. and Baldwin, J. (1972). The ecology and behaviour of squirrel monkeys (*Saimiri oerstedi*) in a natural forest in Western Panama. *Folia primatol.* 18, 161–84.

Bates, B. C. (1970). Territorial behaviour in primates: a review of recent field studies. *Primates* 11, 271–84.

Bearder, S. K. and Doyle, G. A. (1974). Ecology of bushbabies *Galago senegalensis* and *Galago crassicaudatus* with some notes on their behaviour in the field. In *Prosimian biology*, 109–30. (eds. R. D. Martin, G. A. Doyle, and A. C. Walker). Duckworth, London.

Bernstein, I. S. (1967). A field study of the pigtail monkey (*Macaca nemestrina*). *Primates* 8, 217–28.

—— (1968). The lutong of Kuala Selangor. *Behaviour* 32, 1–16.

Bertram, B. (1975). Social factors influencing reproduction in wild lions. *J. Zool., Lond.* 177, 463–82.

Booth, A. H. (1957). Observations on the natural history of the Olive colobus monkey, *Procolobus verus* (Van Beneden). *Proc. zool. Soc. Lond.* 129, 421–30.

Boulière, F., Hunkeler, C., and Bertrand, M. (1970). Ecology and behavior of Lowe's guenon (*Cercopithecus campbelli lowei*) in the Ivory coast. In *Old World monkeys: Evolution, systematics and behavior*, 297–350 (eds. J. R. Napier and P. H. Napier). Academic Press, New York.

Brown, J. L. (1964). The evolution of diversity in avian territorial systems. *Wilson Bull.* 76, 160–9.

—— (1975). *The evolution of behavior*. Norton, New York.

Chalmers, N. R. (1968). Group composition, ecology and daily activities of free living mangabeys in Uganda. *Folia primatol.* 8, 247–62.

Charles-Dominique, P. (1971). Eco-éthologie des Prosimiens du Gabon. *Biologia gabon.* 7, 121–228.

—— (1974). Ecology and feeding behaviour of five sympatric lorisids in Gabon. In *Prosimian biology*, 131–50 (eds. R. D. Martin, G. A. Doyle, and A. C. Walker). Duckworth, London.

—— and Hladik, C. M. (1971). Le lépilémur du sud de Madagascar: écologie, alimentation et vie sociale. *Terre Vie* 25, 3–66.

—— and Martin, R. D. (1972). Behaviour and ecology of nocturnal prosimians. *Z. Tierpsychol.* Supplement 6.

Chivers, D. J. (1969). On the daily behaviour and spacing of howling monkey groups. *Folia primatol.* 10, 48–102.

—— (1972a). An introduction to the socio-ecology of Malayan forest primates. In *Comparative ecology and behaviour of primates*, 101–70 (eds. R. P. Michael and J. H. Crook). Academic Press, London.

—— (1972b). The siamang and the gibbon in the Malay Peninsula. In *Gibbon and siamang* I, 103–35 (ed. D. M. Rumbaugh). Karger, Basel.

—— (1974). The siamang in Malaya. *Contr. Primatol.* 4, 1–335.

——, Raemakers, J. J. and Aldrich-Blake, F. P. G. (1975). Long-term observations of siamang behaviour. *Folia primatol.* 23, 1–49.

Clutton-Brock, T. H. (1972). *Feeding and ranging behaviour of the Red colobus monkey*. Ph.D. thesis. University of Cambridge.

—— (1974a). Why do animals live in groups? *New Scient.* July 1974.

—— (1974b). Primate social organization and ecology. *Nature, Lond.* 250, 539–42.

—— (1975a). Feeding behaviour of red colobus and black and white colobus in East Africa. *Folia primatol.* 23, 165–207.

—— (1975b). Ranging behaviour of red colobus (*Colobus badius tephrosceles*) in the Gombe National Park. *Anim. Behav.* 23, 706–22.

—— (1977). Some aspects of intraspecific variation in feeding and ranging behaviour in primates. In *Primate ecology: Studies of feeding and ranging behaviour in lemurs, monkeys and apes* (ed. T. H. Clutton-Brock). Academic Press, London.

—— and Guiness, F. E. (1975). Behaviour of red deer (*Cervus elaphus* L.) at calving time. *Behaviour* 55, 287–300.

—— and Harvey, P. H. (1976). Evolutionary rules and primate societies. In *Growing points in ethology*: 195–237 (eds. P. P. G. Bateson and R. A. Hinde). Cambridge University Press.

—— (1977). Species differences in feeding and ranging behaviour in primates. In *Primate ecology: Studies of feeding and ranging behaviour in lemurs, monkeys and apes* (ed. T. H. Clutton-Brock). Academic Press, London.

Cody, M. L. (1971). Finch flocks in the Mohave desert. *Theor. Popul. Biol.* 2, 142–58.

—— (1974). *Competition and the structure of bird communities*. Princeton University Press.

Coelho, A. M. (1974). Socio-bioenergetics and sexual dimorphism in primates. *Primates* 15, 263–9.

Coimbra-Filho, A. F. and Mittermeier, R. A. (1973). Distribution and ecology of the genus *Leontipithecus*, Lesson 1840, in Brazil. *Primates* 14, 47–66.

Crook, J. H. (1966). Gelada baboon herd structure and movement: a comparative report. *Symp. zool. Soc. Lond.* No. 18, 237–58.

── (1970). The socio-ecology of primates. In *Social behaviour in birds and mammals*, 103–66 (ed. J. H. Crook). Academic Press, London.

── (1972). Sexual selection, dimorphism and social organization in the primates. In *Sexual selection and the descent of man, 1871–1971:* 231–81 (ed. B. Campbell). Aldine Atherton, Chicago.

──, Ellis, J. E. and Goss-Custard. J. D. (1976). Mammalian social systems: structure and function. *Anim. Behav.* 24, 261–74.

── and Gartlan, J. S. (1966). Evolution of primate societies. *Nature, Lond.* 210, 1200–3.

Darwin, C. (1871). *The descent of man and selection in relation to sex.* Appleton, New York.

Deag, J. M. (1973). Intergroup encounters in the wild barbary macaque, *Macaca sylvanus*. In *Comparative behaviour and ecology of primates*, 315–74 (eds. R. P. Michael and J. H. Crook). Academic Press, London.

── and Crook, J. H. (1971). Social behaviour and 'agonistic buffering' in the wild Barbary macaque, *Macaca sylvana* L. *Folia primatol.* 15, 183–200.

Denham, W. D. (1971). Energy relations and some basic properties of primate social organization. *Am. Anthrop.* 73, 77–95.

Devore, I. and Hall, K. R. L. (1965). Baboon ecology. In *Primate behavior: Field studies of monkeys and apes:* 20–52 (ed. I. Devore). Holt, Rinehart and Winston, New York.

DeVos, A. and Omar, A. (1971). Territories and movements of Sykes monkeys (*Cercopithecus mitis kolbi* Neuman) in Kenya. *Folia primatol.* 16, 196–205.

Dunbar, R. I. M. (in press). Feeding ecology of gelada baboons: a preliminary report. In *Primate ecology: Studies of feeding and ranging behaviour in lemurs, monkeys and apes.* (ed. T. H. Clutton-Brock). Academic Press, London.

── and Dunbar, E. P. (1974). Ecology and population dynamics of *Colobus guereza* in Ethiopia. *Folia primatol.* 21, 188–208.

── and ── (1975). Social dynamics of Gelada baboons. *Contr. Primatol.* 6.

── and Nathan, M. F. (1972). Social organization of the Guinea baboon, *Papio papio*. *Folia primatol.* 17, 321–34.

Durham, N. M. (1971). Effects of altitude differences on group organization of wild black spider monkeys (*Ateles paniscus*). *Int. congr. Primatol.* 3, 32–40.

Eisenberg, J. P. and Kuhn, R. E. (1966). The behavior of *Ateles geoffroyi* and some related species. *Smithson. misc. Collns* 151, 1–63.

── and Lockhart, M. (1972). An ecological reconnaissance of Wilpattu National Park, Ceylon. *Smithson. Contr. Zool.* No. 101, 1–118.

── and McKay, G. M. (1974). Comparison of ungulate adaptations in the New World and Old World tropical forests, with special reference to Ceylon and the rainforests of Central America. In *The behaviour of ungulates and its relation to management*, 586–617 (eds. V. Geist and F. Walther). I.U.C.N., Morges.

──, Muckenhirn, N. A. and Rudran, R. (1972). The relation between ecology and social structure in primates. *Science, Wash.* 176, 863–74.

── and Thorington, R.W. (1973). A preliminary analysis of a neotropical mammal fauna. *Biotropica* 5, 150–61.

Ellefson, J. O. (1974). A natural history of white-handed gibbons in the Malayan Peninsula. *Gibbon and Siamang* 3, 1–136 (ed. D. M. Rumbaugh). Karger, Basel.

Estes, R. D. (1974). Social organization of the African Bovidae. In *The behaviour of ungulates and its relation to management* I 166–205 (eds. V. Geist and F. Walther). I.U.C.N., Morges.

Fooden, J. (1964). Stomach contents and gastro-intestinal proportions in wild-shot Guianan monkeys. *Am. J. Phys. Anthrop.* 22, 227–32.

Fossey, D. (1974). Observations on the home range of one group of mountain gorillas (*Gorilla gorilla beringei*). *Anim. Behav.* 22, 568–81.

── and Harcourt, A. H. (1977). Feeding ecology of free-ranging mountain gorilla (*Gorilla gorilla beringei*). In *Primate ecology: Studies of feeding and ranging behaviour in lemurs, monkeys and apes* (ed. T. H. Clutton-Brock). Academic Press, London.

Galdikas-Brindanour, B. (in press). Orang-utan adaptation at Tanjung Puting Reserve. (Quoted in Rodman, in press *b*).

Gartlan, J. S. (1970). Preliminary notes on the ecology and behavior of the drill, *Mandrillus leucophaeus* Ritgen 1824. In *Behavior of Old World monkeys,* 445-80 (eds. J. H. Napier and P. H. Napier). Academic Press, New York.

—— and Brian, C. K. (1968). Ecology and social variability in *Cercopithecus aethiops* and *C. mitis*. In *Primates, studies in adaptation and variability*, 253-92 (ed. P. C. Jay). Holt, Rinehart and Winston, New York.

Gatinot, B. L. (1975). *Ecologie d'un colobe bai (Colobus badius termincki, Kuhn 1820) dans un milieu marginal au Senegal.* Doctoral thesis, University of Paris VI.

Gautier, J. P. (1971). Etude morphologique et fonctionnelle des annexes extra-larygées des Cercopithecinae: liaison avec les cris d'espacement. *Biologica Gabon.* 7, 229-67.

Gautier-Hion, A. (1970). L'organization sociale d'une bande de talapoins (*Miopithecus talapoin*) dans le nord-est du Gabon. *Folia primatol.* 12, 116-41.

—— (1971). L'écologie du talapoin du Gabon. *Terre Vie* 25, 427-90.

——(1975). Dimorphisme sexuel et organisation sociale chez les cercopithécinés forestiers Africains. *Mammalia* 30, 365-74.

Geist, V. (1971). *Mountain Sheep: A study in behavior and evolution.* University of Chicago Press.

—— (1974*a*). On the relationship of social evolution and ecology in ungulates. *Amer. Zool.* 14, 205-20.

—— (1974*b*). On the relationship of ecology and behaviour in the evolution of ungulates: theoretical consideration. In *The behaviour of ungulates and its relation to management*, 235-46 (eds. V. Geist and F. Walker). I.U.C.N., Morges.

Goss-Custard, J. D. (1970). Feeding dispersion in some over-wintering wading birds. In *Social behaviour in birds and mammals*, 3-35 (ed. J. H. Crook). Academic Press, New York.

——, Dunbar, R. I. M. and Aldrich-Blake, F. P. G. (1972). Survival, mating and rearing strategies in the evolution of primate social structure. *Folia primatol.* 17, 1-19.

Groves, C. P. (1973). Notes on the ecology and behaviour of the Angola colobus (*Colobus angolensis* P. L. Slater 1860) in N.E. Tanzania. *Folia primatol.* 20, 12-26.

Haddow, A. J. (1952). Field and laboratory studies on an African monkey *Cercopithecus ascanius schmidti* Matschie. *Proc. zool. Soc. Lond.* 122, 297-394.

Hall, K. R. L. (1962). Numerical data, maintenance activities and locomotion in the wild chacma baboon, *Papio ursinus. Proc. zool. Soc. Lond.* 139, 181-220.

—— (1966). Behaviour and ecology of the wild Patas monkey, *Erythrocebus patas*, in Uganda. *J. zool., Lond.* 148, 15-87.

—— and Gartlan, J. S. (1965). Ecology and behaviour of the vervet monkey, *Cercopithecus aethiops*, Lolui Island, Lake Victoria. *Proc. zool. Soc. Lond.* 145, 37-56.

Hamilton, W. D. (1971). Selection of selfish and altruistic behavior in some extreme models. In *Man and beast: Comparative social behavior*, 57-91 (eds. J. F. Eisenberg, and W. S. Dillon). Washington, Smithsonian.

Hamilton, W. J. III, Buskirk, R. E. and Buskirk, W. H. (1975). Chacma baboon tactics during intertroop encounters. *J. Mamm.* 56, 857-870.

Hill, W. C. O. (1952). The external and visceral anatomy of the Olive colobus monkey (*Procolobus verus*). *Proc. zool. Soc. Lond.* 122, 127-86.

—— (1966). *Primates* 6, Edinburgh: University Press.

Hinde, R. A. (1956). The biological significance of the territories of birds. *Ibis* 98, 340-69.

Hladik, C. M. (1975). Ecology, diet and social patterning in Old and New World primates. In *Socio-Ecology and psychology of primates*, 3-36 (ed. R. H. Tuttle). Mouton, The Hague.

—— and Charles-Dominique, P. (1974). The behaviour and ecology of the sportive lemur (*Lepilemur mustelinus*) in relation to its dietary pecularities. In *Prosimian biology*, 23-31 (eds. R. D. Martin, G. P. Doyle, and A. C. Walker). Duckworth, London.

—— and Hladik, A. (1972). Disposibilités alimentaires et domains vitaux des primates à Ceylon. *Terre Vie* 2, 149–215.
Horwich, R. H. (1972). Home range and food habits of the Nilgiri langur *Presbytis johnii. J. Bombay nat. Hist. Soc.* 69, 255–67.
Huxley, J. S. (1934). A natural experiment on the territorial instinct. *Br. Birds* 27, 270–77.
—— (1958). Evolutionary processes and taxonomy with special reference to grades. *Uppsala Univ. Arsskr.* 1958, 21–39.
Jarman, P. J. (1974). The social organization of antelope in relation to their ecology. *Behaviour* 48, 215–67.
Jay, P. C. (ed.). (1972). *Primates: Studies in adaptation and variability.* Holt, Rinehart and Winston, New York.
Jewell, P. A. and Oates, J. (1969). Ecological observations on the lorisoid primates of African lowland forest. *Zool. afr.* 4, 231–48.
Jolly, A. (1966). *Lemur behavior.* University Press, Chicago.
—— (1972). *The evolution of primate behavior.* MacMillan, New York.
Jones, C. and Sabater Pi, J. (1968). Comparative ecology of *Cercocebus albigena* (Gray) and *Cercocebus torquatus* (Kerr) in Rio Muni, West Africa. *Folia primatol.* 9, 99–113.
Jorde, L. B. and Spuhler, J. N. (1974). A statistical analysis of selected aspects of primate demography, ecology and social behavior. *J. anthrop. Res.* 30, 199–224.
Kaufmann, J. H. (1974). The ecology and evolution of social organization in the kangaroo family (Macropodidae). *Amer. Zool.* 14, 51–62.
Kavanagh, M. and Dresdale, L. (1975). Observations on the woolly monkey (*Lagothrix lagothricha*) in Northern Colombia. *Primates* 16, 285–94.
Kawabe, M. and Mano, T. (1972). Ecology and behavior of the wild proboscis monkey, *Nasalis larvatus* (Wurmb) in Sabah, Malaysia. *Primates* 13, 213–28.
Kingdon, J. (1971). *East African mammals* 1. Academic Press, London.
Kinzey, W. G. (1977). Diet and feeding behaviour of *Callicebus torquatus.* In *Primate ecology: Studies of feeding and ranging behaviour in lemurs, monkeys and apes* (ed. T. H. Clutton-Brock). London, Academic Press.
—— Rosenberger, A. L., Heisler, P. S., Prowse, D. L. and Trilling, J. S. (in press). A preliminary field investigation of the yellow handed titi monkey, *Callicebus torquatus*, in Northern Peru. *Primates.*
Klein, L. and Klein, B. (1977). Feeding behaviour of the Colombian spider monkey. In *Primate ecology: Studies of feeding and ranging behaviour in lemurs, monkeys and apes* (ed. T. H. Clutton-Brock). Academic Press, London.
Krebs, J. R., MacRoberts, M. H. and Cullen, J. M. (1972). Flocking and feeding in the great tit *Parus major*—an experimental study. *Ibis* 114, 507–30.
Kruuk, H. (1972). *The Spotted hyena.* University of Chicago Press.
—— (1975). Functional aspects of social hunting by carnivores. In *Function and evolution of behaviour* 119–41. (eds. C. Beer and A. Manning). Clarendon Press, Oxford.
Kummer, H. (1968). *Social organization of Hamadryas baboons.* University of Chicago Press.
Kuhn, H. J. (1964). Zur Kenntnis von Bau und Funktion des Magens der Schlankaffen (Colobinae). *Folia primatol.* 2, 193–221.
Lack, D. (1968). *Ecological adaptations for breeding in birds.* Methuen, London.
van Lawick-Goodall, J. (1968). The behaviour of free-living chimpanzees in the Gombe Stream Reserve. *Anim. behav. Monog.* 1, 161–311.
Lindberg, D. G. (1971). The rhesus monkey in North India: an ecological and behavioral study. In *Primate behavior: Developments in field and laboratory research*, 1–106 (ed. L. A. Rosenblum). Academic Press, New York.
—— (1977). Feeding behaviour and diet of rhesus monkeys (*Macaca mulatta*) in a Siwalik forest in North India. In *Primate ecology: Studies of feeding and ranging behaviour in lemurs, monkeys and apes* (ed. T. H. Clutton-Brock). Academic Press, London.

MacKinnon, J. (1974). The behaviour and ecology of wild orang-utans (*Pongo pygmaeus*). *Anim. behav.* 22, 3–74.

McNab, B. W. (1963). Bioenergetics and the determination of home range size. *Amer. Nat.* 97, 133–40.

Maher, W. J. (1970). The Pomarine jaeger as a Brown lemming predator in northern Alaska. *Wilson Bull.* 82, 130–57.

Maples, W. R. (1972). Systematic reconsideration and a revision of the nomenclature of Kenya baboons. *Am. J. Phys. Anthrop.* 36, 9–19.

Marler, P. (1969). *Colobus guereza*: territoriality and group composition. *Science, Wash.* 163, 93–5.

Mason, A. (1968). Use of space by *Callicebus* groups. In *Primates: Studies in adaptation and variability*, 200–16 (ed. P. C. Jay). Holt, Rinehart and Winston, New York.

Milton, K. and May, M. L. (1976). Body weight, diet and home range area in primates. *Nature, Lond.* 259, 459–62.

Moynihan, M. (1964). Some behavior patterns of platyrrhine monkeys I: the night monkey (*Aotus trivirgatus*). *Smithson. misc. Collns.* 146, 1–84.

—— (1970). Some behavior patterns of platyrrhine monkeys II: *Saguinus geoffroyi* and some other tamarins. *Smithson. Contr. Zool.* No. 28, 1–77.

Mucherjee, R. P. and Saha, S. S. (1974). The golden langurs (*Presbytis geei*, Khajuria 1956) of Assam. *Primates* 15, 327–40.

Napier, J. R. and Napier, P. H. (1967). *A handbook of living primates*. Academic Press, London.

Neville, M. K. (1968). Ecology and activity of Himalayan foothill monkeys. *Ecology* 49, 110–23.

—— (1972). The population structure of red howler monkeys (*Alouatta seniculus*) in Trinidad and Venezuela. *Folia primatol.* 17, 56–86.

Nolte, A. (1955). Field observations on the daily routine and social behaviour of common Indian monkeys, with special reference to the bonnet monkey (*Macaca radiata* Geoffroy). *J. Bombay nat. Hist. Soc.* 53, 177–84.

Oates, J. F. (1977). The guereza and its food. In *Primate ecology: Studies of feeding and ranging behaviour in lemurs, monkeys and apes* (ed. T. H. Clutton-Brock). Academic Press, London.

Oppenheimer, J. R. (1969a). Changes in forehead patterns and group composition of the White-faced monkey (*Cebus capucinus*). *Int. congr. Primatol.* 2 (1), 36–42.

—— (1969b). Behavior and ecology of the White-faced monkeys, *Cebus capucinus* on Barro Colorado Island, C.Z. *Diss. Abstr. Int.* 30B, 442–3.

—— and Oppenheimer, E. C. (1973). Preliminary observations of *Cebus nigrivattatus* (Primates: Cebidae) on the Venezuelan Llanos. *Folia primatol.* 19, 409–36.

Orians, G. H. (1961). The ecology of blackbird (*Agelaius*) social systems. *Ecol. Monogs* 31, 285–312.

—— (1969). On the evolution of mating systems in birds and mammals. *Amer. Nat.* 103, 589–603.

Patterson, I. J. (1965). Timing and spacing of broods in the black-headed gull *Larus ridibundus*. *Ibis* 107, 433–59.

Petter, J. J. (1962). Récherches sur l'écologie et l'éthologie des lémuriens malgaches. *Mém. Mus. Hist. nat. Paris* 27A, 1–146.

—— and Hladik, C. M. (1970). Observations sur le domaine vital et la densité de population de *Loris tardigradus* dans les forêts de Ceylon. *Mammalia* 34, 394–409.

Pfeffer, P. (1969). Considérations sur l'écologie des forêts claires du Cambodge oriental. *Terre Vie* 116, 3–24.

Poirier, F. E. (1968). Analysis of a Nilgiri langur (*Presbytis johnii*) home range change. *Primates* 9, 29–44.

—— (1969). The Nilgiri langur (*Presbytis johnii*) troop: its composition, structure, function and change. *Folia primatol.* 10, 20–47.

Pollock, J. I. (1977). The ecology and sociology of feeding in *Indri*. In *Primate eco-*

*logy: studies of feeding and ranging behaviour in lemurs, monkeys and apes* (ed. T. H. Clutton-Brock). Academic Press, London.

Pope, B. L. (1966). The population characteristics of howler monkeys (*Alouatta caraya*) in northern Argentina. *Am. J. Phys. Anthrop.* 24, 361-70.

Rahaman, H. and Parthasarathy, M. D. (1969). Studies of the social behaviour of bonnet monkeys. *Primates* 10, 149-62.

Richard, A. (1977). The feeding behaviour of *Propithecus verreauxi*. In *Primate ecology: Studies of feeding and ranging behaviour in lemurs, monkeys and apes* (ed. T. H. Clutton-Brock). Academic Press, London.

Rodman, P. (1973). Population composition and adaptive organization among orang-utans of the Kutai Reserve. In *Comparative ecology and behaviour of primates*, 171-209 (eds. R. P. Michael and J. H. Crook). Academic Press, London.

— — (in press *a*). Individual activity patterns and the solitary nature of orang-utans. In *The behaviour of the Great apes* (eds. J. Goodall and D. Hamburg). Viking Press, New York.

— — (in press *b*). The feeding ecology of orang-utans. In *Primate ecology: Studies of feeding and ranging behaviour in lemurs, monkeys and apes* (ed. T. H. Clutton-Brock). Academic Press, London.

Rowell, T. E. (1966). Forest living baboons in Uganda. *J. Zool., Lond.* 149, 344-64.

— — (1968). The habits of baboons in Uganda. *Proc. E. Afr. Acad.* 11, 121-7.

— — (1973). Social organization of wild talapoin monkeys. *Am. J. Phys. Anthrop.* 38, 593-7.

Schaller, G. B. (1965). The behavior of the mountain gorilla. In *Primate behavior: Field studies of monkeys and apes*, 324-67 (ed. I. DeVore). Holt, Rinehart and Winston, New York.

— — (1972). *The Serengeti Lion.* University of Chicago Press.

Schenkel, R. and Schenkel-Hulliger, L. (1967). On the sociology of free-ranging colobus (*Colobus guereza caudatus* Thomas, 1885). In *Progress in primatology*, 185-94. Fischer, Stuttgart.

Schoener, T. W. (1968). Sizes of feeding territories among birds. *Ecology* 49, 123-41.

— — (1969). Models of optimal size for solitary predators. *Amer. Nat.* 103, 277-313.

Schultz, A. H. (1956). Post embryonic changes. *Primatologia* 1, 887-964.

Selander, R. K. (1966). Sexual dimorphism and differential niche utilization in birds. *Condor* 68, 113-51.

— — (1972). Sexual selection and dimorphism in birds. In *Sexual selection and the descent of man*, 180-230 (ed. B. Campbell). Aldine, Chicago.

Simonds, P. E. (1965). The bonnet macaque in South India. In *Primate behavior*, 175-96 (ed. I. DeVore). Holt, Rinehart and Winston, New York.

Smith, C. C. (1968). The adaptive nature of the social organization in the genus of tree squirrels, *Tamiasciurus*. *Ecol. Monogr.* 38, 31-63.

Sokal, R. R. and Rohlf, F. J. (1969). *Biometry*. Freeman, San Francisco.

Southwick, C. H., Beg, M. A. and Siddiqi, M. R. (1965). Rhesus monkeys in North India. In *Primate behavior: Field studies of monkeys and apes*, 111-59 (ed. I. DeVore). Holt, Rinehart and Winston, New York.

Stolz, L. P. and Keith, M. E. (1973). A population survey of the chacma baboon in the Northern Transvaal. *J. Hum. Evol.* 2, 195-212.

Struhsaker, R. R. (1967*a*). Social structure among vervet monkeys (*Cercopithecus aethiops*). *Behaviour* 29, 83-121.

— — (1967*b*). Ecology of vervet monkeys (*Cercopithecus aethiops*) in the Masai-Amboseli game reserve, Kenya. *Ecology* 48, 891-904.

— — (1969). Correlates of ecology and social organization among African cercopithecines. *Folia primatol.* 11, 80-118.

— — (1975). *The Red colobus monkey.* Chicago University Press.

Sugiyama, Y. (1965). On the social change of Hanuman langurs (*Presbytis entellus*) in their natural condition. *Primates* 6, 381-418.

—— (1966). Social organizàtion of Hanuman langurs. In *Social communication among primates*, 221-36 (ed. S. A. Altmann). Chicago University Press.

—— (1971). Characteristics of the social life of bonnet macaques (*Macaca radiata*). *Primates* 12, 247-66.

Sussman, R. W. (1975). A preliminary study of the behavior and ecology of *Lemur fulvus rufus* Audubert 1800. In *Lemur biology* (eds. I. Tattersall and R. W. Sussman). Plenum, New York.

—— (1977). Feeding behaviour of *Lemur catta* and *Lemur fulvus*. In *Primate ecology: Studies of feeding and ranging behaviour in lemurs, monkeys and apes* (ed. T. H. Clutton-Brock). Academic Press, New York.

—— and Richard, A. (1974). The role of aggression among diurnal prosimians. In *Primate aggression, territoriality and xenophobia*, 49-76. Academic Press, New York.

Tanaka, J. (1965). Social structure of Nilgiri langurs. *Primates* 6, 107-22.

Tattersall, I. and Sussman, R. W. (1975). Observations on the ecology and behavior of the mongoose lemur, *Lemur mongoz mongoz* Linnaeus (Primates, Lemuriformes) at Ampijoroa, Madagascar. *Anthrop. Pap. Am. Mus. nat. Hist.* 59, 193-216.

Tenaza, R. R. (1975). Territory and monogamy among Kloss' gibbons (*Hylobates klossii*) in Siberut Island, Indonesia. *Folia primatol.* 24, 60-80.

—— and Hamilton, W. J. (1971). Preliminary observations of the Mentawi islands gibbon, *Hylobates klossii. Folia primatol.* 15, 201-11.

Thompson, W. A., Vertinsky, I., and Krebs, J. R. (1974). The survival value of flocking in birds: a simulation model. *J. Anim. Ecol.* 43, 785-820.

Thorington, R. W. (1967). Feeding and activity of *Cebus* and *Saimiri* in a Colombian forest. In *Progress in primatology*, 180-4 (eds. D. Starck, R. Schneider, and H. J. Kuhn). Fischer, Stuttgart.

—— (1968a). Observations of the tamarin, *Saguinas midas. Folia primatol.* 9, 95-8.

—— (1968b). Observations of squirrel monkeys in a Colombian forest. In *The Squirrel monkey*, 69-85 (eds. L. A. Rosenblum and R. W. Cooper). Academic Press, New York.

Tinbergen, N. (1957). The functions of territory. *Bird Study* 4, 14-27.

Treisman, M. (1975a). Predation and the evolution of gregariousness I. Models for concealment and evasion. *Anim. Behav.* 23, 779-800.

—— (1975b). Predation and the evolution of gregariousness. II. An economic model for predator–prey interaction. *Anim. Behav.* 23, 801-5.

Trivers, R. L. (1972). Parental investment and sexual selection. In *Sexual selection and the descent of man*, 156-79 (ed. B. Campbell). Aldine, Chicago.

Verner, J. (1964). Evolution of polygamy in the Long-billed marsh wren. *Evolution* 18, 252-61.

—— and Willson, M. F. (1966). The influence of habitats on mating systems of North American passerine birds. *Ecology* 47, 143-7.

—— and —— (1969). Mating systems, sexual dimorphism and the role of the male North American passerine birds in the nesting cycle. *Orn. Monogr.* 9, 1-76.

Vine, I. (1971). The risk of visual detection and pursuit by a predator and the selective advantage of flocking behaviour. *J. theoret. Biol.* 30, 405-22.

Waser, P. M. (1977). Feeding, ranging and group size in the mangabey, *Cercocebus albigena*. In *Primate ecology: Studies of feeding and ranging behaviour in lemurs, monkeys and apes* (ed. T. H. Clutton-Brock). Academic Press, London.

—— and Floody, O. (1974). Ranging patterns of the mangabey, *Cercocebus albigena*, in the Kibale Forest, Uganda. *Z. Tierpsychol.* 35, 85-107.

Wiley, R. H. (1974). Evolution of social organization and life history patterns among grouse (Aves: Tetraonidae). *Q. Rev. Biol.* 49 201-27.

Wilson, C. C. and Wilson, W. L. (1975). The influence of selective logging on primates and some other animals in East Kalimantan. *Folia primatol.* 23, 245-74.

Wilson, E. O. (1975). *Sociobiology: the new synthesis*. Harvard University Press, Cambridge.

Wrangham, R. W. (1975). *The behavioural ecology of chimpanzees in Gombe National Park*. Ph.D. thesis, University of Cambridge.

Yoshiba, K. (1968). Local and intertroop variability in ecology and social behavior of common Indian langurs. In *Primates: Studies in adaptation and variability*, 217–42 (ed. P. C. Jay). Holt, Rinehart and Winston, New York.

Zimmerman, J. L. (1966). Polygyny in the dickcissel. *Auk* 83, 534–46.

—— (1971). The territory and its density dependent effect in *Spiza americana*. *Auk* 88, 591–612.

# References to the Preface and Introductory sections

## Preface

Bateson, P. P. G. and Hinde, R. A. (eds.) (1976). *Growing points in ethology.* Cambridge University Press.

Dawkins, R. (1976). *The selfish gene.* Oxford University Press.

Gilpin, M. E. (1975). *Group selection in predator-prey communities.* Princeton University Press.

Hamilton, W. D. (1964). The genetical evolution of social behaviour I, II. *J. Theoret. Biol.* **7**, 1—52.

—— (1970). Selfish and spiteful behaviour in an evolutionary model. *Nature, Lond.* **228**, 1218—20.

Hinde, R. A. (1956). The biological significance of the territories of birds. *Ibis* **98**, 340—69.

—— (1970). *Animal behaviour.* McGraw-Hill, New York.

—— (1974). *Biological bases of human social behaviour.* McGraw-Hill, New York.

—— and Stevenson-Hinde, J. (eds.) (1973). *Constraints on learning: limitations and predispositions.* Academic Press, London and New York.

Jarman, P. J. (1974). The social organisation of antelope in relation to their ecology. *Behaviour* **48**, 215—67.

Maynard Smith, J. (1958). *The theory of evolution.* Penguin Book, Harmondsworth.

—— and Parker, G. A. (1976). The logic of asymmetric contests. *Anim. Behav.* **24**, 159—75.

Poirier, F. E. (ed.) (1972). *Primate socialization.* Random House, New York.

Rosenblum, L. A. (ed.) (1970). *Primate behavior*, Vols I and II. Academic Press, New York.

Schoener, T. W. (1968). Sizes of feeding territories among birds. *Ecology* **49**, 123—41.

Selander, R. K. (1966). Sexual dimorphism and differential niche utilisation in birds. *Condor* **68**, 113—51.

—— (1972). Sexual selection and dimorphism in birds. In *Sexual selection and the descent of man, 1871—1971* (ed. B. Campbell) pp. 180—230. Aldine-Atherton, Chicago.

Williams, G. C. (1966). *Adaptation and natural selection.* Princeton University Press.

Wilson, E. O. (1975). *Sociobiology, the new synthesis.* Belknap Press, Harvard.

## 1. Group benefit or individual advantage?

Alexander, R. D. (1974). The evolution of social behaviour. *Ann. Rev. Ecol. Syst.* **5**, 325—83.

Carr-Saunders, A. M. (1922). *The population problem: a study in human evolution.* Clarendon Press, Oxford.

Darwin, C. (1856). *The origin of species.* John Murray, London.

Dawkins, R. (1976). *The selfish gene.* Oxford University Press.

Fisher, R. A. (1930). *The genetical theory of natural selection.* Clarendon Press, Oxford.

Gibb, J. (1956). Food, feeding habits and territory of the Rock Pipit, *Anthus spinoletta.* *Ibis* **98**, 506—30.

Haldane, J. B. S. (1953). Animal populations and their regulation. *Penguin Modern Biology* **15**, 9–24.

Hamilton, W. D. (1963). The evolution of altruistic behavior. *Amer. Natur.* **97**, 354–6,

—— (1964). The genetical theory of social behaviour I, II. *J. Theoret. Biol.* **7**, 1–52.

—— (1972). Altruism and related phenomena, mainly in social insects. *Ann. Rev. Ecol. Syst.* **3**, 193–232.

Jacquard, A. (1974). *Genetic structure of populations.* Springer-Verlag, Berlin.

Lack, D. (1954). *The natural regulation of animal numbers.* Clarendon Press, Oxford.

—— (1966). *Population studies of birds.* Oxford University Press.

LeBoeuf, B. J. (1974). Male-male competition and reproductive success in elephant seals. *Amer. Zool.* **14**, 163–76.

Maynard Smith, J. (1964). Group selection and kin selection: a rejoinder. *Nature, Lond.* **201**, 1145–7.

—— (1976). Group selection. *Quart. Rev. Biol.* **51**, 277-83.

Mayr, E. (1963). *Animal species and evolution.* Belknap Press, Harvard.

Orlove, M. J. (1975). A model of kin selection not invoking coefficients of relationships. *J. Theoret. Biol.* **49**, 289–310.

Trivers, R. L. and Hare, H. (1976). Haplo-diploidy and the evolution of the social insects. *Science* **191**, 249–63.

Watson, A. and Jenkins, D. (1968). Experiments on population control in red grouse. *J. animal Ecol.* **37**, 595–614.

West-Eberhard, M. J. (1975). The evolution of social behavior by kin selection. *Quart. Rev. Biol.* **50**, 1–34.

Wilson, D. S. (1974). A theory of group selection. *Proc. natn. Acad. Sci., U.S.A.* **72**, 143–6.

Wilson, E. O. (1975). *Sociobiology, the new synthesis.* Belknap Press, Harvard.

Wynne-Edwards, V. C. (1962). *Animal dispersion in relation to social behaviour.* Oliver and Boyd, Edinburgh.

## 2. Reproductive strategies

Bourlière, F. (1975). Mammals, small and large: the ecological implications of size. In Golley, F. B., Petrusewicz, K., and Ryszkowski, L. (eds). *Small mammals: their productivity and population dynamics*, pp. 1–9. Cambridge University Press.

Breder, C. M. and Rosen, D. E. (1966). *Modes of reproduction in fishes.* Natural History Press, New York.

Clutton-Brock, T. H. and Harvey, P. H. (1977). Primate ecology and social organisation. *J. Zool.* **183**, 1–39.

Crow, J. F. and Kimura, M. (1965). Evolution in sexual and asexual populations. *Amer. Natur.* **99**, 439–50.

Darwin, C. (1871). *The descent of man, and selection in relation to sex.* Appleton, New York.

Davis, J. W. F. and O'Donald, P. (1976). Sexual selection for a handicap: a critical analysis of Zahavi's model. *J. Theoret. Biol.* **57**, 345–54.

Dawkins, R. (1976). *The selfish gene.* Oxford University Press.

—— and Carlisle, T. R. (1976). Parental investment, mate desertion and a fallacy. *Nature, Lond.* **262**, 131–3.

Emlen, J. M. (1970). Age specificity and ecological theory. *Ecology* **51**, 588–601.

Fagen, R. M. (1972). An optimal life history strategy in which reproductive effort decreases with age. *Amer. Natur.* **106**, 258–61.

Fisher, R. A. (1930). *The genetical theory of natural selection.* Clarendon Press, Oxford.

French, N. R., Stoddart, D. M., and Bobek, B. (1975). Patterns of demography in small mammal populations. In Golley, F. B., Petrusewicz, K. and Ryszkowski, L. (eds.). *Small mammals: their productivity and population dynamics*, pp. 73–103. Cambridge University Press.

Gadgil, M. and Bossert, W. H. (1970). Life history consequences of natural selection. *Amer. Natur.* **104**, 1–24.

Hamilton, W. D. (1966). The moulding of senescence by natural selection. *J. Theoret. Biol.* **12**, 12–45.

—— (1967). Extraordinary sex ratios. *Science* **156**, 477–88.

Holgate, P. (1967). Population survival and life history phenomena. *J. Theoret. Biol.* **14**, 1–10.

Huxley, J. S. (1938). *Evolution: essays on aspects of evolutionary biology*, pp. 11–42. Clarendon Press, Oxford.

Jenni, D. A. (1974). Evolution of polyandry in birds. *Amer. Zool.* **14**, 129–44.

Lack, D. (1968). *Ecological adaptations for breeding in birds*. Methuen, London.

Leigh, E. G. (1970). Sex ratios and differential mortality between the sexes. *Amer. Natur.* **104**, 205–10.

Lewontin, R. C. (1965). Selection for colonizing ability. In Baker, H. G. and Stebbins, G. L. (eds.). *The genetics of colonising species*, pp. 77–94. Academic Press, New York.

MacArthur, R. H. (1965). Ecological consequences of natural selection. In Waterman, T. T. H. and Morowitz, H. J. (eds.) *Theoretical and mathematical biology*, pp. 388–97. Blaisdall, New York.

—— and Wilson, E. O. (1967). *The theory of island biogeography*. Princeton University Press.

Maynard Smith, J. (1956). Fertility, mating behaviour, and sexual selection in *Drosophila subobscura*. *J. Genet.* **54**, 261–79.

—— (1968). Evolution in sexual and asexual populations. *Amer. Natur.* **102**, 469–73.

—— (1975). *The theory of evolution*. Penguin Books, Harmondsworth.

—— (1976). Sexual selection and the handicap principle. *J. Theoret. Biol.* **57**, 239–42.

—— (1977). Parental investment: a prospective analysis. *Anim. Behav.* **25**, 1–9.

—— and Ridpath, M. G. (1972). Wife sharing in the Tasmanian native hen, *Tribonyx mortierii*: a case of kin selection? *Amer. Natur.* **106**, 447–52.

Medawar, P. B. (1957). *The uniqueness of the individual*. Methuen, London.

Murphy, G. I. (1968). Patterns in life history. *Amer. Natur.* **102**, 391–403.

Orians, G. H. (1969). On the evolution of mating systems in birds and mammals. *Amer. Natur.* **103**, 589–603.

Pianka, E. R. (1970). On *r* and *K* selection. *Amer. Natur.* **104**, 592–7.

Ralls, K. (1976). Mammals in which females are larger than males. *Quart. Rev. Biol.* **51**, 245–76.

Schoener, T. W. (1969). Models of optimal size for solitary predators. *Amer. Natur.* **103**, 277–313.

Selander, R. K. (1966). Sexual dimorphism and differential niche utilisation in birds. *Condor* **68**, 113–51.

—— (1972). Sexual selection and dimorphism in birds. In Campbell, B. (ed.) *Sexual selection and the descent of man, 1871–1971*, pp. 180–230. Aldine-Atherton, Chicago.

Trivers, R. L. (1972). Parental investment and sexual selection. In Campbell, B. (ed.) *Sexual selection and the descent of man, 1871–1971*, pp. 136–79. Aldine-Atherton, Chicago.

—— and Willard, D. E. (1973). Natural selection and parental ability to vary the sex ratio of offspring. *Science* **179**, 90–2.

Wiley, R. H. (1974). Evolution of social organisation and life history patterns among grouse (Aves: Tetranidae). *Quart. Rev. Biol.* **49**, 201–27.

Williams, G. C. (1966a). *Adaptation and natural selection: a critique of some current evolutionary thought*. Princeton University Press.

—— (1966b). Natural selection, the costs of reproduction, and a refinement of Lack's principle. *Amer. Natur.* **100**, 687–90.

—— (1975). *Sex and evolution*. Princeton University Press.

Wilson, E. O. (1971). Competitive and aggressive behavior. In Eisenberg, J. F. and Dillon, W. (eds.) *Man and beast: comparative social behavior*, pp. 183—217. Smithsonian Institution Press, Washington D.C.

—— (1975). *Sociobiology, the new synthesis*. Belknap Press, Harvard.

—— and Bossert, W. H. (1971). *A primer of population biology*. Sinauer, Stamford.

Zahavi, A. (1975). Mate selection—a selection for a handicap. *J. Theoret. Biol.* **53**, 205—14.

## 3. Co-operation and disruption

Alcock, J. (1975). *Animal behavior, an evolutionary approach*. Sinauer, Stamford.

Alexander, R. D. (1974). The evolution of social behavior. *Ann. Rev. Ecol. Syst.* **5**, 325—83.

Barrington, E. J. W. (1967). *Invertebrate structure and function*. Nelson, London.

Bertram, B. R. (1976). Kin selection in lions and in evolution. In Bateson, P. P. G. and Hinde, R. A. (eds.), *Growing points in ethology*, pp. 281—301. Cambridge University Press.

Brown, J. L. (1975). *The evolution of behavior*. Norton, New York.

Charnov, E. L. and Krebs, J. R. (1975). The evolution of alarm calls: altruism or manipulation. *Amer. Natur.* **109**, 107—12.

Clapham, W. B. (1973). *Natural ecosystems*. Macmillan, New York.

Clutton-Brock, T. H. (1975). Feeding behaviour of red colobus and black and white colobus in East Africa. *Folia primatologica* **23**, 165—207.

—— and Harvey, P. H. (1976). Evolutionary rules and primate societies. In Bateson, P. P. G. and Hinde, R. A. (eds.), *Growing points in ethology*, pp. 195—237. Cambridge University Press.

Cody, M. L. (1971). Finch flocks in the Mohave Desert. *Theoret. Popul. Biol.* **2**, 142—58.

Crook, J. H. (1970). The socio-ecology of primates. In Crook, J. H. (ed.), *Social behaviour in birds and mammals: essays on the social ethology of animals and man*, pp. 103—166. Academic Press, New York.

Dawkins, R. (1976). *The selfish gene*. Oxford University Press.

Eibl-Eibesfeldt, I. (1970). *Ethology: the biology of behavior*. Holt, Rinehart and Winston, New York.

Emlen, J. M. (1973). *Ecology, an evolutionary approach*. Addison Wesley, New York.

Epple, G. von (1967). Verleichunde Untersuchungen über Sexual- und Social-verhalten der Krallenhaffen (Hapalidae). *Folia primatologica* **7**, 36—65.

Fry, C. H. (1972). The social organisation of bee eaters (Meropidae) and co-operative breeding in hot-climate birds. *Ibis* **114**, 1—14.

Gale, H. S. and Eaves, L. J. (1975). Logic of animal conflict. *Nature, Lond.* **254**, 463—4.

Geist, V. (1966). The evolution of horn-like organs. *Behaviour* **27**, 377—416.

—— (1971). *Mountain sheep: a study in behavior and evolution*. University of Chicago Press.

—— (1974). On the relationship of social evolution and ecology in Ungulates. *Amer. Zool.* **14**, 205—20.

Hamilton, W. D. (1970). Selfish and spiteful behaviour in an evolutionary model. *Nature, Lond.* **228**, 1218—20.

—— (1971a). Geometry for the selfish herd. *J. Theoret. Biol.* **31**, 295—311.

—— (1971b). Selection of selfish and altruistic behavior in some extreme models. In Eisenberg, J. F. and Dillon, W. S. (eds.), *Man and beast: comparative social behavior*, pp. 55—91. Smithsonian Institution Press, Washington D.C.

Hinde, R. A. (1974). *Biological bases of human behaviour*. McGraw-Hill, New York.

Hutchinson, G. E. and MacArthur, R. H. (1959). On the theoretical significance of aggressive neglect in interspecific competition. *Amer. Natur.* **93**, 133—4.

Krebs, J. R. (1974). Colonial nesting and social feeding as strategies for exploiting food resources in the great blue heron (*Ardea herodias*). *Behaviour* **51**, 99—134.

# 388 Comparative social behaviour

MacRoberts, M. H., and Cullen, J. M. (1972). Flocking and feeding in the great tit *Parus major*—an experimental study. *Ibis* **114**, 507–30.

Kruuk, H. (1972). *The spotted hyena: a study of predation and social behavior.* University of Chicago Press.

Kühme, W. (1965). Communal food distribution and division of labour in African hunting dogs. *Nature, Lond.* **205**, 443–4.

Lack, D. (1968). *Ecological adaptations for breeding in birds.* Methuen, London.

Lorenz, K. Z. (1966). *On aggression.* Methuen, London.

Maynard Smith, J. (1965). The evolution of alarm calls. *Amer. Natur.* **99**, 59–63.

—— (1974). The theory of games and the evolution of animal conflict. *J. Theoret. Biol.* **47**, 209–21.

—— (1976). Evolution and the theory of games. *Amer. Scient.* **64**, 41–5.

—— and Parker, G. A. (1976). The logic of asymmetric contests. *Anim. Behav.* **24**, 159–75.

—— and Price, G. R. (1973). The logic of animal conflict. *Nature, Lond.* **246**, 15–18.

—— and Ridpath, M. G. (1972). Wife sharing in the Tasmanian native hen, *Tribonyx mortierii*: a case of kin selection? *Amer. Natur.* **106**, 447–52.

Packer, C. (1977). Reciprocal altruism in olive baboons. *Nature, Lond.* **265**.

Parker, G. A. (1970). The reproductive behaviour and the nature of sexual selection in *Scatophaga stercoraria* L. (Diptera: Scatophagidae), VII The origin and evolution of the passive phase. *Evolution* **24**, 774–88.

—— (1974). Assessment strategy and the evolution of fighting behaviour. *J. Theoret. Biol.* **47**, 223–43.

Patterson, I. J. (1965). Timing and spacing of broods in the black-headed gull *Larus ridibundus*. *Ibis* **107**, 433–59.

Schaller, G. B. (1972). *The Serengeti lion: a study of predator-prey relations.* University of Chicago Press.

Skutch, A. F. (1961). Helpers among birds. *Condor* **63**, 198–226.

Smythe, N. (1970). On the existence of 'pursuit invitation' signals in mammals. *Amer. Natur.* **104**, 491–4.

Thompson, W. A., Vertinsky, I., and Krebs, J. R. (1974). The survival value of flocking in birds: a simulation model. *J. Anim. Ecol.* **43**, 785–820.

Tinbergen, N. (1951). *The study of instinct.* Clarendon Press, Oxford.

Trivers, R. L. (1971). The evolution of reciprocal altruism. *Quart. Rev. Biol.* **46**, 35–57.

—— (1974). Parent-offspring conflict. *Amer. Zool.* **14**, 249–64.

Ward, P. (1965). Feeding ecology of the black-faced dioch *Quelea quelea* in Nigeria. *Ibis* **107**, 173–214.

Watts, C. R. and Stokes, A. W. (1971). The social order of turkeys. *Scient. Amer.* **224**, 112–18.

Wilkinson, P. F. and Shank, C. C. (1976). Rutting-fight mortality among musk oxen on Banks Island, Northwest Territories, Canada. *Anim. Behav.* **24**, 756–8.

Wilson, E. O. (1971). Competitive and aggressive behavior. In Eisenberg, J. F. and Dillon, W. (eds.), *Man and beast: comparative social behavior*, pp. 183–217. Smithsonian Institution Press, Washington D.C.

Wrangham, R. W. (1975). *The behavioural ecology of chimpanzees in Gombe National Park.* Ph.D. thesis, University of Cambridge, U.K.

Vine, A. A. (1971). The risk of visual detection and pursuit by a predator and the selective advantage of flocking behaviour. *J. Theoret. Biol.* **30**, 405–22.

Zahavi, A. (1971). The function of pre-roost gatherings and communal roosts. *Ibis* **113**, 106–9.

## 4. Comparative social behaviour

Barlow, G. W. (1974). Contrasts in social behavior between Central American cichlid fishes and coral-reef surgeon fishes. *Amer. Zool.* **14**, 9–34.

Brattstrom, B. H. (1974). The evolution of reptilian social behavior. *Amer. Zool.* **14**, 35—49.

Eisenberg, J. F. Muckenhirn, N. A., and Rudran, R. (1972). The relation between ecology and social structure in primates. *Science* **176**, 863—74.

Estes, R. D. (1974). Social organisation of the African Bovidae. In *Behaviour of ungulates and its relation to management I*, Geist, V. and Walther, F. (eds.), pp. 166—205, Morgues, I.U.C.N.

French, N. R., Stoddart, D. M., and Bobek, B. (1975). Patterns of demography in small mammal populations. In *Small mammals: their productivity and population dynamics*, Golley, F. B., Petrusewicz, K. and Ryszkowski, L. (eds.), pp. 73—103. Cambridge University Press.

Goss-Custard, J. D. (1970). Feeding dispersion in some overwintering wading birds. In *Social behaviour in birds and mammals: essays on the social ethology of animals and man*, Crook, J. H. (ed.), pp. 3—35. Academic Press, New York.

Hinde, R. A. (1956). The biological significance of territories of birds. *Ibis* **98**, 340—69.

Jarman, P. J. (1974). The social organisation of antelope in relation to their ecology. *Behaviour* **48**, 215—67.

Jolly, A. (1972). *The evolution of primate behavior.* Macmillan, New York.

Kaufmann, J. H. (1974). The ecology and evolution of social organisation in the kangaroo family (Macropodidae). *Amer. Zool.* **14**, 51—62.

Kruuk, H. (1972). *The spotted hyena.* University of Chicago Press.

— (1975). Functional aspects of social hunting by carnivores. In *Function and evolution of behaviour*, Beer, C. and Manning, A. (eds.). Clarendon Press, Oxford.

Lack, D. (1968). *Ecological adaptations for breeding in birds.* Methuen, London.

McNab, B. K. (1963). Bioenergetics and the determination of home range size. *Amer. Natur.* **97**, 133—40.

Milton, K. and May, M. L. (1976). Body weight, diet and home range size in primates. *Nature, Lond.* **259**, 459—62.

Newton, I. (1972). *Finches.* Collins, London.

Schoener, T. W. (1968). Sizes of feeding territories among birds. *Ecology* **49**, 123—41.

Stokes, A. W. (ed.) (1974). *Territory.* Dowden, Hutchinson, and Ross, Pennsylvania.

Tinbergen, N. (1957). The functions of territory. *Bird Study* **4**, 14—27.

Wiley, R. H. (1974). Evolution of social organisation and life history patterns among grouse (Aves: Tetraonidae). *Quart. Rev. Biol.* **49**, 201—27.

# Index